S0-CKV-528

CONTRIBUTIONS TO PROBABILITY AND STATISTICS

APPLICATIONS AND CHALLENGES

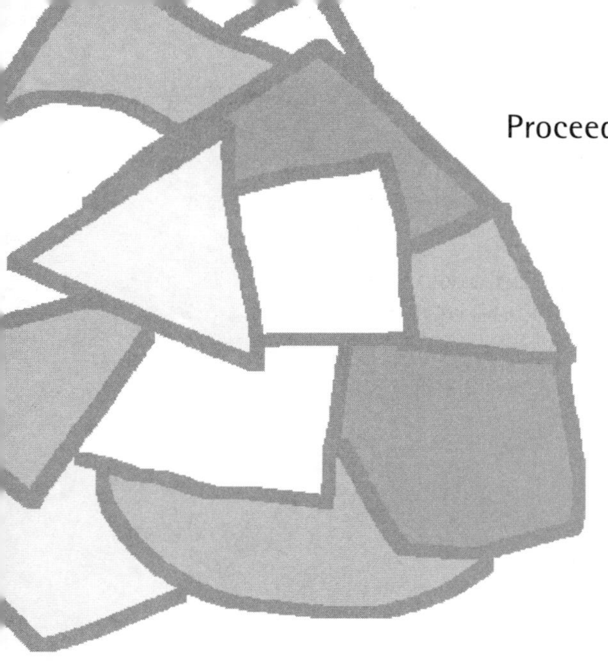

Proceedings of the International Statistics Workshop

University of Canberra

4 – 5 April 2005

CONTRIBUTIONS TO PROBABILITY AND STATISTICS

APPLICATIONS AND CHALLENGES

PETER BROWN ▪ SHUANGZHE LIU ▪ DHARMENDRA SHARMA

University of Canberra

World Scientific

NEW JERSEY • LONDON • SINGAPORE • BEIJING • SHANGHAI • HONG KONG • TAIPEI • CHENNAI

Published by

World Scientific Publishing Co. Pte. Ltd.
5 Toh Tuck Link, Singapore 596224
USA office: 27 Warren Street, Suite 401-402, Hackensack, NJ 07601
UK office: 57 Shelton Street, Covent Garden, London WC2H 9HE

British Library Cataloguing-in-Publication Data
A catalogue record for this book is available from the British Library.

CONTRIBUTIONS TO PROBABILITY AND STATISTICS: APPLICATIONS AND CHALLENGES
Proceedings of the International Statistics Workshop

ISBN 981-270-391-8

Printed in Singapore by World Scientific Printers (S) Pte Ltd

Preface

The International Statistics Workshop was held at the University of Canberra, Canberra, Australia on 4–5 April 2005. It was timed to fit in with the 55th Session of the International Statistical Institute (ISI) held in Sydney, Australia, from 5 to 12 April 2005.

Attending were a number of international academics, academics from around Canberra (Australian National University (ANU), Australian Defense Force Academy (attached to University of New South Wales) (ADFA), Commonwealth Scientific and Industrial Research Organisation (CSIRO) and various Commonwealth Government Departments), academics from within the University of Canberra with research interests in the area, and members of the Mathematics and Statistics Discipline at the University of Canberra.

It was a pleasure to welcome the many international and local speakers, participants and friends to the workshop. Speakers addressed topics in probability and statistics, with applications, both specialised and general. All contributions were strongly research and/or application oriented. The Workshop was arranged to include especially the areas of interest at the University of Canberra: applications in economic and financial areas, ecology and psychology, teaching and education, health and sports studies, and computer and IT-data mining, in addition to statistical theory and methodology.

We wish to thank the many contributors, referees and participants, without whose efforts the Workshop and these Proceedings would not have been possible. In particular, we would like to thank Joe Gani, Chris Heyde and George Styan for their thoughtful advice and assistance in the organisation of the workshop and in the editing process during the last two years. We also wish to thank Daryl Daley, Peter Hall, Mary Hewett, Ian Lisle, Mike Osborne and Alan Welsh for their support on various matters. Finally, we thank the staff members of the School of Information Sciences and Engineering and of the University of Canberra more widely who have been involved in arranging and supporting the Workshop and Proceedings.

Peter Brown, Shuangzhe Liu & Dharmendra Sharma
Editors.

Organising group: Shuangzhe Liu, Dharmendra Sharma and Peter Brown;
Editorial group: Peter Brown, Shuangzhe Liu and Dharmendra Sharma;
all at
School of Information Sciences and Engineering,
University of Canberra.

CONTENTS

PART A

Mathematics and Statistics in Society

Estimating Internet Access for Welfare Recipients in Australia

Anne Daly

School of Business and Government, University of Canberra
Canberra ACT 2601, Australia
E-mail: anne.daly@canberra.edu.au

Rachel Lloyd

Centre for Labour Market Research and NATSEM, University of Canberra,
Canberra ACT 2601, Australia
E-mail: rachel.lloyd@canberra.edu.au

The internet offers a quick and cheap method for government agencies to contact their clients. Many agencies are now exploring ways in which they can utilise new technologies to improve the efficiency of their communication with clients. Centrelink is currently responsible for the administration of the Australian welfare system and the agency is keen to know whether the use of the internet as a vehicle for transmitting information to clients would be a feasible option. This paper builds on earlier work by Lloyd and Bill (2004) based on data from the 2001 Census of Population. In this earlier study, the researchers estimated an equation for the determinants of internet usage for the Australian adult population. The Census does not identify welfare recipients. In this paper the earlier estimates are applied to data from the Household Expenditure Survey (HES) to provide estimates of the level of internet usage among those identified in the HES as welfare recipients. This involves using variables that are available on both data sets to estimate the probability of internet usage for welfare recipients.

Keywords: welfare recipients; internet usage; Australia.

1. Introduction

Technological developments in computers and the internet have opened new opportunities for government in providing services to clients. Centrelink, as one of the largest Commonwealth agencies dealing directly with individual clients, has been keen to explore possibilities for maintaining and improving services while reducing costs. As the Commonwealth agency currently responsible for the payment of all pensions and benefits, for example the Old Age and Disability pensions, Parenting Payment Single and Partnered, and

Newstart Allowance, there is considerable scope to improve service delivery using new technologies. There has been a continuing concern that a digital divide should not develop in Australia on the basis of socio-economic characteristics, age and location of residence (see for example Lloyd and Hellwig 2000, Daly 2002), and a number of government programs have been designed to address this issue.

However the question remains: do Centrelink clients have access to computers and the internet? Would a reliance on communications using these methods disadvantage clients? Data to answer these questions directly are not available. The purpose of this study has therefore been to estimate the likely use of a home computer and of the internet for individuals identified in the 1998/99 Household Expenditure Survey (HES), conducted by the Australian Bureau of Statistics (ABS), as receiving a payment from the Commonwealth government. These are mainly pension and benefit recipients but Centrelink is also responsible for administering family payments that are available to a much wider group in the community. Estimates are made by applying the results of logistic regression equations from the 2001 Census to data from the HES. The results reported here are of work-in-progress.

2. Results

In 2001, for the first time, the Australian Census of Population and Housing, which is conducted every five years by the ABS, included questions on computer and internet use. People were asked whether they had used a home computer in the week prior to Census night and were also asked if they had accessed the internet at all during the week prior to Census night. Respondents were asked to indicate whether access to the internet had taken place at home, at work or elsewhere and were given the opportunity to indicate if it had been at a combination of these locations. None of these questions asked about intensity of use, so an individual who used a home computer once in the preceding week is counted in the same way as a person who had used it for long periods every day.

Lloyd and Bill (2004) and Bill and Lloyd (2003) provide a detailed description of the Census results. They found that certain characteristics were associated with not having used a home computer or the internet in the week before the Census. Over two-thirds of those in each of the following categories reported that they had not used a home computer or accessed the internet: those who did not speak English well, did not currently go to school, attended school to year 8 or below, were aged over 65 years, had family incomes $300–399/week, were Indigenous, were born in Southern or Eastern Europe, were not in the labour force or were occupied as labourers (Lloyd and Bill 2004). They used the data to formalise these relationships by estimating logistic regressions for home computer and internet usage.

While the 2001 Census contained information on home computer and internet usage, it did not contain information on sources of income so it was not possible to investigate computer and internet usage for Centrelink clients using the Census. The HES 1998/99 includes information on sources of income and therefore has been used as the data set for estimating computer and internet usage for Centrelink clients. Table 1 presents the independent variables used in the estimation of results that form the basis of this study. Tables 2 and 3 include the logistic regression results from Bill and Lloyd (2003) using Census data on variables that are common between the Census and the HES. The full equations reported in Lloyd and Bill (2004) also include variables for Indigenous status, speaking English not well or not at all and regions and remote location of residence. These variables were all significant in the full equation so there will be some omitted variable bias from the equations excluding them. However, these additional variables are not available in the HES.

The first column of Tables 2 and 3 present the estimated coefficient from the logistic equations. Each of the coefficients is significant at the usual levels as they were estimated using the full Census file. The second column reports the odds ratio relative to the base case. The base case is a married man aged 25–44, with no post-secondary education but not currently studying, employed in an occupation other than trades or labouring, with a household income of $500–999/week, no dependent children and living in New South Wales. Bill and Lloyd estimate that the probability of such a person using a home computer was 43.8% and of using the internet was 51.2%. The odds ratio shows the effect of a change in one variable on the base-case probability. For example, a male with a degree was over three times as likely to use a home computer and four times as likely to use the internet as the base-case male with no post-secondary education holding all the other base case characteristics constant. A married man living in a household with a weekly income above $1500 was 1.7 times as likely to have used a home computer and twice as likely to have used the internet as was an identical man in a household with a weekly income of between $500 and $999.

The logistic regression results reported in Tables 2 and 3 below show the large effects of education and income on home omputer and internet usage.

Table 1. Explanatory variables used in regression model

Explanatory variable	Classes (Base Class*)
Gender & marital status	Male & not married
	Female & not married
	Male & married*
	Female & married
Age	15–24 years
	25–44 years*
	45–64 years
	65+ years
Educational qualifications & study status	Bachelor degree or higher
	Advanced diploma, diploma or certificate
	No post school qualification*
	Still at school
	Other still studying
Labour force status & occupation	Employed as Tradesperson or Labourer
	Employed in other occupations*
	Unemployed
	Not in the labour force
Household income	Household income under $500 per week
	Household income $500–$999*
	Household income $1,000–$1,499
	Household income $1,500 per week and over
Dependents (in household)	Dependents
	No dependents (& not applicable)*
State	New South Wales*
	Victoria
	Queensland
	South Australia
	Western Australia
	Tasmania
	Australian Capital Territory
	Northern Territory & other territories

Source: Bill & Lloyd (2003)

Table 4 presents some descriptive statistics from the HES for those receiving government payments and those who did not. The first column shows the proportion of those receiving payments who were in each category. The second column shows the proportion of those who did not receive payments in each category. The results show that payment recipients are more likely to be female, be aged 65+, have no post-secondary qualification and be outside the labour force. The regression results reported in Tables 2 and 3 show that each of these characteristics is likely to be associated with lower probabilities of home computer and internet usage.

Table 2. Centrelink regression model for people using a home computer, 2001

	Coefficient estimate	Odds ratio
Intercept	−0.3407	
Male and not married	−0.2976	0.743
Female and not married	−0.5828	0.558
Male and married (base)		1
Female and married	−0.2054	0.814
15–24 years	0.1149	1.122
25–44 years (base)		1
45–64 years	−0.2627	0.769
65+years	−1.1263	0.324
Degree level	1.1457	3.145
Diploma/certificate	0.5275	1.695
No post-school (base)		1
Still at school	1.9454	6.996
Still studying other	1.7393	5.693
Employed as tradesperson or labourer	−0.8739	0.417
Employed in other occupation (base)		1
Unemployed	−0.3875	0.679
Not in the labour force	−0.6549	0.519
Household income under $500 per week	−0.2628	0.769
$500–$999 (base)		1
$1,000–$1,499	0.2485	1.282
$1,500 or more	0.5371	1.711
Dependent children	0.4466	1.563
No dependent children (base)		1
New South Wales (base)		1
Victoria	0.0661	1.068
Queensland	0.1581	1.171
South Australia	0.0943	1.099
Western Australia	0.1563	1.169
Tasmania	−0.0536	0.948
Australian Capital Territory	0.3517	1.421
NT and other territories	−0.3933	0.675

Source: Bill and Lloyd (2003)

3. Concluding Remarks

These results have been used to predict home computer and internet usage for the population aged 15+ in the HES. This population has been divided into those who received government payments and those who did not. The preliminary estimates show that the probability of those receiving government payments making use of home computers and the internet was about half that for the non-recipients in the sample. Sensitivity and benchmarking analysis of these results is still to be completed.

There are a number of outstanding issues arising from this project before

Table 3. Centrelink regression model for people using the Internet, 2001

Parameter	Coefficient estimate	Odds ratio
Constant	0.0462	
Male and not married	−0.212	0.809
Female and not married	−0.398	0.671
Male and married (base)		1
Female and married	−0.454	0.635
15–24 years	0.1071	1.113
25–44 years (base)		1
45–64 years	−0.57	0.565
65+ years	−1.685	0.185
Degree or higher	1.4504	4.265
Diploma/certificate	0.5434	1.722
No post-school (base)		1
Still at school	1.9125	6.77
Still studying other	1.8703	6.49
Employed as tradesperson or labourer	−1.35	0.259
Employed other occupation (base)		1
Unemployed	−0.76	0.468
Not in labour force	−1.053	0.349
Household income under $500 per week	−0.338	0.713
$500–$999 (base)		1
$1,000–$1,499	0.3039	1.355
$1,500 or more	0.7164	2.047
Dependent children	0.0092	1.009
No dependent children (base)		1
New South Wales (base)		1
Victoria	0.0734	1.076
Queensland	0.1278	1.136
South Australia	0.0629	1.065
Western Australia	0.1559	1.169
Tasmania	0.0301	1.031
Australian Capital Territory	0.5721	1.772
NT and other territories	−0.195	0.823

Source: Bill and Lloyd (2003)

we can fully answer the question of whether the use of the internet as a vehicle to communicate with customers is a feasible option. Firstly, the geographical breakdown available in the HES is limited to identification of the State or Territory where the respondent lives. Evidence from the 2001 Census shows that there is a consistent pattern of higher levels of home computer and internet use in the capital cities than in regional and rural areas. If the distribution of the population between the capital cities and outside those cities of those receiving income from the government is very different from that of the population as a whole, then the ability of these predictions to estimate access will be reduced. In addition, there

Table 4. Descriptive statistics for those receiving Government payments compared with those who did not; Australia, 1998/99

Characteristic	Recipient of Payment Proportion	Non-recipient of Payment Proportion
Male and not married	0.14	0.22
Female and not married	0.26	0.16
Male and married	0.20	0.63
Female and married	0.40	0.25
15–24 years	0.11	0.22
25–44 years (base)	0.35	0.42
45–64 years	0.22	0.32
65+ years	0.32	0.04
Degree level	0.06	0.18
Diploma/certificate	0.26	0.33
No post-school (base)	0.62	0.40
Still at school	0.02	0.07
Still studying other	0.08	0.12
Employed as tradesperson or labourer	0.05	0.20
Employed in other occupation (base)	0.19	0.63
Unemployed	0.09	0.02
Not in the labour force	0.67	0.15
Household income < $500/week	0.52	0.11
$500–$999/week (base)	0.30	0.30
$1000–$1499/week	0.13	0.27
$1500+/week	0.06	0.32
Dependent children	0.43	0.40
No dependent children (base)	0.57	0.60
New South Wales	0.33	0.35
Victoria	0.25	0.25
Queensland	0.20	0.17
South Australia	0.09	0.07
Western Australia	0.09	0.10
Tasmania	0.03	0.02
ACT	0.01	0.02
NT and other territories	0.00	0.01

Source: HES Unit record file

have also been some significant policy changes with respect to eligibility for government payments since 1998/99. These include welfare reforms and the introduction of a new tax system. The characteristics of those receiving Centrelink payments may have changed as a result of these policy changes. Finally, the 2001 Census is now almost four years old. A more recent ABS survey shows that 66 per cent of households (as distinct from persons as discussed above) in 2003 had access to a home computer compared with 58 per cent in 2001 (ABS 2004). The percentage of households with internet access had risen from 42 per cent in 2001 to 53 per cent in 2003. If Centrelink were to use estimates from the above analysis to estimate the computer and internet usage of individual clients, it would be important to adjust these estimates in the light of the rapid growth in use of these technologies among the Australian population. Finally it may be worthwhile developing the capacity to communicate with clients through the internet as a way of encouraging the development of computer and internet skills among this group. However, at this stage, our preliminary estimates suggest that it would be inappropriate to rely completely on this form of communication with Centrelink clients.

References

1. Australian Bureau of Statistics (2004) *Household Use of Information Technology*, ABS cat. no. 8146.0, Canberra.
2. Bill, A. and Lloyd, R. (2003) *Modelling Internet and Computer Use for Centrelink Customers*, Draft paper, NATSEM, University of Canberra.
3. Daly, A. (2002) Telecommunications services in rural and remote Indigenous communities in Australia, *Economic Papers*, 21 (1): 18–31.
4. Lloyd, R. and Hellwig, O. (2000) The digital divide, *Agenda*, 17 (4): 345–358.
5. Lloyd, R. and Bill, A. (2004) *Australia Online: How Australians are Using Computers and the Internet*, ABS cat. no. 2056.0, Canberra.

Two Classification Methods of Individuals for Educational Data and an Application

Atsuhiro Hayashi
Research Division, The National Center for University Entrance Examinations, Tokyo 153-8501, Japan
E-mail: hayashi@rd.dnc.ac.jp

Both methods, Rule Space Method (RSM) and Neural Network Model (NNM), are techniques of statistical pattern recognition and classification approaches developed from different fields — one is for behavioural sciences and the other is for neural sciences.

RSM is developed in the domain of educational statistics. It starts from the use of an incidence matrix Q that characterises the underlying cognitive processes and knowledge (Attribute) involved in each Item. It is a grasping method for each examinee's mastered/non-mastered learning level (Knowledge State) from item response patterns. RSM uses multivariate decision theory to classify individuals, and NNM, considered as a nonlinear regression method, uses the middle layer of the network structure as classification results. We have found some similarities and differences between the results from the two approaches, and moreover both methods have characteristics supplemental to each other when applied to the practice.

In this paper, we compare both approaches by focusing on the structures of NNM and on knowledge States in RSM. Finally, we show an application result of RSM for a reasoning test in Japan.

Keywords: Rule Space Method; Neural Network Model; educational statistics; cognitive science.

1. Introduction

A Neural Network model was proposed for the purpose of modelling the information processing in a person's brain in the 1940s. Neurons (nerve cell elements) are considered as the minimum composition unit of cerebral functions that can entangle in a complicated and organic manner. The model shows that all the logical reasoning can be described in a finite size of the number of neurons and connections [2]. The model enables us to express acquisition of new knowledge from learned examples in the past; therefore it can be used to help to solve one of the weaknesses in constructing an

AI (Artificial Intelligence) system. It is known that expressing knowledge acquisition in an AI system is extremely difficult.

On the other hand, the Rule Space Method (RSM) is a technique of clustering examinees into one of the predetermined latent Knowledge States (KS) that are derived logically from an expert's hypotheses about how students learn. The method can be considered as a statistical testing technique of the expert's hypotheses. These hypotheses are expressed by an item-attribute matrix (incidence matrix) Q where attributes are representing underlying knowledge and cognitive processing skills required in addressing problems [1]. A Knowledge State consists of attributes of the type mastered/non-mastered, and a list of all the possible Knowledge States can be generated algorithmically by applying Boolean Algebra to the incidence matrix Q. This method is fairly new but has lately started getting some attention because it is possible to provide diagnostic scoring reports for a large-scale assessment [3]. We have found that there are similarities between the results from the two approaches, and moreover they have complementary characteristics when applied in practice. In this paper, we discuss the comparisons of both approaches by focusing on the structure of the Neural Network Model (NNM) and of Knowledge States in the RSM. We show an application result for a reasoning test.

2. Rule Space Method

RSM is a technique developed in the domain of educational statistics [6]. It starts from the use of an incidence matrix Q that characterises the underlying cognitive processes and knowledge (Attribute) involved in each Item. It is a grasping method for each examinee's mastered/non-mastered learning level (Knowledge State, KS) from item response patterns. Up to now, the results of examinees' performances on a test are reported by total scores or scaled scores. However, if this technique is used in educational practices, it is possible to report which attributes each student mastered or did not master, in addition to his/her total scores. It is often true that several different Knowledge States may arise from the same total score. By reporting detailed information of his/her Knowledge State, learning can be facilitated more effectively than by just providing total scores.

3. Feed-Forward Neural Network Model

In spite of that the mathematical formulation of the Feed-Forward NNM is simple. Almost any nonlinear function can be approximated by selecting different numbers of middle layers and connections between neurons. When we apply this technique to existing data obtained from learning processes, we can use this model to search for a strategy of any joint intensity between units.

From a statistical point of view, NNM is a nonlinear regression model. In this paper Feed-Forward NNM is considered as a model-fitting procedure to estimate the optimum values of the parameters in the regression model [4].

This procedure is called parameter estimation in statistics, but is called a learning algorithm in NNM. One of the learning algorithms commonly used is Back Propagation (BP), that is a method of learning by passing on errors to previous layers. BP is an adaptation of the steepest descent method to the NNM field. This method has a reducible faculty of convergence to the local minimum point.

4. Science Reasoning Test

The Science Reasoning Test (SR Test) is an entrance examination that measures the student's interpretation, analysis, evaluation, reasoning and problem-solving skills required in the natural sciences [5].

Since we got the ACT's (American College Testing, Inc.) cooperation, we used one open-form of their ACT Assessment tests for our experimentation. The test is based on units containing scientific information and a set of multiple choice questions about the scientific information. Calculators are not permitted to be used for the test. The scientific information for the test is provided in one of three types of formats.

The first format, data representation, presents graphic and tabular material similar to that found in science journals and texts. The questions associated with this format measure skills such as graph reading, interpretation of scatter plots, and interpretation of information presented in tables. The second format, research summaries, provides students with descriptions of one or more related experiments. The questions focus upon the design of experiments and interpretation of experimental results. The third format, conflicting viewpoints, presents students with expressions of several hypotheses or views that, being based on differing premises or on incomplete data, are inconsistent with one another. The questions focus upon the understanding, analysis and comparison of alternative viewpoints or hypotheses.

The Science Reasoning Test questions require students to use scientific reasoning to answer the questions. The students are required to recognise and understand the basic features of, and concepts related to, the provided information; to critically examine the relationships between the information provided and the conclusions drawn or hypotheses developed; and to generalise from given information to gain new information, draw conclusions, or make predictions.

5. Numerical Examples

We applied the RSM to data from fraction addition problems, and got a tree structure for the Knowledge State. We related RSM that derives the Knowledge State from an incidence matrix Q, to the Feed-Forward NNM. For that, we designed the network of a three-layer structure in which items were assigned to the input layer and Attributes to the output layer. The Knowledge States in the RSM were considered to correspond to the middle layers of NNM. We applied several numerical examples to both methods and found close similarities in their results, although they were not identical.

Also we applied the RSM to data from Science Reasoning Test results of 286 Japanese students. The number of attributes and items are 12 and 18, respectively. Figure 1 is the tree representation of the Knowledge States that shows the examinee's mastered/non-mastered learning level. In this figure, each circle is a Knowledge State, and the numbers in the circle are the IDs of non-mastered attributes. The number in parentheses is the number of examinees classified in this Knowledge State. We observe that the main solving attribute IDs are $6, 8$ and 9, and secondary attribute IDs are 2 and 5. The total examinees classified in these Knowledge States is 225, which is about 80% of all. The main streams to reach the fully-mastered state are the three Knowledge States on the left-hand side in the third layer from the top.

6. Discussion and Conclusions

We investigated the relationship between the characteristics of the middle layer of NNM and the Knowledge States in the RSM, and discussed their similarities and usefulness for the weaknesses existing in the RSM.

It is well known that the composition of an incidence matrix Q in the RSM is a very laborious task, and requires experts' intense cooperation. The experts identify attributes involved in each item and express them in an incidence matrix Q. Multiple solution strategies for each item need to be investigated. This is extremely hard work. If an examinee's mastering level (cluster) is known to some extent from past experiences, it is also possible to

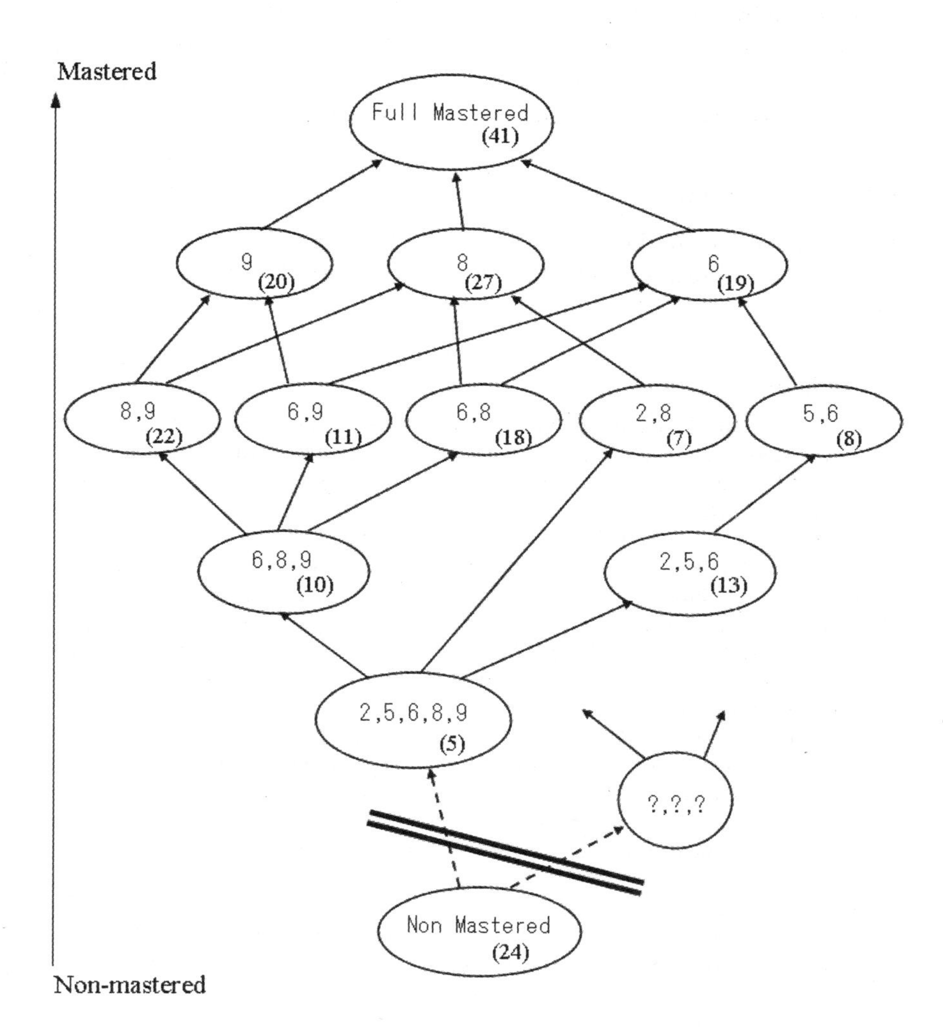

Fig. 1. A tree representation of Knowledge States for the SR-Test data

construct a network in which these clusters are assigned to the output layer of NNM. The middle layer drawn from this model is expected to correspond to Attributes. It may be possible to use this result for replacing the task analysis required in making an incidence matrix Q in RSM.

We plan to clarify the differences and similarities of the two models with numerical examples, and should get useful results by applying these methods to the SR-Test data and our real examination data.

Acknowledgements

We are grateful to Dr. James Maxey and ACT who gave us their permission to use one open-form in their ACT Assessment tests for our experimentation.

References

1. Mary F. Klein, Menucha Birenbaum *et al.* (1981). Logical Error Analysis and Construction of Tests to Diagnose Student "Bugs" in Addition and Subtraction of Fractions, University of Illinois, Computer-based Education Research Laboratory, Research Report 81–86.
2. Takio Kurita and Yoichi Motomura (1993). Feed-Forward Neural Networks and their Related Topics, Japanese Journal of Applied Statistics, Vol. 22, No. 3, 99–114 (in Japanese).
3. Kikumi K. Tatsuoka (1995). Architecture of Knowledge Structures and Cognitive Diagnosis: A Statistical Pattern Recognition and Classification Approach, Paul D. Nichols *et al.* (Eds.), Cognitively Diagonostic Assesment, 327–359, Lawrence Erlbaum Associates.
4. Atsuhiro Hayashi and Yasumasa Baba (1998). An Analysis of University Entrance Examination Data by using Neural Network Models, Compstat 1998, Physica-Verlag, Bristol, Short Communications, 45–46.
5. James Maxey (2000). Introduction to the ACT Assessment, International comparison study for university entrance examinations, M. Huzii *et al.* (Eds.), 42–55.
6. Kikumi K. Tatsuoka and Atsuhiro Hayashi (2001). Statistical method for individual cognitive diagnosis based on latent knowledge state, Journal of The Society of Instrument and Control Engineers, Vol. 40, No. 8, 561–567 (in Japanese).

Measurement of Skill and Skill Change

Ross Kelly and Philip E. T. Lewis

Centre for Labour Market Research, University of Canberra,
Canberra ACT 2601, Australia
E-mail (1): ross.kelly@canberra.edu.au
E-mail (2): phil.lewis@canberra.edu.au

The measurement of skill and skill change is important for much empirical research on labour markets, education and technological change. This paper presents a novel way of measuring skills and provides an example of its use in investigating the degree and nature of change in the Australian labour market over a period of rapid technological and structural change.

Keywords: structural change; technological change; measurement; motor skills; interactive skills; cognitive skills; educational skills.

1. Introduction

The issue of measurement will confront any analysis of skills at an aggregate level. Measures typically favoured by economists in studies of human capital rely on years or level of education as a measure of skill attainment. The obvious shortfalls are that these measures do not necessarily capture the actual skill requirements of jobs — the rapid growth in educational attainment may have as much to do with credentialism as with skill attainment (Attewell 1990). An alternative favoured by sociologists focuses on the skill attributes required of jobs, as defined in the US Department of Labour's Dictionary of Occupational Titles (DOT). It provides a convenient basis for the analysis of skills independent of productivity measures and knowledge of individuals or workplaces and so is used for the following analyses. A brief overview of how skill scores are assigned to an occupation or industry follows. The full details can be found in Kelly and Lewis (2003). New results generated using this method are presented to illustrate how skill changes can be decomposed into those due to technological and structural change as well as differences in the movement from full-time to part-time employment.

Mean skill scores (i.e. average skill per hour worked in the whole economy) for four skill dimensions for 1991, 1996 and 2001 were calculated. Census data for 1991, 1996 and 2001 were used to compile the skill indices for each of the skill dimensions being considered. Measures of skill were constructed using data and information from Australian occupational task descriptions contained in the Australian Standard Classification of Occupations (ASCO), 2nd edition. These were then combined with occupation by industry employment matrices showing total hours worked for part-time and full-time workers and scales of skill complexity for four skill dimensions developed by the United States Department of Labor (USDOL).

The Dictionary of Occupational Titles (DOT), 4th edition (1991), used in the US provides a schema for rating skills at the finest level of occupational detail, as shown in Table 1. In DOT jobs are classified as requiring workers to function to some degree in relation to data, people and things. The scale for each skill dimension shown in Table 1 is in descending order.

Those tasks that involve more complex responsibility and judgment are assigned lower numbers for each category and the less complicated have higher numbers. For example, for the data skill dimension (see Table 1) 'compiling' would be considered a more complex task than 'copying'. The same applies for the other dimensions. Each dimension is considered separately. The scale relates to an ordering of the complexity of tasks normally undertaken in an occupation; it does not signal anything about the intensity of use of those skills. At an industry level, this is determined by the hours of employment, or utilisation, of the skills embodied in an occupation. The occupation, in turn, tells us something about the tasks undertaken and how they relate to the scale of complexity shown in Table 1.

Table 1. Scale of complexity for skill categories.

Data	People	Things
0 Synthesising	0 Mentoring	0 Setting Up
1 Coordinating	1 Negotiating	1 Precision Working
2 Analysing	2 Instructing	2 Operating-Controlling
3 Compiling	3 Supervising	3 Driving-Operating
4 Computing	4 Diverting	4 Manipulating
5 Copying	5 Persuading	5 Tending
6 Comparing	6 Comparing	6 Feeding-Off bearing
	7 Serving	7 Handling
	8 Taking Instructions-Helping	

Source: USDOL (2000)

Four types of skill are analysed: motor skills, education, interactive skills and cognitive skills. The 'data' category in Table 1 provides a measure of cognitive skills, the 'people' category aligns with interactive skills and the 'things' category provides a good indicator of motor skills. The education category used for this study comes from the education requirement listed for each occupation in the ASCO (2nd edition). The levels of education, based on the Australian Qualifications Framework (AQF), were grouped into six levels, with masters and doctoral degree the highest and AQF I & II the lowest, the measure being made complete by the addition of a 'no qualification required' level. AQF I & II are the most basic of qualifications requiring a narrow range of elementary competencies, such as demonstrating "...basic practical skills such as the use of relevant hand tools ..." (AQF 2002). All other measures were inverted, that is, the least complex tasks were given the lowest score. The scale was converted to a common scale of 0 to 10. Finally, the scores were assigned to a given occupation for each skill dimension at the finest level of information on occupations, the ASCO (2nd edition) six-digit level. The most complex task undertaken in an occupation for each skill dimension, as identified from the ASCO, provided the basis for applying the scores.

Thus, the hourly weighted mean skill score for a given dimension in industry k is as follows:

$$S_k = \frac{\sum\limits_{m,n} s_m O_{kmn}}{\sum\limits_{m,n} O_{kmn}} \tag{1}$$

with the mean skill for a given skill dimension for the economy defined as:

$$S = \frac{\sum\limits_{k,m,n} s_m O_{kmn}}{O_T}, \tag{2}$$

where S is the mean skill score, S_k is the mean skill score across industry k, s_m is the skill score of occupation m and is constant across time, O_{kmn} is the number of hours worked in the appropriate triplet, and $O_T = \sum_{k,m,n} O_{kmn}$ is the total hours worked across the economy. The subscripts $k = (1, \ldots, q)$, $m = (1, \ldots, r)$ and $n = (1, \ldots, u)$ denote industry, occupation and part-time or full-time employment status, respectively.

Given that the skill score for a given skill dimension and occupation is held constant for each time period, it is changes to the occupational composition of employment that determines changes in the economy-wide mean skill level. This can be represented as:

$$\Delta S = \sum_{k,m,n} s_m \Delta \left(\frac{O_{kmn}}{O_T} \right) \tag{3}$$

with the subscripts as described above.

From (2) it is apparent that changes in mean skills in the economy can arise from changes in the share of an occupation in an industry and changes in industry shares of total hours employed in the economy.

To simplify exposition we denote the occupation-m share for industry k as:

$$b_{km} = \frac{\sum_{n} O_{kmn}}{\sum_{m,n} O_{kmn}} \tag{4}$$

and industry k's share of total hours employed in the economy as:

$$h_k = \frac{\sum_{m,n} O_{kmn}}{O_T}. \tag{5}$$

Thus, the change in mean skill for the economy as a whole for any skill dimension is:

$$S_t - S_{t-1} = \sum_{k,m} s_m \left((b_{km} h_k)_t - (b_{km} h_k)_{t-1} \right). \tag{6}$$

An exact decomposition is provided by:

$$\Delta S = \sum_{k,m} s_m \Delta b_{km} \bar{h}_k + \sum_{k,m} s_m \bar{b}_{km} \Delta h_k \tag{7}$$

with change expressions identified by the delta symbol and the bar over expressions indicating the inter-temporal mean.

The first term on the right of equation (7) provides the within-industry effect, the second expression the between-industry effect. Both of these are further decomposed to show the contribution of the part-time and full-time workforce to changes in mean skill.

The way changing industry shares of total employment affect economy-wide mean skill scores can be explained as follows. If an industry with a relatively high proportion of skilled workers increases its share of overall employment, then there will be an increase in the economy-wide average skill level. This is the inter-industry effect and can be split into the contributions from part-time and full-time employment by applying the respective weights for part and full-time hours employed.

Intra-industry changes to occupation composition work the same way. When an occupation that is relatively highly skilled increases its share of employment within a given industry, that industry experiences an increase in its mean skill level. This can be further decomposed into the contributions from part-time and full-time employment. This enables an examination of whether the large shift towards part-time employment over the last decade has resulted in de-skilling. If the occupational composition of part-time employment is different from that of full-time employment, then a change in emphasis within an industry toward one or the other will influence the economy-wide mean skill score. The sum of such changes across the economy shows the within-industry effect on the change in the economy-wide average skill level.

2. Results

Table 2 shows the percentage change for each of these dimensions between 1991 and 2001. The mean skill levels for full-time workers for interactive, cognitive and education skills increased by about 9.9, 9.2 and 6.4 per cent respectively between 1991 and 2001. Motor skills per hour employed for full-time workers declined by 10.8 per cent. The decline in motor skills for part-time workers was 8.1 per cent. Mean education skills for part-time workers also showed a small decline. This does not necessarily mean that the part-time workforce became less educated over the period in question, but rather, that the educational level required (as represented by the hours employed) was less intensive. The increases for part-time workers in interactive and cognitive skills were quite modest.

Table 2. Change in average skill levels, Australia, 1991-2001, percentage.

	motor	interactive	cognitive	education
All	−12.4	6.9	5.7	2.6
Part-time	−8.1	3.3	1.7	−0.8
Full-time	−10.8	9.9	9.2	6.4

Overall the increase in mean skills was highest for interactive and cognitive skills and relatively modest for education skills (or educational attainment). Motor skills dropped significantly. It is clear that the changes in total mean skills mask the differing outcomes between the part-time and full-time workforce.

Table 3. Decomposition of economy-wide change in average skill levels, Australia, 1991-2001.

1991–2001	motor	interactive	cognitive	education
total change	−0.317	0.248	0.162	0.088
total within industry	−0.192	0.148	0.118	0.058
p-t mean skill	−0.003	0.035	0.025	0.014
f-t mean skill	−0.160	0.170	0.135	0.105
p-t share of industry employment	0.093	0.159	0.124	0.127
f-t share of industry employment	−0.122	−0.216	−0.167	−0.187
total between industry	−0.125	0.101	0.045	0.030
industry share of total employment p-t	0.098	0.264	0.192	0.208
industry share of total employment f-t	−0.223	−0.164	−0.147	−0.178

It can be seen from Table 3 that motor skills declined substantially over the decade from 1991 to 2001. Interactive, cognitive and education skills increased, although the change for education was relatively small. The within-industry effect for all skill dimensions was dominant (see Table 4).

The data in Table 3 also decomposes the total skill change between 1991 and 2001 into contributions from part-time and full-time workers. From Table 3 it can also be seen that the mean skill of both part-time and full-time employment within industries for the motor skill dimension decreased, although the part-time decrease was fairly minor. The net effect of the changing status of employment (i.e. the changing shares of hours worked by part-time and full-time workers) reinforced the decline in the mean skill of both the part-time and the full-time workforce. Thus the switch to part-time employment was de-skilling — the −0.122 in mean skills given up from the drop in full-time employment within industries was greater than the 0.093 contribution to skills from the increased use of the part-time workforce.

Other skill dimensions showed increases in the mean skill level of both part-time and full-time employment. The common thread for all skill dimensions was that the contribution from the change in mean part-time skill levels was relatively small; the full-time workforce was the main contributor to mean skill levels. The status effect, the impact of changing part-time and full-time shares of employment within industries, pulled the skill level down. One way of viewing the changes taking place is that the growth in full-time work has been relatively skilled in nature, but less important in terms of its contribution to total employment. Growth in part-time employment has been substantial, but continues to be in occupations that are relatively lower skilled than are occupations for full-time jobs. This has tended to moderate the overall increase in the non-manual skill dimensions between 1991 and 2001. In the case of motor skills it had a reinforcing effect.

In all cases the within-industry effects were larger than the between-industry effects. The within-industry effect is usually interpreted as resulting from technological change, changes in the capital intensity within an industry and changes in the relative price of labour skills (Pappas 1998). Disentangling capital intensity effects over time from technological change is not a straightforward matter, since much of the technological development we observe is embodied in new capital.

The other effect on change in mean skill levels comes from the changing composition of industry shares of total employment in the economy. The between-industry effect captures the impact of changing product demands, trade and other structural change (Pappas 1998). Around 0.1 of the increase in interactive skills was due to the between-industry effect, while for motor skills the effect was to reduce the mean skill level for the economy by 0.125, reinforcing the within-industry effect. Table 4 shows the relative contributions of within- and between-industry effects for each of the skill dimensions for the period 1991 to 2001.

Table 4. Contribution to change, percentage.

1991–2001	motor	interactive	cognitive	education
Total within industry	60.6	59.5	72.5	66.3
Total between industry	39.4	40.5	27.5	33.7
Total	100	100	100	100

In summary, the within-industry effects were the main contributor to change for all skill dimensions over the 1991–2001 period. These were, in relative terms, most pronounced for cognitive skills. The conclusion to draw from these results is that technological change has been the dominant influence. It has allowed for, or driven, a restructuring of occupations within industries. Although greater emphasis on part-time employment has been de-skilling (suggesting technology-skill substitution), this has been outweighed by the changing occupational contribution within full-time employment and, to a lesser extent, part-time employment (suggesting that technology exhibits skill complementarity).

The full-time workforce (other than for motor skills) has become more skilled, but is numerically less important in production. The increased share of part-time employment has not been as highly skilled as the full-time employment it displaced. The balance of the two effects has still seen increases in mean industry skill levels for the economy for the non-manual skills, particularly for interactive skills. This is consistent with the nature of technological change that has taken place. This has been in the form of information systems and transactional processing technologies. It has been shown elsewhere (for example, Autor *et al.*, 2000; Caroli 1999) that these enable better management technologies to be implemented, allowing tighter scheduling of labour, flatter management structures and smaller workforces for a given output.

Changes in mean skills have also been decomposed for the relevant sub-periods. Coinciding with the census periods for Australia, the decomposition is for 1996–2001 and 1991–1996. The weights applied are for the whole period, that is, 1991–2001. Only the change variables shown in equation (7) vary for the sub-periods.

The data in Tables 5 and 6 are interpreted in the same way as before. Table 7 summarises these changes into the percentage contribution from each effect for each sub-period and skill respectively. It can be seen that the change that occurred between 1996 and 2001 was predominantly from within-industry changes in occupational composition. Although less pronounced, the opposite was the case for the period 1991 to 1996.

Table 5. Decomposition of economy-wide change in mean skill levels, Australia, 1996–2001.

1996–2001	motor	interactive	cognitive	education
total change	−0.150	0.122	0.094	0.043
total within industry	−0.112	0.110	0.092	0.048
mean skill for p-t	−0.012	0.038	0.030	0.023
mean skill for f-t	−0.088	0.093	0.078	0.048
p-t share of industry employment	0.042	0.069	0.054	0.059
f-t share of industry employment	−0.055	−0.090	−0.071	−0.081
total between industry	−0.037	0.012	0.002	−0.005
industry share of total employment p-t	0.049	0.101	0.076	0.083
industry share of total employment f-t	−0.086	−0.089	−0.073	−0.087

With the exception of cognitive skills the contributions to total change for each of the sub-periods was about the same. For cognitive skills 58 per cent of the change in mean skill levels occurred in the latter half of the decade. The individual effects are strikingly different for the sub-periods. The within-industry effects for the later period account for 74, 78 and 83

Table 6. Decomposition of economy-wide change in mean skill levels, Australia, 1991–1996.

1991–1996	motor	interactive	cognitive	education
total change	−0.167	0.126	0.068	0.044
total within industry	−0.080	0.038	0.026	0.010
mean skill for p-t	0.009	−0.003	−0.005	−0.009
mean skill for f-t	−0.072	0.076	0.057	0.057
p-t share of industry employment	0.050	0.090	0.070	0.068
f-t share of industry employment	−0.066	−0.125	−0.096	−0.106
total between industry	−0.088	0.089	0.042	0.034
industry share of total employment p-t	0.049	0.163	0.116	0.125
industry share of total employment f-t	−0.137	−0.075	−0.074	−0.091

Table 7. Contribution to mean skill change, Australia, 1991–2001, percent.

	total change		*total within industry*		*total between industry*	
	1991–1996	1996–2001	1991–1996	1996–2001	1991–1996	1996–2001
motor	53	47	42	58	70	30
interactive	51	49	26	74	88	12
cognitive	42	58	22	78	95	5
education	51	49	17	83	116	−16

per cent of the total within-industry effect for interactive, cognitive and education skill changes respectively that occurred between 1991 and 2001. Most of the occupational adjustment that has occurred has coincided with the rapid increase in Information and Communication Technology (ICT) capital expenditure observed from 1996–7 onwards and has been due to within-industry changes (ABS 2003).

The between-industry changes that took place and the impact that these had on changes in mean skill levels are nearly entirely attributable to the post-recession period, 1991–1996. In total the effects played a much smaller part in the change observed over the decade. One possible explanation is that the effects of the recession were not evenly spread across industries, or that the normal pattern of recovery sees some industries grow more quickly than others. Thus, the changes in skills mix over the first part of the decade can be attributed largely to structural change as the economy became more open to the global economy. In the second half of the decade the changes can be largely attributed to technological change.

3. Conclusion

Significant de-skilling of the part-time workforce occurred in the first half of the 1990s, although the effect of this on the overall skill level of the workforce was more than compensated for by the increasing skill level of the full-time workforce. Most of the impact on skill demand from the changing industry structure of the economy occurred prior to 1996, with the change after this time having very little impact. The latter half of the decade was characterised by the changing structure of occupations within industies. There was further intensification of higher skill occupations among the full-time workforce, while the part-time workforce experienced only modest (positive) change to its mean skill level. These changes coincided with the rapid increase in ICT investment and, significantly, with the increasing share of ICT in the capital stock.

The increasing importance of ICT in the capital stock is clearly having an impact on the type of skills demanded in the economy. This is most likely occurring due to direct demand for ICT-related skills and indirectly through the enabling characteristics of ICTs. Importantly, it is not only the increasing emphasis of computers in the workplace and industry, but the rapid increase in the uptake of software applications by industry. It appears that ICTs have allowed a substantial re-ordering of occupations within industries; that is, they are enabling a reorganisation of workplaces that places greater emphasis on skills, particularly interactive and cognitive skills. The extent to which these skills are able to be diffused through formal training and education needs to be explored.

The implications for policy are clear — traditional 'blue-collar' skills will stagnate or continue to decline, and this will test the ability of the labour market to adjust and absorb the existing supply of these skills. The inability of many individuals to adjust to the current and expected skill demands of industry will continue to see a large component of unemployment in Australia being structural in nature. When capital becomes technologically obsolete, the social consequences will be relatively benign. When the skills of workers become obsolete, the social consequences are much more serious, with unemployment, financial hardship and marginalisation the likely outcome. The vocational education and training (VET) sector should be taking a lead role on this issue, as equipping displaced workers with relevant skills will be critical to successful re-adjustment.

In the longer term, the balance between the VET and university education sectors in Australia may need revisiting. Within vocational education training the move away from traditional skilled manual trades and low skill employment needs to be acknowledged and greater emphasis placed on skills to meet the needs of the service sector. The university sector will continue to be the primary source of skills as the knowledge intensity of production in the economy increases.

References

1. Attewell, P. (1990), What is Skill?, *Work and Occupations*, Vol. 17, No. 4, pp. 422–448.
2. Australian Bureau of Statistics (ABS 1997), *Australian Standard Classification of Occupations (ASCO)*, Second Edition, Cat. no. 1220.0
3. Australian Bureau of Statistics (2003), *Labour Force*, Australia Catalogue No. 6291.0.55.001, Australian Bureau of Statistics, Canberra. (accessed online on 30/05/2003: `http://www.abs.gov.au/ausstats/`)
4. Australian Qualifications Framework Advisory Board (2002), *Australian Qualifications Framework Implementation Handbook, 3rd Ed.*, Australian Qualifications Framework (AQF) Advisory Board, Melbourne.
5. Autor, D., Levy, F. and Murnane, R. (2000), *Upstairs-Downstairs: Complementarity and Computer-Labour Substitution on Two Floors of a Large Bank*, NBER Working Paper No. 7890, National Bureau of Economic Research, Cambridge.
6. Caroli, E. (1999), New Technologies, Organisational Change and the Skill Bias: a go into the Black Triangle, in *Employment and Economic Integration*, P. Petit and L. Soete (eds), Edward Elgar, London.
7. Kelly, R. and Lewis, P.E.T. (2003), The New Economy and Demand for Skills, *Australian Journal of Labour Economics*, Vol. 6, No. 1, pp. 135–152.
8. Pappas, N. (1998), Changes in the Demand for Skilled Labour in Australia, in *Working for the Future: Technology and Employment in the Global Knowledge Economy*, P. Sheehan and G. Teggart (eds), Victoria University Press, Melbourne.
9. United States Department of Labor (2000), *Dictionary of Occupational Titles, 4th Ed.* — Appendix B.
(accessed online 19 September, 2000:
`http://www.oalj.dol.gov/public/dot/refrnc/dotappb.htm`)

Some Comments about Issai Schur (1875–1941) and the Early History of Schur Complements

Simo Puntanen

Department of Mathematics, Statistics and Philosophy
FI-33014 University of Tampere, Finland
E-mail: sjp@uta.fi

George P. H. Styan

Department of Mathematics and Statistics, McGill University
Burnside Hall Room 1005
805 rue Sherbrooke Street West
Montréal (Québec), Canada H3A 2K6
E-mail: styan@math.mcgill.ca

After some biographical remarks about Issai Schur (1875–1941), we note that he attended the Nicolai Gymnasium in Libau (Kurland) from 1888 to 1894 and suggest that the adjectival noun Nicolai is used here in honour of Saint Nicholas of Bari, Bishop of Myra, who is widely associated with Christmas and after whom Santa Claus is named. We also comment on the fact that Issai Schur published under I. Schur and under J. Schur and propose an explanation for this.

The term *Schur complement* was introduced in 1968 by Emilie Virginia Haynsworth (1916–1985) in view of a lemma (Hilfssatz) in the 1917 paper[132] by Issai Schur. We continue with some useful early results involving Schur complements by Alexander Craig Aitken (1895–1967), Tadeusz Banachiewicz (1882–1954) and William Jolly Duncan (1894–1960), along with some biographical remarks. The article ends with proofs of the Cauchy–Schwarz Inequality and the Frisch–Waugh–Lovell and Gauß–Markov Theorems using Schur complements.

Keywords: Aitken block-diagonalisation formula; Banachiewicz inversion formula; biography; Duncan inversion formula; Gothic alphabet; Guttman rank-additivity formula; Haynsworth inertia-additivity formula; myrrh; Old Cemetery in Tel Aviv; old postcards; Schur determinant lemma.

AMS Classification: 01A55, 01A60, 01A70, 15A09, 15A15, 15A99, 19C09, 20Cxx, 26Dxx, 62J05.

1. Issai Schur (1875–1941)

Joseph, Melnikov & Rentschler dedicate their recent book entitled *Studies in Memory of Issai Schur*[84] to the "Memory of the Great Mathematician Issai Schur" and as Joseph & Melnikov[84] point out, Issai Schur "is popularly known for his lemma on Group Representations[a] familiar to mathematicians, physicists and chemists alike. It is an essential component of even the most elementary course on the subject ... Yet for all its celebrity it remains one of the most modest of Schur's achievements." Our interest in Issai Schur, however, was prompted by the linear-algebraic concept known as the *Schur complement.*

Fig. 1. Issai Schur *c.* 1917.

In 1968 Emilie Virginia Haynsworth (1916–1985) called the matrix

$$\mathbf{H} = \mathbf{S} - \mathbf{R}\mathbf{P}^{-1}\mathbf{Q}$$

the *Schur complement* of the invertible matrix $\mathbf{P}$ in the block-partitioned matrix

$$\mathbf{G} = \begin{pmatrix} \mathbf{P} & \mathbf{Q} \\ \mathbf{R} & \mathbf{S} \end{pmatrix}$$

in two papers[72,73] published, respectively, in the *Basel Mathematical Notes*[b] and in *Linear Algebra and its Applications.* "Schur" in "Schur complement" was chosen by Haynsworth in view of a lemma (Hilfssatz) in the 1917 paper[132] by Issai Schur published in the *Journal für die reine und ange-*

---

[a]In German: Darstellungstheorie, see, e.g., Refs. 18, 41, 98, 135, and 150.

[b]The *Basel Mathematical Notes* (BMN) reported research through the support and sponsorship of the U.S. Department of Army with the Institute of Mathematics, University of Basle (Universität Basel), in Basel, Switzerland. (Basle is the English and an older name for Basel, or Bâle in French.) Copies were distributed by the Clearinghouse for Federal Scientific & Technical Information, Springfield, Virginia, USA. The only library which has copies of the BMN seems to be the Zürich Zentralbibliothek (Zürich, Switzerland), where the serial title is catalogued with the spelling *Basel* and the holdings are BMN 1–BMN 50 (1960–1977), apparently a complete set. The editor is given as A. M. Ostrowski and the publisher, the Mathematisches Institut der Universität Basel. Issues numbered BMN 3, 4, 5, 18, 20, 21, and 22 have *Basle Mathematical Notes* on the cover page, all the others have *Basel Mathematical Notes* on the cover page.

wandte Mathematik[c] in which he showed that when **P** and **R** commute, then $\det \mathbf{G} = \det(\mathbf{PS} - \mathbf{RQ})$; here **P** need not be invertible. As Chandler & Magnus[34] point out, "The coining of new technical terms is an absolute necessity for the evolution of mathematics."

Issai Schur was born in Mogilev on the Dnieper, Russia, on 10 January 1875,[26,27] and died in Tel Aviv, Palestine, on his 66th birthday 10 January 1941.[26,27] In October 1894, Schur enrolled in the University of Berlin, studying mathematics and physics; on 27 November 1901 he passed his doctoral examination *summa cum laude* with the thesis entitled "Über eine Klasse von Matrizen, die sich einer gegebenen Matrix zuordnen lassen":[131] his thesis adviser was Ferdinand Georg Frobenius (1849–1917). According to Vogt,[162] in this thesis Schur used his first name "Issai" for the first time.[d]

Feeling that he "had no chance whatsoever of sustaining himself as a mathematician in czarist Russia"[34] and since he now wrote and spoke German so perfectly that one would guess that German was his native language, Schur stayed on in Germany. According to Ref. 134, he was *Privatdozent* at the University of Berlin from 1903 till 1913, and *ausserordentlicher Professor* (associate professor) at the University of Bonn from 21 April 1913 till 1 April 1916,[146] as successor to Felix Hausdorff (1868–1942). In 1916 Schur returned to Berlin where in 1919 he was appointed full professor; in 1922 he was elected a member of the Prussian Academy of Sciences to fill the vacancy following the death of Frobenius in 1917. We believe that our portrait of Issai Schur in Figure 1 above was made at the Atelier[e] Hanni Schwarz, N.W. Dorotheenstraße 73 (in Berlin), *c.* 1917, the year in which Schur's seminal determinantal lemma was published.

Schur lived in Berlin as a highly respected member of the academic community and was a quiet unassuming scholar who took no part in the fierce struggles that preceded the downfall of the Weimar Republic. "A leading mathematician and an outstanding and highly successful teacher, [Schur] occupied the very prestigious chair at the University of Berlin for 16 years".[34] Until 1933 Schur's algebraic school at the University of Berlin was, without any doubt, the single most coherent and influential group of

[c] Also known as "Crelle's Journal" after August Leopold Crelle (1780–1855), who founded the *Journal für die reine und angewandte Mathematik* in 1826 and edited it until his death in 1855; see Frei.[61]

[d] Apparently Issai Schur used the first name "Schaia" rather than "Issai" until his mid-20s.[162]

[e] The caption in the photograph in Figure 1 has "Atelieir", which we suppose should be "Atelier" (same spelling in German and in English), meaning "an artist's or designer's studio or workroom".[103]

mathematicians in Berlin and among the most important in all of Germany. With Schur as its charismatic leader, the school centred around his research on group representations, which was extended by his students in various directions (soluble groups, combinatorics, matrix theory).[31] "Schur made fundamental contributions to algebra and group theory which, according to Hermann Weyl (1885–1955), were comparable in scope and depth to those of Emmy Amalie Noether (1882–1935)".[112]

When Schur's lectures were cancelled in 1933 there was an outcry among the students and professors for he was respected and very well liked.[31] Thanks to his colleague Erhard Schmidt (1876–1959), Schur was able to continue his lectures till the end of September 1935,[112] Schur being the last Jewish professor to lose his job at the University of Berlin at that time.[146] Schur's "lectures on number theory, algebra, group theory and the theory of invariants" attracted large audiences. On 10 January 1935 some of the senior postgraduates congratulated Schur in the lecture theatre on his sixtieth birthday. Replying in mathematical language, Schur said he "hoped that the good relationship between himself and his student audience would remain invariant under all the transformations to come".[112]

Indeed Schur was a superb lecturer. His lectures were meticulously prepared and were exceedingly popular. Walter Ledermann (b. 1911) remembers attending Schur's algebra course which was held in a lecture theatre filled with about 400 students:[93] "Sometimes, when I had to be content with a seat at the back of the lecture theatre, I used a pair of opera glasses to get a glimpse of the speaker." In 1938 Schur was pressed to resign from the Prussian Academy of Sciences and on 7 April 1938 he resigned "voluntarily" from the Commissions of the Academy. Half a year later, he had to resign from the Academy altogether.[31]

The names of the 22 persons who completed their dissertations from 1917 to 1936 under Schur, together with the date in which the Ph.D. degree was awarded and the dissertation title, are listed in the *Issai Schur Gesammelte Abhandlungen*;[28] see also Refs. 31 and 84. One of these 22 persons was Alfred Theodor Brauer (1894–1985), who completed his Ph.D. dissertation under Schur on 19 December 1928 and who with Hans Rohrbach edited the *Issai Schur Gesammelte Abhandlungen*.[28] Alfred Brauer was a faculty member in the Department of Mathematics at The University of North Carolina at Chapel Hill for 24 years and directed 21 Ph.D. dissertations,[f] including that of Emilie Haynsworth, who in 1968 introduced the

[f]The Mathematics Genealogy Project website[38] indicates that Alfred Brauer has 20

term "Schur complement".

A comment by Alfred Brauer,[26,27] see also Ref. 31, sheds light on Schur's situation after he finally left Germany in 1939: "When Schur could not sleep at night, he read the *Jahrbuch über die Fortschritte der Mathematik.*[g] When he came to Tel Aviv (then in the British Mandate of Palestine) and for financial reasons offered his library for sale to the Institute for Advanced Study in Princeton, he finally excluded the *Jahrbuch* in a telegram only weeks before his death."

2. Nicolai Gymnasium, Libau (Kurland), and Riga Technical University at Liepāja (Latvia)

Issai Schur was born on 10 January 1875, the son of Golde Schur (née Landau) and the *Kaufmann* Moses Schur, according to Schur's *Biographische Mitteilungen.*[134] Brauer[26] identifies Moses Schur as a *Großkaufmann* — in English: wholesale merchant.[27]

Writing in German in his *Biographische Mitteilungen*,[134] Schur gives his place of birth as Mohilew am Dnjepr (Russland) — in English: Mogilev on the Dnieper, Russia. Founded in the 13th century, Mogilev changed hands frequently among Lithuania, Poland, Sweden and Russia, and was finally annexed to Russia in 1772 in the first partition of Poland.[12] By the late 19th century, almost half of the population of Mogilev was Jewish.[80] About 200 km east of Minsk, Mogilev is in the eastern part of the country now known as Belarus (Belorussia, White Russia) and called Mahilyow in (transliterated) Belarusian.[104]

From Brauer,[26,27] see also Refs. 84, 109, and 146, we learn that in 1888 when he was 13, Schaia Schur, as he was then known,[162] went to live with his older sister and brother-in-law in Libau (Kurland), about 640 km northwest of Mogilev. Also founded in the 13th century, Libau (Liepāja in Latvian or Lettish) is on the Baltic coast of what is now Latvia[h] in the region of Courland (Kurland in German, Kurzeme in Latvian), which from 1562 to 1795 was a semi-independent duchy linked to Poland but with a prevailing German influence.[21,145] Indeed the German way of life was dominant in Courland in 1888, with mostly German (rather than Yiddish) being the

(rather than 21) students and 87 descendants and that Issai Schur has 25 (rather than 22) students and 1209 descendants.

[g]Published from 1871 to 1944 and then merged into the *Zentralblatt für Mathematik und ihre Grenzgebiete*, now the *Zentralblatt MATH* website `www.emis.de/ZMATH/`

[h]The Baltic republic of Latvia lies between Estonia (on the north), Russia (on the east), Belarus (on the southeast) and Lithuania (on the south).

spoken language of the Jewish community until 1939.[14] At that time there were many synagogues in Libau, the Great Synagogue in Babylonian style with three cupolas being a landmark.[21]

According to his *Biographische Mitteilungen*,[134] Schur attended the German-language "Nicolai Gymnasium in Libau (Kurland)", apparently from 1888 to 1894, and at graduation he received the highest mark on the final examination and a gold medal.[84,109,134,146,162] It was in the German-language Nicolai Gymnasium in Libau that he became fluent in German (we believe that his first language was probably Yiddish). In Germany the Gymnasium is a "state-maintained secondary school that prepares pupils for higher academic education".[54]

We have found three postcards that depict the Nicolai Gymnasium in Libau in the early 20th century and have purchased two[i] of these postcards from Bartko-Reher-GbR in Berlin in March 2005. The first of the two Bartko postcards, see Figure 2 below, has the heading "LIBAU — Nikolai-Gymnasium" (the other two postcards use the spelling Nicolai rather than Nikolai) and was issued by the "Verlag Typ. Victoria, W. Sarchi, Libau" and sent through the German army postal service (Feldpost) in October 1916[j] by "Ldstr. Hugo Hofmann I König. I Bat. Landst. I.R. 19. [unidentified word] 6. Res. Div. im Osten" to "Herrn Paul Schölzel Bäkerei [Bäckerei], Schönbach b/Sebnitz i. Sa."[k]

The second Bartko postcard has the heading "Nicolai-Gymnasium Libau" (and also in Russian) and was issued by A. Dunkert, Libau/Otto Henjes & Co., Hannover, but was never mailed. The Muser postcard, which has the heading "Gruss aus Libau: Nicolai Gymnasium" was mailed to St. Petersburg (apparently from Libau in December 1909, by "Union Postale Universelle Russie" with a Russian postage stamp).

---

[i]The third postcard was available from Catherine Muser on the `heimatsammlung.de` website on 14 February 2004 but seems not to be available on 24 January 2006.

[j]Written on 15 October 1916, and postmarked "18 October 1916: I. Bataillon Landsturm-Inf-Rgt. Nr. 19" and "K.B. Feldpostexp. der 6. Res. Div." We believe that "Ldstr." is an abbreviation for "Landstürmer" or member of the veteran reserve (Landsturm) and that "I König. I Bat. Landst. I.R. 19" is an abbreviation for "I Königliche I Batallion, Landsturm-Infanterie-Regiment Nr. 19". The unidentified word may be some word that qualifies the "Landsturm-Infanterie-Regiment Nr. 19". Moreover, "6. Res. Div. im Osten" is an abbreviation for "6. Reserve Division im Osten", with "im Osten" here apparently meaning in the German-occupied territory "in the east" (we suppose that the postcard was sent from Libau, which was then in German-occupied Courland).

[k]Schönbach bei Sebnitz im Sachsen is in the part of Germany known as the Sächsische Schweiz (Saxon Switzerland) near Dresden and the border today with the Czech Republic (in 1916 the border with Austria–Hungary).

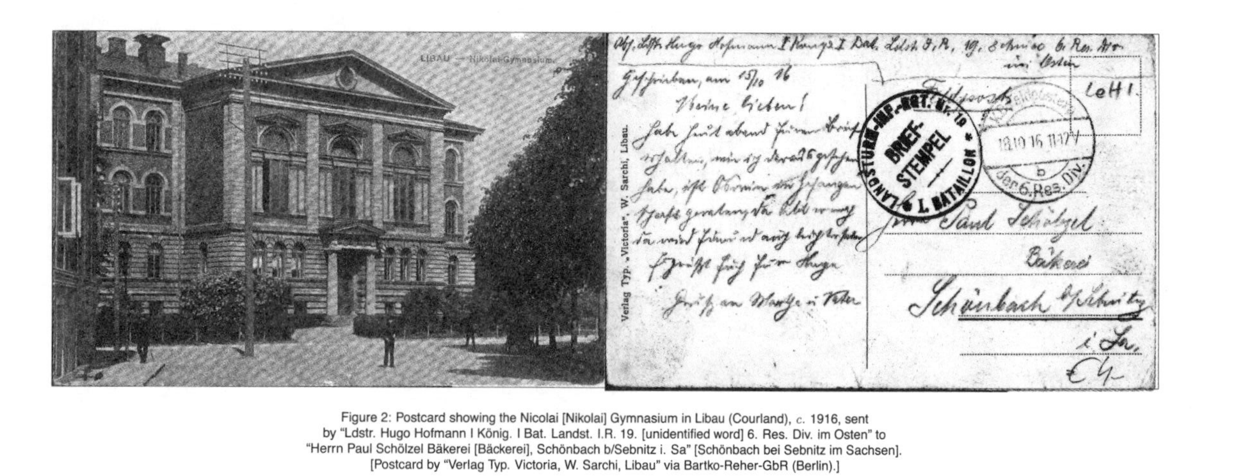

Figure 2: Postcard showing the Nicolai [Nikolai] Gymnasium in Libau (Courland), c. 1916, sent by "Ldstr. Hugo Hofmann I König. I Bat. Landst. I.R. 19. [unidentified word] 6. Res. Div. im Osten" to "Herrn Paul Schölzel Bäkerei [Bäckerei], Schönbach b/Sebnitz i. Sa" [Schönbach bei Sebnitz im Sachsen]. [Postcard by "Verlag Typ. Victoria, W. Sarchi, Libau" via Bartko-Reher-GbR (Berlin).]

Figure 3: (left panels) Rīgas Tehniskā Universitāte Liepājas Filiāle on Krišjānis Valdemārs Street, Liepāja, Latvia, May 2004, with (left upper panel) plaque on the building façade for Gabriel Narutowicz (1865–1922) and (right) his portrait.

In May 2004 our friend Timo Mäkeläinen (University of Helsinki) and his son Kari Tapio Mäkeläinen visited Liepāja and found the building that used to house the Nicolai Gymnasium on (what is now called) Krišjānis Valdemārs Street and that it now (see Figure 3 above) houses the Rīgas Tehniskā Universtāte Liepājas Filiāle (Riga Technical University, Liepāja branch). The façade of this building is embellished with two memorial plaques, one of which, see our Figure 3 above, is in honour of Gabriel Narutowicz (1865–1922), who studied at the Nicolai Gymnasium in Libau from 1873 to 1883; Schur studied there from 1888 to 1894.[1]

From the *Wikipedia* website[163] we learn that Gabriel Narutowicz had been a professor at the Zürich Polytechnic from 1908 and had directed the construction of many hydroelectric plants in western Europe. After Poland regained independence in 1918, Narutowicz became involved in national politics and served as minister of public works, 1920–1921, and as minister of foreign affairs in 1922. On December 9, 1922, Narutowicz was elected by the Polish parliament as the first president of Poland. He was sworn in on December 11, 1922.

Gabriel Narutowicz was a sympathiser (though not a member) of the "Liberation" peasant party, the more radical of the peasant parties, and so was considered a leftist. Since he was elected by left, center, peasant, and minorities deputies, the right, particularly the National Democrats, were strongly opposed to him. On December 16, 1922, five days after his inauguration, while attending the opening of an art exhibition at Warsaw's Zacheta Gallery, Gabriel Narutowicz was shot to death by a National Democrat sympathiser, painter, art professor, and critic, Eligiusz Niewiadomski (1869–1923), for which he was sentenced to death and executed. So Poland lost not only its first President but also a passionate art professor who had strong political interests.

Although Schur was an outstanding student excelling in mathematics, he pursued a much quieter career path than Narutowicz, but unlike Narutowicz, Schur apparently had no interest in politics.

3. Nicolai Gymnasium, Libau (Kurland), and Saint Nicholas of Bari, Bishop of Myra

We do not know why the adjectival noun Nicolai was used for the Nicolai Gymnasium in Libau but we do know that there are apparently more buildings dedicated to the Saint Nicholas who is also known as Saint Nicholas

[1]There is apparently no memorial plaque in this building for Issai Schur.

of Bari, Bishop[m] of Myra, than to any other Nicholas. This Saint Nicholas is widely associated with Christmas and it is after him that Santa Claus is named; see, e.g., the detailed scholarly book entitled *Saint Nicholas of Myra, Bari, and Manhattan: Biography of a Legend* by Charles W. Jones.[82] "Holy St. Nicholas is also a favorite subject of iconography throughout art history";[78] see, e.g., Figure 4. According to Fournier[59] Saint Nicholas is regarded as the "special patron of children" and Jones[82] says he is the "patron of all scholars". We suggest, therefore, that the adjectival noun Nicolai used for the Nicolai Gymnasium in Libau was done so in honour of Saint Nicholas of Bari, Bishop of Myra.

Jones[82] notes that "Professor [Karl] Meisen[102] of Bonn mapped the public monuments dedicated to Saint Nicholas erected before the year 1500 only in France, Germany, and the Low Countries. He tabulated 2,137 monuments. Yet his list was notably incomplete. Professor [Heinrich] Börsting[23] immediately named 50 in the single diocese of Münster, where Meisen listed but 9." Jones[82] finds it "likely that all occurrences of the name Nicholas from the time of the Roman Emperor Justinian (527–565 AD) find an inspiration, direct or indirect, in Saint Nicholas."

There is a *Nikolaischule* in Leipzig, Germany. In fact in July 1653 the mathematician and philosopher Gottfried Wilhelm von Leibniz (1646–1716) "entered the Nicolai School in Leipzig where he remained until Easter 1661";[6] see also Ref. 57.[n] According to Hocquél-Schneider[77] the *Alte Nikolaischule Leipzig* dates from 1395, with the initial record[o] for a *Privatschule* under the name *Schola Nicolaitana* in 1490. The picture[p] of the Nicolai School in Leipzig in 1553[77] shows the "Niclas Schule" adjacent to the "Kirche zu Sanct Niclas" (now known as the *Nikolaikirche*), which was founded in about 1165.[q,71] It seems very clear, therefore, that the Nikolaischule in Leipzig is named[77] after Saint Nicholas.

The legend of Saint Nicholas has evolved from the Bishop Nicholas who

---

[m]Or Archbishop; see, e.g., *The Book of Saints*.[22]

[n]The article by Farebrother, Styan & Tee[57] includes images of the 8 postage stamps issued in honour of Leibniz; see also Miller[106] and Wilson.[169]

[o]In German: Aktenkundliche Erwähnung.

[p]"Kupferstich [copperplate engraving] von Christian Romsted, 1702."

[q]The first church to ber named after Saint Nicholas seems to be the church built by Emperor Justinian in Constantinople (now Istanbul, Turkey) in the suburb of Blachernae (or Blaquernae), *c.* 430 AD.[51,55,59] Jones[82] observes that a Church of Saint Nicholas was built almost 300 years earlier than that in Leipzig, in 882 in Kiev (Kyïv), Ukraine, and from the *Lonely Planet Guide*,[168] we find that there is a "magnificent St. Nicholas' Russian Orthodox Cathedral (1900–1903)" in Libau, apparently located about 5 km north of the building which housed the Nicolai Gymnasium.

was born in Patara and became Bishop of Myra at the beginning of the 4th century AD. Myra and Patara were then the main cities of the ancient district of Lycia in Asia Minor, now in southwestern Turkey.[r]

The exact years of the birth and death of Bishop Nicholas of Myra are not known. For the year of his birth, e.g., the St. Nicholas Center website[149] says "between 260 and 280" while Nes[108] gives "between 240 and 245" and *Wikipedia*[163] *c.* 270. For the year of his death, Cruz[40] says "Traditions are unanimous ... that St. Nicholas died in 342" but Nes[108] says "probably in 326", while *Wikipedia*[163] gives 343 and 345/352. Jones[82] says that according to *The Golden Legend*,[65] Bishop Nicholas died in 343 AD, but also mentions as possibilities 287, 312, 341, 342, 345, 351 and says that "Réau's *Iconographie*[124] unhesitatingly says 342, but papal commissions of modern times have stated that no date is certain."

Fig. 4. Handmade Byzantine icon of Saint Nicholas (from Crete).

Saint Nicholas is known as "Saint Nicholas of Bari", since "a shipload of Italian sailors rescued (or stole) the bones of St. Nicholas and took them to Bari, Italy, arriving on 9 May 1087";[59] see also Cruz.[40] A postage stamp commemorating this event was issued by Vatican City in 1987 (see Figure 5[s]), and a pilgrimage church was erected on the site in Bari and at this shrine in Bari, a sweet smell is often reported.

This smell may be of myrrh:[t] "an aromatic, bitter-tasting gum resin obtained principally from a small thorny tree, *Commiphora myrrha* (family *Burseraceae*) (or dindin tree).[46,49,120] The Egyptians used myrrh in em-

---

[r]The small adjacent towns of Demre (or Dembre) and Kale on the Mediterranean coast of Turkey (about 50 km east of Kaş and 150 km southwest of Antalya) now occupy the site of Myra, but "many Lycian rock tombs, some dating from the 4th century" remain; part of the ancient city of Patara, also located on the Turkish Mediterranean coast (but about 50 km west of Kaş and so about 100 km west of Myra), remains today with a splendid 20 km long beach and with the 2nd century theatre excellently preserved and partly buried in sand.[113]

[s]For other postage stamps depicting Saint Nicholas, see Klimchalk.[89,90]

[t]From Latin: *myrrha*, Middle English: *myrre*, Arabic: *murr*, meaning "bitter".[103] Cruz[40] says that the smell comes from "The oil that exuded from the saint's bones ... has been called unction, myrrh, medicinal liquor, balm, manna or ... distilled bone oil ... a combination of hydrogen and oxygen, and because of its extremely low content of bacteria the product was declared biologically pure."

balming, filling body cavities with powdered myrrh; and, along with frankincense and gold, it was a gift of the Magi to the infant Jesus".[46] The name of the Lycian city Myra may be derived from myrrh.[29]

Fig. 5. Saint Nicholas of Bari: Vatican City 1987 (Scott #803).

We find it interesting that there was and possibly still is a Nicolai Gymnasium in nearby Reval, Russia (now Tallinn, Estonia,[u]) with a library, in which in 1893 was found, according to Coomber,[36,37] the only surviving copy of the 1547 English translation of the seminal book[172,v] on accounting and bookkeeping. Moreover, from the *Lonely Planet Guide*,[168] we learn that there is also a St. Nicholas Church (Niguliste Kirik) in Tallinn, and that it was named after "St. Nicholas of Bari".

In view of all this, we conclude that it is very probable that the adjectival noun Nicolai was used for the Nicolai Gymnasium in Libau in honour of Saint Nicholas of Bari, Bishop of Myra.

4. Publications under ℑ. 𝔖𝔠𝔥𝔲𝔯 and 𝔍. 𝔖𝔠𝔥𝔲𝔯

Issai Schur published under "I. Schur" (ℑ. 𝔖𝔠𝔥𝔲𝔯) and under "J. Schur" (𝔍. 𝔖𝔠𝔥𝔲𝔯). As is pointed out by Ledermann in his biographical article[93] on Schur, this has caused some confusion: "For example I have a scholarly work on analysis which lists amongst the authors cited both J. Schur and I. Schur, and an author on number theory attributes one of the key results to I. J. Schur."

We have identified 81 publications by Issai Schur which were published before he died in 1941; several further publications by Schur were, however, published posthumously, including the book[136] published in 1968. On the title page of (the original versions of) the articles 132 and 133, the author is given as "J. Schur"; indeed for all but one of the other 11 papers by Issai

[u]The Baltic republic of Estonia lies between Latvia (to the south), Russia (to the east), and the Gulf of Finland (to the north).

[v]Apparently first written in Italian by Giovanni Paolo di Bianchi, this book was first translated into Flemish by the Antwerp merchant into French by his widow Anne Ympyn Christoffels (née Swinters), and finally into English; see also Gordon,[66] Kats,[85] and Yamey & Kojima.[171] This English translation was removed from Reval in 1917 and "taken to Luckyanov" in Nijni Novgorod (about 400 km east of Moscow) and then to the Lenin Library in Moscow. It was later microfilmed and a copy deposited in the British Museum.

Schur that we found published in the *Journal für die reine und angewandte Mathematik* the author is given as "J. Schur". Brauer[27] observes that "In the first of these papers Schur's first name had been abbreviated wrongly and Schur considered it proper not to change this for the subsequent papers in the journal. But in fact he always attached great importance to being quoted as I. Schur. Nevertheless even today his first name is sometimes wrongly abbreviated."

For the lecture notes[135] published in Zürich in 1936, the author is given as J. Schur on the title page[w] and so cited in the preface. For all other publications by Issai Schur that we have found, however, the author is given as "I. Schur", and posthumously as "Issai Schur"; moreover Schur edited the *Mathematische Zeitschrift* from 1917 to 1938 and he is listed there on the journal title pages as I. Schur.

In the complete German Gothic alphabets from Bentz[17] and *Ahn's Second German Book* by Peter Henn,[76] we see, the "I" and "J" are very similar but not quite the same. In the Gothic Print Alphabet given in Ref. 153, the "I" and "J" are almost the same except that the vertical part of the "J" is slightly longer than that of the "I"; see Figure 6.

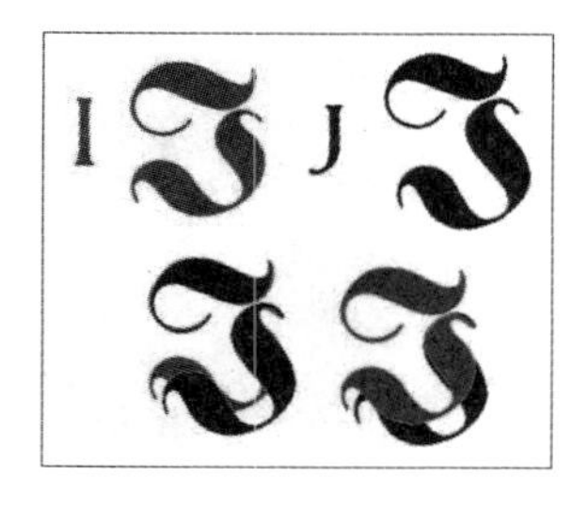

Fig. 6. Gothic "I" and "J".

The confusion here between "I" and "J" probably stems from there being two major styles of writing German: *Fraktur script*, also known as *black letter script* or *Gothic script*, in use since the ninth century and prevailing until 1941,[39] and *Roman* or *Latin*, which is common today.[79] According to Mashey,[99] "it is a defect of most styles of German type that the same character $\mathfrak{I}$ is used for the capitals I (i) and J (j)"; when followed by a vowel it is the consonant "J" and when followed by a consonant, it is "I", see also Refs. 17, 76, and 160.

The way Schur wrote and signed his name, as in his *Biographische Mitteilungen*[134] (see also Figure 1 above), his first name could easily be interpreted as "Jssai" rather than "Issai"; see also the signature at the bottom of the photograph in the *Issai Schur Gesammelte Abhandlungen*.[28]

The official letter, reprinted in Soifer,[146] dated 28 September 1935 and signed by Kunisch,[92] relieving Issai Schur of his duties at the University of Berlin, is addressed to "Jssai Schur"; the second paragraph starts with "Jch

[w]See also Stammbach.[150]

übersende Jhnen ..." which would now be written as "Ich übersende Ihnen ..."; see also Ref. 84. Included in the article by Ledermann & Neumann[94] are copies of many documents associated with Issai Schur. These are presented there in chronological order, with a transcription first, followed by a translation. It is noted there[94] that "Schur used Roman script" but "sometimes, particularly in typed official letters after 1933, initial letters I are rendered as J."

5. The Old Cemetery in Tel Aviv

Fig. 7. The gravestone of Issai Schur and Regina Schur in The Old Cemetery, Tel Aviv, Israel, December 2003.

Issai Schur died of a heart attack in Tel Aviv on his 66th birthday, 10 January 1941,[26,27] and is buried in Tel Aviv in the Old Cemetery[x] on Trumpeldor Street,[y] which was "reserved for the Founders' families and persons of special note". On Schur's tombstone in Tel Aviv, see Figure 7,[z] the lettering is entirely in Hebrew. The first two lines on the gravestone translate as "Issai[aa] Schur, Professor of Mathematics" and the third line as "Regina Schur".

The dates at the bottom of the gravestone are the corresponding birth and death dates in the Hebrew calendar for Issai Schur (left-hand side): 4th of Sh'vat,

---

[x]For aerial photographs of the cemetery taken in 1917 and 1997, see Kedar.[88]

[y]Trumpeldor Street in Tel Aviv is named after the soldier and early pioneer-settler Joseph Trumpeldor (1880–1920), who joined the Russian army in 1902 and served in the Russian–Japanese war, losing his left arm and being taken prisoner; he received a high Tsarist order of merit for his gallantry and zeal. In 1912 he settled in "Eretz Yisrael" but following the outbreak of the First World War and his refusal to take Ottoman citizenship, Trumpeldor was expelled from the country and joined the Allied war effort. He was a founder of the Zion Mule Corps in 1915. In 1918 he established He-Halutz, the pioneering youth organisation that prepared youngsters for settlement in Eretz Yisrael. Following his return to Eretz Yisrael and his involvement in the defense of Tel Hai (a settlement in the Galilee) against the Arabs, he was killed together with seven other defenders and it is claimed that as he lay on his death bed, one of his final utterances was, "Never mind, it is good to die for one's country". The town of Kiryat Shmona ("City of Eight") is named after Trumpeldor and the seven others who died defending Tel Hai.[163]

[z]and also Refs. 31, 84, 88, and 107.

[aa]The first name on the gravestone transliterates as "Yisha'yahu" or "Ishayahu".

5635 [10 January 1875] and 12th of Tevet, 5701 [11 January 1941] and for Regina Schur (right-hand side): 8th of Sh'vat, 5641 [8 January 1881] and 20th of Adar I, 5725 [22 February 1965]. The dates in the Gregorian calendar appearing here in square brackets are computed using the Hebrew Date Converter website;[123] see also Ref. 125. The Hebrew calendar starts at sunset (on the previous day) in the Gregorian calendar and so the 12th of Tevet, 5701, the day in the Hebrew calendar on which Issai Schur died started at sunset on 10 January 1941 (Schur's 66th birthday).

Issai Schur was survived by his wife, medical doctor Regina (née Frumkin), their son Georg (born 1907 and named after Frobenius), and daughter Hilde (born 1911, later Hilda Abelin-Schur), who in "A story about father"[1] writes:

> One day when our family was having tea with some friends, [my father] was enthusiastically talking about his work. He said: "I feel like I am somehow moving through outer space. A particular idea leads me to a nearby star on which I decide to land. Upon my arrival I realize that somebody already lives there. Am I disappointed? Of course not. The inhabitant and I are cordially welcoming each other, and we are happy about our common discovery." This was typical of my father; he was never envious.

6. The Schur Determinant Lemma: 1917, and Emilie Virginia Haynsworth (1916–1985)

The adjectival noun "Schur" in "Schur complement" was chosen[ab] by Emilie Haynsworth because of the following lemma (Hilfssatz), first published in the paper[132] by Issai Schur in 1917.

Lemma. *Let* $\mathbf{P}$, $\mathbf{Q}$, $\mathbf{R}$ *and* $\mathbf{S}$ *denote four* $n \times n$ *matrices and suppose that* $\mathbf{P}$ *and* $\mathbf{R}$ *commute. Then the determinant* $\det \mathbf{G}$ *of the* $2n \times 2n$ *matrix*

$$\mathbf{G} = \begin{pmatrix} \mathbf{P} & \mathbf{Q} \\ \mathbf{R} & \mathbf{S} \end{pmatrix} \tag{1}$$

is equal to the determinant of the matrix $\mathbf{PS} - \mathbf{RQ}$.

This lemma, which we refer to as the *Schur determinant lemma*, appears in Ref. 132; see also Refs. 28, 63, and 117. In general when the submatrix $\mathbf{P}$

[ab]The Schur complement may just be a counterexample to Stigler's Law of Eponymy,[152] which "in its simplest form" states that "no scientific discovery is named after its original discoverer".[152]

is nonsingular, then the *Schur complement* of $\mathbf{P}$ in the partitioned matrix $\mathbf{G}$ in (1) is defined to be[ac]

$$\mathbf{G}/\mathbf{P} = \mathbf{S} - \mathbf{R}\mathbf{P}^{-1}\mathbf{Q} \tag{2}$$

and when the submatrix $\mathbf{S}$ is nonsingular, then the Schur complement of $\mathbf{S}$ in the partitioned matrix $\mathbf{G}$ is defined to be

$$\mathbf{G}/\mathbf{S} = \mathbf{P} - \mathbf{Q}\mathbf{S}^{-1}\mathbf{R}. \tag{3}$$

The Schur complement $\mathbf{G}/\mathbf{P} = \mathbf{S} - \mathbf{R}\mathbf{P}^{-1}\mathbf{Q}$ is used in the proof of the Schur determinant lemma but the lemma holds even if the square matrix $\mathbf{P}$ is singular. Moreover, the Schur complements defined in (2) and (3) remain well defined even when none of the matrices $\mathbf{Q}$, $\mathbf{R}$ and $\mathbf{S}$ are square (but they must be conformable so that $\mathbf{G}$ is well defined).

Fig. 8. Emilie Virginia Haynsworth *c.* 1968.

Emilie Virginia Haynsworth[ad] was born on 1 June 1916 and died on 4 May 1985, both at home in Sumter, South Carolina. As observed in the obituary article[32] by Carlson, Markham & Uhlig, "In her family there have been Virginia Emilies or Emilie Virginias for over 200 years. From childhood on, Emilie had a strong and independent mind, so that her intellectual pursuits soon gained her the respect and awe of all her relatives and friends". Throughout her life Emilie Haynsworth was eager to discuss any issue whatsoever.

From Carlson, Markham & Uhlig,[32] we quote Philip J. Davis (b. 1923): "She was a strong mixture of the traditional and the unconventional and for years I could not tell beforehand on what side of the line she would locate a given action". In *The Education of a Mathematician*,[45] Davis observes that Emilie Haynsworth

[ac]We may also write the Schur complements $(\mathbf{G}/\mathbf{P})$ and $(\mathbf{G}/\mathbf{S})$ with parentheses. For further remarks concerning the notation used for the Schur complement, see Puntanen & Styan.[117]

[ad]Our portrait here of Emilie Haynsworth is reproduced, with permission, from the Auburn University website.[10] See also Davis[45] and Zhang.[174] We conjecture that the photograph was taken *c.* 1968, the year in which the term Schur complement was introduced by Haynsworth.[72,73]

"had a fine sense of mathematical elegance — a quality not easily defined. Her research can be found in a number of books on advanced matrix theory under the topic: 'Schur complement'. Emilie taught me many things about matrix theory."

In 1952 Emilie Haynsworth received her Ph.D. degree in mathematics at The University of North Carolina at Chapel Hill with Alfred Brauer as her dissertation adviser. We recall that Issai Schur was Alfred Brauer's Ph.D. dissertation adviser and that the topic of Haynsworth's dissertation was determinantal bounds for diagonally dominant matrices.

From 1960 until retirement in 1983, Emilie Haynsworth taught at Auburn University (Auburn, Alabama) "with a dedication which honors the teaching profession"[32] and supervised 18 Ph.D. students.[ae] The mathematician Alexander Markowich Ostrowski (1893–1986), with whom Haynsworth co-authored the paper[74,af] on the inertia formula for the apparently not-then-yet-publicly-named Schur complement, wrote the following upon her death: "I lost a very good, life-long friend and mathematics [lost] an excellent scientist. I remember how on many occasions I had to admire the way in which she found a formulation of absolute originality."

7. Aitken Block-Diagonalisation Formulas: 1939, and Alexander Craig Aitken (1895–1967)

The Schur complement is a very useful matrix function; see, e.g., Zhang.[174] In particular, we find the *Aitken block-diagonalisation formulas* to be particularly useful:[2,117]

$$\begin{pmatrix} \mathbf{I} & \mathbf{0} \\ -\mathbf{R}\mathbf{P}^{-1} & \mathbf{I} \end{pmatrix} \begin{pmatrix} \mathbf{P} & \mathbf{Q} \\ \mathbf{R} & \mathbf{S} \end{pmatrix} \begin{pmatrix} \mathbf{I} & -\mathbf{P}^{-1}\mathbf{Q} \\ \mathbf{0} & \mathbf{I} \end{pmatrix} = \begin{pmatrix} \mathbf{P} & \mathbf{0} \\ \mathbf{0} & \mathbf{G}/\mathbf{P} \end{pmatrix} \tag{4}$$

and

$$\begin{pmatrix} \mathbf{I} & -\mathbf{Q}\mathbf{S}^{-1} \\ \mathbf{0} & \mathbf{I} \end{pmatrix} \begin{pmatrix} \mathbf{P} & \mathbf{Q} \\ \mathbf{R} & \mathbf{S} \end{pmatrix} \begin{pmatrix} \mathbf{I} & \mathbf{0} \\ -\mathbf{S}^{-1}\mathbf{R} & \mathbf{I} \end{pmatrix} = \begin{pmatrix} \mathbf{G}/\mathbf{S} & \mathbf{0} \\ \mathbf{0} & \mathbf{S} \end{pmatrix}. \tag{5}$$

The mathematician and statistician Alexander Craig Aitken (1895–1967) was elected FRS, Fellow of the Royal Society, and from the Royal Society website[127] we find his election citation for fellowship as follows: "Dis-

[ae]The Mathematics Genealogy Project website[38] indicates that Emilie Haynsworth has 17 (rather than 18) students and 19 descendants.
[af]See also Haynsworth & Ostrowski.[75]

tinguished for his researches in mathematics. Author of 43 papers, chiefly in algebra and statistics. In particular, Aitken discovered (a) a theory of duality which links determinantal theory with the combinatory partition-theory and with group-character theory, (b) the transformation of the rational canonical form of any matrix into classical canonical form, (c) the concept of minimal vectors associated with a singular pencil of matrices, furnishing a method of great value, (d) a theorem of which most of the fundamental expansions of function theory and interpolation theory are special cases, and (e) a final solution, theoretical and practical, of the statistical problem of polynomial representations."

Fig. 9. Alexander Craig Aitken.[109]

There is much published biographical information on Aitken.[ag] As Silverstone[143] points out in his obituary on Aitken, "A brief biographical note must fail to do justice to one who has been described by an eminent mathematician as the 'greatest algebraist since Cayley' and by a prominent psychologist as being endowed with the ability to calculate mentally 'with a skill which possibly exceeds that of any person for whom precise authenticated records exist'. ... A man who never ceased to wonder at the beauty revealed by a mathematical discovery, a man of great enthusiasm but with great humility, a humanist, a lover of nature, generous with the ideas he gave to his students and a constant source of inspiration to his fellow workers."

8. The Guttman Rank-Additivity Formula: 1946, and Louis Guttman (1916–1987)

From (4) and (5), we obtain immediately the Schur determinant formulas

$$\det \mathbf{G} = \det \begin{pmatrix} \mathbf{P} & \mathbf{Q} \\ \mathbf{R} & \mathbf{S} \end{pmatrix} = \det \mathbf{P} \cdot \det(\mathbf{G}/\mathbf{P}) = \det(\mathbf{G}/\mathbf{S}) \cdot \det \mathbf{S}, \qquad (6)$$

from which we see at once that when $\mathbf{P}$ is nonsingular, then $\mathbf{G}$ is nonsingular if and only if $\mathbf{G}/\mathbf{P}$ is nonsingular, and when $\mathbf{S}$ is nonsingular, then $\mathbf{G}$ is

[ag]Another portrait photograph of Aitken is in the biographical memoir by Whittaker & Bartlett[167] and on The University of York website;[95] see also Refs. 3, 4, 5, 56, 87, 109, 144, and 170.

nonsingular if and only if $\mathbf{G}/\mathbf{S}$ is nonsingular. The determinant formulas (6) hold only when $\mathbf{G}$ and $\mathbf{P}$ (and hence $\mathbf{S}$) are square. But the *Guttman rank-additivity formula*[69,117] holds even when none of the matrices in $\mathbf{G}$ are square:

$$\begin{aligned} \operatorname{rank}\mathbf{G} = \operatorname{rank}\begin{pmatrix} \mathbf{P} & \mathbf{Q} \\ \mathbf{R} & \mathbf{S} \end{pmatrix} &= \operatorname{rank}\mathbf{P} + \operatorname{rank}(\mathbf{G}/\mathbf{P}) \\ &= \operatorname{rank}(\mathbf{G}/\mathbf{S}) + \operatorname{rank}\mathbf{S}, \end{aligned} \tag{7}$$

which we believe was first established in 1946 by the social scientist and statistician Louis Guttman (1916–1987) in Ref. 69.

Louis Guttman[ah] was born in Brooklyn, New York, on 10 February 1916 and obtained his B.A., M.A. and Ph.D. degrees from the University of Minnesota. In 1947 Guttman became the Founder and Scientific Director of The Israel Institute of Applied Social Research in Jerusalem, one year before the establishment of the state of Israel. He died on 25 October 1987 in Minneapolis.

Fig. 10. Louis Guttman.[95]

According to Katz,[86] "The development of scaling theory by Guttman and the mathematical psychologist Clyde Coombs (1912–1988) is one of the 62 major advances in social science identified and analyzed in *Science* by Deutsch, Platt & Senghaas.[48]"

Guttman's research in statistics and matrix theory included factor analysis, inequalities for eigenvalues of matrices, the Ising model in statistical mechanics, least-squares image analysis, and matrix factorisations. Guttman (1954) proposed an alternative approach to factor analysis for the analysis of correlation matrices with positive elements, introducing the *simplex*, *circumplex* and *radex* models; see, e.g., Anderson[8] and Browne.[30]

[ah] Our biographical comments on Louis Guttman come from Shye[142] in *Leading Personalities in Statistical Sciences*, and the obituaries by Katz[86] and Styan.[156]

9. Banachiewicz Inversion Formulas: 1937, and Tadeusz Banachiewicz (1882–1954)

When $\mathbf{G}$ and $\mathbf{P}$ are nonsingular, then inversion of the Aitken block-diagonalisation formula (4) yields

$$\begin{pmatrix} \mathbf{P} & \mathbf{Q} \\ \mathbf{R} & \mathbf{S} \end{pmatrix}^{-1} = \begin{pmatrix} \mathbf{I} & -\mathbf{P}^{-1}\mathbf{Q} \\ \mathbf{0} & \mathbf{I} \end{pmatrix} \begin{pmatrix} \mathbf{P}^{-1} & \mathbf{0} \\ \mathbf{0} & (\mathbf{G}/\mathbf{P})^{-1} \end{pmatrix} \begin{pmatrix} \mathbf{I} & \mathbf{0} \\ -\mathbf{R}\mathbf{P}^{-1} & \mathbf{I} \end{pmatrix} \tag{8}$$

$$= \begin{pmatrix} \mathbf{P}^{-1} + \mathbf{P}^{-1}\mathbf{Q}(\mathbf{G}/\mathbf{P})^{-1}\mathbf{R}\mathbf{P}^{-1} & -\mathbf{P}^{-1}\mathbf{Q}(\mathbf{G}/\mathbf{P})^{-1} \\ -(\mathbf{G}/\mathbf{P})^{-1}\mathbf{R}\mathbf{P}^{-1} & (\mathbf{G}/\mathbf{P})^{-1} \end{pmatrix} \tag{9}$$

$$= \begin{pmatrix} \mathbf{P}^{-1} & \mathbf{0} \\ \mathbf{0} & \mathbf{0} \end{pmatrix} + \begin{pmatrix} \mathbf{P}^{-1}\mathbf{Q} \\ -\mathbf{I} \end{pmatrix} (\mathbf{G}/\mathbf{P})^{-1} \begin{pmatrix} \mathbf{R}\mathbf{P}^{-1} & -\mathbf{I} \end{pmatrix} \tag{10}$$

and when $\mathbf{G}$ and $\mathbf{S}$ are nonsingular, then inversion of (5) yields

$$\begin{pmatrix} \mathbf{P} & \mathbf{Q} \\ \mathbf{R} & \mathbf{S} \end{pmatrix}^{-1} = \begin{pmatrix} \mathbf{I} & \mathbf{0} \\ -\mathbf{S}^{-1}\mathbf{R} & \mathbf{I} \end{pmatrix} \begin{pmatrix} (\mathbf{G}/\mathbf{S})^{-1} & \mathbf{0} \\ \mathbf{0} & \mathbf{S}^{-1} \end{pmatrix} \begin{pmatrix} \mathbf{I} & -\mathbf{Q}\mathbf{S}^{-1} \\ \mathbf{0} & \mathbf{I} \end{pmatrix} \tag{11}$$

$$= \begin{pmatrix} (\mathbf{G}/\mathbf{S})^{-1} & -(\mathbf{G}/\mathbf{S})^{-1}\mathbf{Q}\mathbf{S}^{-1} \\ -\mathbf{S}^{-1}\mathbf{R}(\mathbf{G}/\mathbf{S})^{-1} & \mathbf{S}^{-1} + \mathbf{S}^{-1}\mathbf{R}(\mathbf{G}/\mathbf{S})^{-1}\mathbf{Q}\mathbf{S}^{-1} \end{pmatrix} \tag{12}$$

$$= \begin{pmatrix} \mathbf{0} & \mathbf{0} \\ \mathbf{0} & \mathbf{S}^{-1} \end{pmatrix} + \begin{pmatrix} -\mathbf{I} \\ \mathbf{S}^{-1}\mathbf{R} \end{pmatrix} (\mathbf{G}/\mathbf{S})^{-1} \begin{pmatrix} -\mathbf{I} & \mathbf{Q}\mathbf{S}^{-1} \end{pmatrix}. \tag{13}$$

The formulas (8)–(13) are known as the *Banachiewicz inversion formulas*, see, e.g., Puntanen & Styan,[117] and are due to the astronomer and mathematician Tadeusz Banachiewicz.[11] See also the classic book by the three aeronautical engineers Robert Alexander Frazer (1891–1959), William Jolly Duncan (1894–1960) and Arthur Roderick Collar (1908–1986) entitled *Elementary Matrices and Some Applications to Dynamics and Differential Equations*,[60] first published in 1938, just one year after Banachiewicz;[11] the appearance in Ref. 60 of the Banachiewicz inversion formula is almost surely its first appearance in a book.

Tadeusz Banachiewicz was born on 13 February 1882 in Warsaw (then in Russian occupied Poland) and died on 17 November 1954 in Cracow (Kraków in Polish). According to Prominent Poles website,[114] see also Zawada,[173] "his parents, Artur Banachiewicz and Zofia Rzeszotarska, owned the Cychry estate near Warsaw. Banachiewicz published his first article in the *Astronomische Nachrichten* in 1903 as a student at Warsaw University." At that time he sent a telegram to the Central Astronomical Bureau in

Kilonia (now Kiel, Germany) stating that on 19 September 1903 the planet Jupiter would cover the changeable star catalogued as BD +6° 6191.

After passing the habilitation exam in 1910, he obtained work in Kazan, Russia, conducting heliometrical studies of the moon. He was appointed professor of astronomy at the University in Dorpat, Russia (now Tartu, Estonia) in 1915. In 1919, after Poland regained her independence, Banachiewicz moved to Cracow, becoming a professor at the Jagiellonian University and the director of Cracow Observatory.[105] He modified the method of determining parabolic orbits and published approximately 200 research papers. In 1925, he invented a theory of "Cracovians" (a special kind of matrix algebra) which brought him international recognition. This theory was used to solve several astronomical, geodesic, mechanical and mathematical problems.

A postage stamp in honor of Tadeusz Banachiewicz, featuring a portrait of him, see Figure 11, was issued by Poland on 25 March 1983; a copy appears on the *Images of Mathematicians on Postage Stamps* website[106] and in print in the biographical article by Grala, Markiewicz & Styan.[67] The booklet by Zawada[173] contains a detailed biography and many photographs of Tadeusz Banachiewicz.

Fig. 11. Tadeusz Banachiewicz: Poland 1983 (Scott #2565).

Tadeusz Banachiewicz was the recipient of Doctor Honoris Causa titles from the University of Warsaw, the University of Poznań and the University of Sofia (Bulgaria). He was also the founder of the journal *Acta Astronomica*. Moreover, Banachiewicz invented a chronocinematograph, and one of the lunar craters is named after him. He was married to Laura (Larissa) Solohub, a Ukrainian painter and poet.

10. The Duncan Inversion Formula: 1944, and William Jolly Duncan (1894–1960)

If we equate the top-left hand corners of (12) and (9), we obtain the *Duncan inversion formula*; see, e.g., Puntanen & Styan:[117]

$$(\mathbf{G}/\mathbf{S})^{-1} = \mathbf{P}^{-1} + \mathbf{P}^{-1}\mathbf{Q}(\mathbf{G}/\mathbf{P})^{-1}\mathbf{R}\mathbf{P}^{-1} \qquad (14)$$

or explicitly

$$(\mathbf{P} - \mathbf{Q}\mathbf{S}^{-1}\mathbf{R})^{-1} = \mathbf{P}^{-1} + \mathbf{P}^{-1}\mathbf{Q}(\mathbf{S} - \mathbf{R}\mathbf{P}^{-1}\mathbf{Q})^{-1}\mathbf{R}\mathbf{P}^{-1}, \tag{15}$$

which we believe was first established by Duncan in 1944.[52]

The aeronautical engineer William Jolly Duncan (1894–1960) was elected FRS, Fellow of the Royal Society. From the Royal Society website,[127] we have the 1947 election citation for this fellowship[ai] as follows:

> "Distinguished for researches in aero-elasticity (flutter), elasticity of materials, and other problems of mathematical physics. With Dr. R. A. Frazer he laid the foundations of the new subject of the flutter of aeroplane wings, tail surfaces, or airscrew blades. He combines a high degree of mathematical ability with a thorough knowledge of engineering, and has been conspicuously successful in his application of analysis to practical engineering problems. He has contributed much to the subject of the stability of aircraft and is at present in charge of the flight research section of the Royal Aircraft Establishment."

We may use the Duncan inversion formula (15) to obtain the inverse of the $n \times n$ matrix $\mathbf{A}$ for $n \geq 2$:

$$\mathbf{A} = \begin{pmatrix} a & b & b & \dots & b \\ b & a & b & \dots & b \\ \vdots & \vdots & \vdots & \ddots & \vdots \\ b & b & b & \dots & a \end{pmatrix} = (a-b)\mathbf{I}_n + b\mathbf{e}\mathbf{e}' = (a-b)(\mathbf{I}_n + c\mathbf{e}\mathbf{e}'), \tag{16}$$

where $\mathbf{I}_n$ is the $n \times n$ identity matrix, $\mathbf{e} = (1, 1, \dots, 1)'$ and $c = b/(a-b)$, with $b \neq 0$ and $a - b \neq 0$.

When $a = b$, then $\mathbf{A}$ has rank equal to 1 and so is singular. When $b = 0$, then $\mathbf{A} = a\mathbf{I}_n$ and $\mathbf{A}^{-1} = (1/a)\mathbf{I}_n$ when $a \neq 0$. When $b \neq 0$ and $a - b \neq 0$, then $c = b/(a-b)$ is well defined and nonzero. When $a = 1$ and $\mathbf{A}$ is nonnegative definite, then $\mathbf{A}$ is called an *intraclass correlation matrix* or *equicorrelation matrix*.

[ai] For more biographical information on William Jolly Duncan and a portrait photograph see the memoir by Relf.[126] Duncan's coauthors for the famous book,[60] Robert Alexander Frazer and Arthur Roderick Collar, were also elected FRS; see the biographical memoirs, respectively by Pugsley[115] and Bishop,[20] which include portrait photographs.

We set

$$\mathbf{G} = \begin{pmatrix} \mathbf{P} & \mathbf{Q} \\ \mathbf{R} & \mathbf{S} \end{pmatrix} = \begin{pmatrix} \mathbf{I}_n & \mathbf{e} \\ \mathbf{e}' & -\frac{1}{c} \end{pmatrix}, \tag{17}$$

and so

$$\mathbf{G}/\mathbf{S} = \mathbf{I}_n + c\mathbf{e}\mathbf{e}' \quad \text{and} \quad \mathbf{G}/\mathbf{P} = -\frac{1}{c} - n = -\frac{cn+1}{c}, \tag{18}$$

and hence with $a - b \neq 0$ and $a + b(n-1) \neq 0$,

$$(\mathbf{I}_n + c\mathbf{e}\mathbf{e}')^{-1} = \mathbf{I}_n - \frac{c}{cn+1}\mathbf{e}\mathbf{e}',$$

$$\big((a-b)\mathbf{I}_n + b\mathbf{e}\mathbf{e}'\big)^{-1} = \frac{1}{a-b}\left(\mathbf{I}_n - \frac{b}{a+b(n-1)}\mathbf{e}\mathbf{e}'\right), \tag{19}$$

$$\mathbf{A}^{-1} = \begin{pmatrix} a & b & b & \cdots & b \\ b & a & b & \cdots & b \\ \vdots & \vdots & \vdots & \ddots & \vdots \\ b & b & b & \cdots & a \end{pmatrix}^{-1}$$

$$= \frac{1}{(a-b)\big(a+b(n-1)\big)} \begin{pmatrix} a+b(n-2) & -b & -b & \cdots & -b \\ -b & a+b(n-2) & -b & \cdots & -b \\ \vdots & \vdots & \vdots & \ddots & \vdots \\ -b & -b & -b & \cdots & a+b(n-2) \end{pmatrix}.$$

11. The Haynsworth Inertia Formula: 1968, and the Cauchy–Schwarz Inequality

Now suppose that $\mathbf{G}$ is symmetric, and so $\mathbf{R} = \mathbf{Q}'$ and the Schur complement $\mathbf{G}/\mathbf{P} = \mathbf{S} - \mathbf{Q}'\mathbf{P}^{-1}\mathbf{Q}$. Then the Aitken diagonalisation formula (4) becomes

$$\begin{pmatrix} \mathbf{I} & \mathbf{0} \\ -\mathbf{Q}'\mathbf{P}^{-1} & \mathbf{I} \end{pmatrix} \begin{pmatrix} \mathbf{P} & \mathbf{Q} \\ \mathbf{Q}' & \mathbf{S} \end{pmatrix} \begin{pmatrix} \mathbf{I} & -\mathbf{P}^{-1}\mathbf{Q} \\ \mathbf{0} & \mathbf{I} \end{pmatrix} = \begin{pmatrix} \mathbf{P} & \mathbf{0} \\ \mathbf{0} & \mathbf{G}/\mathbf{P} \end{pmatrix} \tag{20}$$

and so if $\mathbf{P}$ is positive definite (and hence also nonsingular) then

$$\mathbf{G} = \begin{pmatrix} \mathbf{P} & \mathbf{Q} \\ \mathbf{Q}' & \mathbf{S} \end{pmatrix} \geq_{\mathsf{L}} \mathbf{0} \;\Leftrightarrow\; \mathbf{G}/\mathbf{P} = \mathbf{S} - \mathbf{Q}'\mathbf{P}^{-1}\mathbf{Q} \geq_{\mathsf{L}} \mathbf{0}, \tag{21}$$

where the Löwner partial ordering $\mathbf{A} \geq_{\mathsf{L}} \mathbf{B}$ means $\mathbf{A} - \mathbf{B}$ is nonnegative definite (positive semi-definite, possibly singular). The result (21) also follows from the *Haynsworth inertia additivity formula*

$$\operatorname{In} \mathbf{G} = \operatorname{In} \mathbf{P} + \operatorname{In}(\mathbf{G}/\mathbf{P}), \tag{22}$$

where the *inertia* (or *inertia triple*) of the symmetric matrix $\mathbf{A}$ is defined to be the ordered integer triple

$$\operatorname{In} \mathbf{A} = \{\pi,\, \nu,\, \delta\},$$

with the nonnegative integers $\pi = \pi(\mathbf{A})$, $\nu = \nu(\mathbf{A})$ and $\delta = \delta(\mathbf{A})$ giving the numbers, respectively, of positive, negative and zero eigenvalues of $\mathbf{A}$. The result (22) was proved in 1968, apparently for the first time, by Haynsworth.[72,73] From (22), it follows at once that rank is additive on the Schur complement in a symmetric matrix. As Guttman showed, see (7) above, this rank additivity holds more generally: $\mathbf{G}$ need not even be square — we need only that $\mathbf{P}$ be square and nonsingular.

We may use the Schur complement inequality (21) to prove the *Cauchy–Schwarz Inequality*:

$$\mathbf{a}'\mathbf{a} \cdot \mathbf{b}'\mathbf{b} \geq (\mathbf{a}'\mathbf{b})^2, \tag{23}$$

where $\mathbf{a}$ and $\mathbf{b}$ are both $k \times 1$ vectors. Equality holds in (23) if and only if $\mathbf{a} = f\mathbf{b}$ for some scalar f.

The Cauchy–Schwarz Inequality is also known as the Cauchy–Bouniakowsky–Schwarz Inequality[164] (or CBS-inequality[129]) and is named after [Baron] Augustin-Louis Cauchy (1789–1857), Viktor Yakovlevich Bouniakowsky [Buniakovski, Bunyakovsky] (1804–1899), and [Karl] Hermann Amandus Schwarz (1843–1921). See Cauchy,[33] Bouniakowsky,[25] and Schwarz,[140] and the recent book by Steele.[151]

Schreiber[129] cites Schwarz,[139] published originally in 1874,[138] but we could not find the inequality there. The corresponding inequality for integrals is often called the Schwarz Inequality; see Steele[151] and the new book by Dragomir.[50] For a postage stamp depicting Cauchy, see Wilson[169] and Miller.[106]

We prove a more general matrix version of the Cauchy–Schwarz Inequality. Let $\mathbf{A}$ be $k \times m$ and let $\mathbf{B}$ be $k \times n$, and suppose that $\mathbf{B}$ has full column rank equal to n and so $\mathbf{B}'\mathbf{B}$ is positive definite. Then

$$\mathbf{G} = \begin{pmatrix} \mathbf{A}'\mathbf{A} & \mathbf{A}'\mathbf{B} \\ \mathbf{B}'\mathbf{A} & \mathbf{B}'\mathbf{B} \end{pmatrix} = \begin{pmatrix} \mathbf{A}' \\ \mathbf{B}' \end{pmatrix} \begin{pmatrix} \mathbf{A} & \mathbf{B} \end{pmatrix} \geq_{\mathsf{L}} \mathbf{0} \tag{24}$$

Fig. 12. Augustin-Louis Cauchy (left), Viktor Yakovlevich Bouniakowsky and Hermann Amandus Schwarz.[109]

implies that the Schur complement

$$\mathbf{G}/(\mathbf{B}'\mathbf{B}) = \mathbf{A}'\mathbf{A} - \mathbf{A}'\mathbf{B}(\mathbf{B}'\mathbf{B})^{-1}\mathbf{B}'\mathbf{A} \geq_{\mathsf{L}} \mathbf{0}, \tag{25}$$

or equivalently

$$\mathbf{A}'\mathbf{A} \geq_{\mathsf{L}} \mathbf{A}'\mathbf{B}(\mathbf{B}'\mathbf{B})^{-1}\mathbf{B}'\mathbf{A}, \tag{26}$$

and so when $m = n = 1$ and $\mathbf{A} = \mathbf{a}$ and $\mathbf{B} = \mathbf{b}$ then (23) follows at once.

Equality holds in (26) if and only if $\operatorname{rank}(\mathbf{A}\ \ \mathbf{B}) = \operatorname{rank}(\mathbf{B})$, which holds if and only if $\mathbf{A} = \mathbf{BF}$ for some matrix $\mathbf{F}$. When $m = n = 1$ then $\mathbf{F}$ becomes a scalar f, say.

12. Gauß–Markov & Frisch–Waugh–Lovell Theorems

We may use the Schur complement inequality (21) to prove the well-known *Gauß–Markov Theorem* (or *Gauss–Markov Theorem*), which is named after the two well-known mathematicians [Johann] Carl Friedrich Gauß (1777–1855) and Andrei Andreyevich Markov (1856–1922).[aj]

Let $\mathbf{A}'\mathbf{y}$ be a linear unbiased estimator of $\boldsymbol{\gamma}$ in the *general linear model with white noise:* $\mathrm{E}(\mathbf{y}) = \mathbf{X}\boldsymbol{\gamma}$ and $\mathrm{D}(\mathbf{y}) = \sigma^2\mathbf{I}$; here $\mathrm{E}(\cdot)$ denotes expectation and $\mathrm{D}(\cdot)$ dispersion matrix. We assume that $\mathbf{X}$ has full column rank and so it follows that $\mathbf{A}'\mathbf{X} = \mathbf{I}$. Then the Gauß–Markov Theorem states that the *ordinary least-squares estimator* (OLSE) $\hat{\boldsymbol{\gamma}} = (\mathbf{X}'\mathbf{X})^{-1}\mathbf{X}'\mathbf{y}$ is the *best linear*

[aj]For 3 postage stamps depicting Gauß, see Wilson[169] and Miller.[106] Odell[110] states that the Gauß–Markov Theorem was first proved by Gauß in the first part of his *Theoria Combinationis* (1821) and that the name "Gauss–Markov Theorem" was first used in 1938 by David & Neyman.[42]

Fig. 13. Carl Friedrich Gauß (left) and Andrei Andreyevich Markov.[109]

unbiased estimator (BLUE) of $\boldsymbol{\gamma}$ in that $\mathrm{D}(\mathbf{A}'\mathbf{y}) \geq_{\mathsf{L}} \mathrm{D}(\hat{\boldsymbol{\gamma}})$, i.e., $\mathbf{A}'\mathbf{A} \geq_{\mathsf{L}} (\mathbf{X}'\mathbf{X})^{-1}$ with $\mathbf{A}'\mathbf{X} = \mathbf{I}$. We note that

$$\begin{pmatrix} \mathbf{A}'\mathbf{A} & \mathbf{I} \\ \mathbf{I} & \mathbf{X}'\mathbf{X} \end{pmatrix} = \begin{pmatrix} \mathbf{A}'\mathbf{A} & \mathbf{A}'\mathbf{X} \\ \mathbf{X}'\mathbf{A} & \mathbf{X}'\mathbf{X} \end{pmatrix} = \begin{pmatrix} \mathbf{A}' \\ \mathbf{X}' \end{pmatrix} \begin{pmatrix} \mathbf{A} & \mathbf{X} \end{pmatrix} \geq_{\mathsf{L}} \mathbf{0}\,, \tag{27}$$

and so the Schur complement $\mathbf{A}'\mathbf{A} - (\mathbf{X}'\mathbf{X})^{-1} \geq_{\mathsf{L}} \mathbf{0}$, or equivalently $\mathbf{A}'\mathbf{A} \geq_{\mathsf{L}} (\mathbf{X}'\mathbf{X})^{-1}$.

We may use the Banchiewicz inversion formula (10) to prove the *Frisch–Waugh–Lovell Theorem*, which is named after the Nobel laureate Ragnar Frisch (1895–1973),[ak] the American agricultural economist Frederick Vail Waugh (1898–1974),[al] and the American economist Michael C. Lovell (b. 1930),[am] in view of results in Frisch & Waugh[62] and Lovell;[96] see, e.g., Davidson & MacKinnon,[44] Fiebig, Krämer & Bartels,[58] and Puntanen & Styan.[118]

Let us consider the two linear models:

$$\mathrm{E}(\mathbf{y}) = \mathbf{X}_1\boldsymbol{\gamma}_1 + \mathbf{X}_2\boldsymbol{\gamma}_2 \tag{28}$$

and

$$\mathrm{E}(\mathbf{M}_1\mathbf{y}) = \mathbf{M}_1\mathbf{X}_2\boldsymbol{\gamma}_2\,, \tag{29}$$

[ak]The Norwegian economist Ragnar Frisch (1895–1973) won the 1969 Bank of Sweden Prize in Economic Sciences in Memory of Alfred Nobel [Nobel Memorial Prize in Economic Sciences] jointly with the Dutch economist Jan Tinbergen (1903–1994) "for having developed and applied dynamic models for the analysis of economic processes".[158,163] The year 1969 was the first year that this prize was awarded.

[al]For an unsigned obituary of Frederick Vail Waugh (1898–1974) see Ref. 165.

[am]See also the home page: Lovell.[97]

where the partitioned matrix $(\mathbf{X}_1\ \mathbf{X}_2)$ has full column rank and $\mathbf{M}_1 = \mathbf{I} - \mathbf{X}_1(\mathbf{X}_1'\mathbf{X}_1)^{-1}\mathbf{X}_1' = \mathbf{I} - \mathbf{H}_1$, where $\mathbf{H}_1$ is the *hat matrix* associated with $\mathbf{X}_1$. Then the Frisch–Waugh–Lovell Theorem says that both (1) the OLS estimate $\hat{\gamma}_2$ of γ_2 and (2) the residual vector $\mathbf{r}$, say, in the two models (28) and (29) are numerically identical.

For the model (29) it is easy to see that

$$\hat{\gamma}_2^{(29)} = (\mathbf{X}_2'\mathbf{M}_1\mathbf{X}_2)^{-1}\mathbf{X}_2'\mathbf{M}_1\mathbf{y}, \tag{30}$$

$$\mathbf{r}^{(29)} = \mathbf{M}_1\mathbf{y} - \mathbf{M}_1\mathbf{X}_2(\mathbf{X}_2'\mathbf{M}_1\mathbf{X}_2)^{-1}\mathbf{X}_2'\mathbf{M}_1\mathbf{y}. \tag{31}$$

From the Banachiewicz inversion formula (10) we find that

$$\begin{pmatrix} \mathbf{X}_1'\mathbf{X}_1 & \mathbf{X}_1'\mathbf{X}_2 \\ \mathbf{X}_2'\mathbf{X}_1 & \mathbf{X}_2'\mathbf{X}_2 \end{pmatrix}^{-1} = \begin{pmatrix} (\mathbf{X}_1'\mathbf{X}_1)^{-1} & \mathbf{0} \\ \mathbf{0} & \mathbf{0} \end{pmatrix}$$
$$+ \begin{pmatrix} (\mathbf{X}_1'\mathbf{X}_1)^{-1}\mathbf{X}_1'\mathbf{X}_2 \\ -\mathbf{I} \end{pmatrix} (\mathbf{X}_2'\mathbf{M}_1\mathbf{X}_2)^{-1}\big(\mathbf{X}_2'\mathbf{X}_1(\mathbf{X}_1'\mathbf{X}_1)^{-1}, -\mathbf{I}\big)$$

and so $\hat{\gamma}_2^{(28)} = \hat{\gamma}_2^{(29)}$ as given in (30). Moreover the residual vector

$$\begin{aligned} \mathbf{r}^{(28)} &= \mathbf{y} - \mathbf{X}_1\hat{\gamma}_1^{(28)} - \mathbf{X}_2\hat{\gamma}_2^{(28)} = \mathbf{y} - \mathbf{X}_1\hat{\gamma}_1^{(28)} - \mathbf{X}_2\hat{\gamma}_2^{(29)} \\ &= \mathbf{y} - \mathbf{H}_1\mathbf{y} + \mathbf{H}_1\mathbf{X}_2(\mathbf{X}_2'\mathbf{M}_1\mathbf{X}_2)^{-1}\mathbf{X}_2'\mathbf{M}_1\mathbf{y} \\ &\qquad - \mathbf{X}_2(\mathbf{X}_2'\mathbf{M}_1\mathbf{X}_2)^{-1}\mathbf{X}_2'\mathbf{M}_1\mathbf{y} \\ &= \mathbf{M}_1\mathbf{y} - \mathbf{M}_1\mathbf{X}_2(\mathbf{X}_2'\mathbf{M}_1\mathbf{X}_2)^{-1}\mathbf{X}_2'\mathbf{M}_1\mathbf{y} = \mathbf{r}^{(29)}, \end{aligned} \tag{32}$$

as given in (31). For extensions of the Frisch–Waugh–Lovell Theorem, see Gross & Puntanen[68] and Puntanen & Styan.[118]

Acknowledgements

This article is based on the invited talk "Issai Schur (1875–1941) and the early development of the Schur complement: photographs, documents and biographical remarks" presented at the International Statistics Workshop, School of Information Sciences and Engineering, University of Canberra (Canberra, Australia), 4 April 2005, and expands on our reports.[119,120] This article also builds on the chapters by Simo Puntanen & George P. H. Styan entitled "Historical introduction: Issai Schur and the early development of the Schur complement"[117] and "Schur complements in statistics and probability"[118] in the book *The Schur Complement and Its Applications*, edited by Fuzhen Zhang;[174] see also Ouellette[111] and Styan.[155]

The portrait photograph of Issai Schur in Figure 1 appears in Refs. 63 and 174 and that of Gabriel Narutowicz in Figure 3 is in *Wikipedia*;[163] the other three pictures in Figure 3 were taken by Kari Tapio Mäkeläinen. The picture of the Schur gravestone in Figure 7 was taken by Geva Maimon Reid. The portrait photograph of Emilie Virginia Haynsworth in Figure 8 appears on the Auburn University website,[10] and in Davis[45] and Zhang.[174] The portrait photographs of Alexander Craig Aitken in Figure 9, Augustin-Louis Cauchy, Viktor Yakovlevich Bouniakowsky, and Hermann Amandus Schwarz in Figure 12, and Carl Friedrich Gauß and Andrei Andreyevich Markov in Figure 13 are on the MacTutor website.[109] The portrait photograph of Louis Guttman in Figure 10 is on the Portraits of Statisticians website.[95]

We are particularly grateful to Google for leading us to many of the websites accessed for this article. Our thanks go also to Oskar Maria Baksalary, Bartko-Reher-GbR (Berlin), Nathan Beit-Aharon, David R. Bellhouse, Adi Ben-Israel, Abraham Berman, Torsten Bernhardt, Nora Bohossian, Eva Brune, Marco Carone, Ka Lok Chu, Richard William Farebrother, Daniel Hershkowitz, Roger A. Horn, Jarkko Isotalo, Jörg Kaufmann, Bernd Kirstein, Miriam C. Klein, Sabina Klein, Tõnu Kollo, Shuangzhe Liu, Geva Maimon Reid, Kari Tapio Mäkeläinen, Timo Mäkeläinen, Augustyn Markiewicz, Jarmo Niemelä, Tina Reid, Michelle E. Ross, Vera Rosta, Annelise Schmidt, Klaus Schmidt, Hans Schneider, Evelyn Matheson Styan, Gerald E. Subak-Sharpe, Garry J. Tee, Götz Trenkler, Frank Uhlig, Kimmo Vehkalahti, and Fuzhen Zhang for their help. This research was supported in part by the Natural Sciences and Engineering Research Council of Canada.

References and Further Reading

1. Abelin-Schur, Hilda. A story about my father. In Joseph, Melnikov & Rentschler.[84]
2. Aitken, A. C. *Determinants and Matrices.* University Mathematical Texts, Oliver & Boyd, Edinburgh, 1939. (2nd–9th editions, 1942–1956; 9th edition, reset & reprinted, 1967. Reprint edition: Greenwood Press, Westport, Connecticut, 1983.)
3. Aitken, A. C. The art of mental calculation: with demonstrations. *Transactions of the Society of Engineers*, 44 (1954), 295–309.
4. Alexander Craig Aitken: unsigned obituary. *Proceedings of the Edinburgh Mathematical Society, Series 2*, 16 (1968/1969), 151–176.
5. Aitken, A. C. *To Catch the Spirit: The Memoir of A. C. Aitken.* With a biographical introduction by Peter C. Fenton. University of Otago Press, Dunedin, New Zealand, 1995.
6. Aiton, E. J. *Leibniz: A Biography.* Adam Hilger, Bristol, 1985.

7. Albanese, Catherine L. Review of Jones.[82] *Winterthur Portfolio*, 15, 75–78 (1980).
8. Anderson, T. W. Some stochastic process models for intelligence test scores. In Arrow, Karlin & Suppes;[9] reprinted in Ref. 157.
9. Arrow, Kenneth J.; Karlin, Samuel; Suppes, Patrick, eds. *Mathematical Methods in the Social Sciences, 1959: Proceedings of the First Stanford Symposium.* Stanford University Press, 1960.
10. Auburn University. `www.auburn.edu/~fitzpjd/ben/images/Emilie.gif` website: Portrait photograph of Emilie Virginia Haynsworth. Department of Mathematics, Auburn University, Auburn, Alabama, accessed 28 December 2003. (Portrait photograph is also in Davis[45] and in Zhang.[174])
11. Banachiewicz, T. Zur Berechnung der Determinanten, wie auch der Inversen, und zur darauf basierten Auflösung der Systeme lineare Gleichungen. *Acta Astronomica, Série C*, 3 (1937), 41–67.
12. Barnavi, Eli, ed. *A Historical Atlas of the Jewish People, from the Time of the Patriarchs to the Present.* First American edition (translated from the French: *Juifs, une histoire universelle*), Eli Barnavi, General editor; Miriam Eliav-Feldon, English edition editor. Alfred A. Knopf, New York, 1992.
13. Beare, Arlene. *A Guide to Jewish Genealogy in Latvia and Estonia.* The Jewish Genealogical Society of Great Britain, London, 2001.
14. Beare, Arlene, ed. `www.jewishgen.org/Latvia/SIG_History_of_Latvia_and_Courland.html` website: *History of Latvia and Courland*, accessed 1 March 2004. (This history is derived from a few sources including [13] but mainly edited from the presentation made by Ruvin Ferber at the 21st International Conference of Jewish Genealogy held in London in July 2001.)
15. Begehr, Heinrich G. W. *Mathematik in Berlin: Geschichte und Dokumentation, Erster Halbband, Zweiter Halbband.* Berichte aus der Geschichtswissenschaft, Shaker Verlag, Aachen, 1998.
16. Begehr, H. G. W.; Koch, H.; Kramer, J.; Schappacher, N.; Thiele, E.-J.; on behalf of the Berliner Mathematische Gesellschaft, eds. *Mathematics in Berlin.* Birkhäuser Verlag, Berlin, 1998.
17. Bentz, Edna M. *If I Can, You Can Decipher Germanic Records.* Published by the author: Edna M. Bentz [San Diego, California], 1982. (Second printing 1983.)
18. Beyl, F. Rudolf; Tappe, Jürgen. *Group Extensions, Representations, and the Schur Multiplicator.* Springer-Verlag, Berlin, 1982.
19. Bierbaum, Max; Börsting, Heinrich. *Liudger und sein Erbe: dem 70. Nachfolger des heiligen Liudger, Clemens August Kardinal von Galen, Bischof von Münster, zum Gedächtnis.* Westfalia sacra 1, Regensberg, Münster, 1948.
20. Bishop. R. E. D. Arthur Roderick Collar: 22 February 1908–12 February 1986. *Biographical Memoirs of Fellows of the Royal Society*, 33 (1987), 165–185.
21. Blaushild, Immanuel. Libau. In Snyder.[145]
22. *The Book of Saints, A Dictionary of Servants of God Canonised by the Catholic Church: Extracted from the Roman & Other Martyrologies.* Com-

piled by The Benedictine Monks of St. Augustine's Abbey, Ramsgate. A. & C. Black, Ltd., London, 1921. (7th edition, entirely revised and reset: *The Book of Saints: A Comprehensive Biographical Dictionary* edited by Dom Basil Watkins OSB, on behalf of the Benedictine monks of St. Augustine's Abbey, Ramsgate. Continuum International Publishing Group, New York.)

23. Börsting, Heinrich. Liudger-träger des Nikolaikultes im Abendland. In Bierbaum & Börsting.[19]
24. Börsting, Heinrich. *Sankt Liudger: Gedenkschrift zum 1150. Todestage des Heiligen.* Pfarramt St. Ludgerus, Essen-Werden, 1959.
25. Bouniakowsky, V. Sur quelques inégalités concernant les intégrales ordinaires et les intégrales aux différences finies. *Mémoires de l'Académie Impériale des Sciences de St.-Pétersbourg, Septième Série*, vol. 1, no. 9, pp. 1–18 (1859).
26. Brauer, Alfred. Gedenkrede auf Issai Schur. In Brauer & Rohrbach.[28] (Translated into English[27] in Ledermann & Neumann.[94])
27. Brauer, Alfred. Memorial address on Issai Schur. In Ledermann & Neumann.[94] (English translation of Brauer.[26])
28. Brauer, Alfred; Rohrbach, Hans, eds. *Issai Schur Gesammelte Abhandlungen: Band I; Band II; Band III.* Springer-Verlag, Berlin, 1973.
29. Brosnahan, Tom. `www.turkeytravelplanner.com/WhereToGo/med/demre/` website: Turkey Travel Planner, accessed 4 February 2006.
30. Browne, Michael W. On the T. W. Anderson–Herman Rubin contribution to statistical inference in factor analysis and On T. W. Anderson's stochastic process models for Guttman's simplex, circumplex and radex. In Styan.[157]
31. Brüning, Jochen; Ferus, Dirk; Siegmund-Schultze, Reinhard. *Terror and Exile: Persecution and Expulsion of Mathematicians from Berlin between 1933 and 1945.* An Exhibition on the Occasion of the International Congress of Mathematicians, Technische Universität Berlin, August 19 to 27, 1998, Deutsche Mathematiker-Vereinigung, Berlin, 1998.
32. Carlson, David; Markham, Thomas L.; Uhlig, Frank. Emilie Haynsworth, 1916–1985. *Linear Algebra and its Applications*, 75 (1986), 269–276.
33. Cauchy, Augustin. Mémoire sur les fonctions qui ne peuvent obtenir que deux valeurs égales et de signes contraires par suites des transpositions opérées entre les variables qu'elles renferment. *Journal de l'École Polytechnique, Paris*, vol. 10, cahier 17, pp. 29–112 (1812). (Reprinted in *Œuvres Complètes d'Augustin Cauchy*, 2e série, tome 1: Mémoires extraits du Journal de l'École Polytechnique, pp. 91–169, Gauthier-Villars, Paris, 1905.)
34. Chandler, Bruce; Magnus, Wilhelm. *The History of Combinatorial Group Theory: A Case Study in the History of Ideas.* Studies in the History of Mathematics and Physical Sciences 9. Springer-Verlag, New York, 1982.
35. Cioffari, Gerardo. *Saint Nicholas: His Life, the Translation of his Relics and his Basilica in Bari*, translated by Philip L. Barnes, Centro Studi Nicolaiani, Bari, Italy, 1994.
36. Coomber, R. Robert. The English translation of Ympyn. *Accounting Research*, 5 (1954), 363. (Reprinted in Solomons & Zeff.[147])

37. Coomber, R. R. The English translation of Ympyn Christoffels's *Nouvelle Instruction* of 1547. *Accounting Research*, 6 (1955), 281–284; reprinted in Solomons & Zeff.[147]
38. Coonce, Harry, ed. `www.genealogy.ams.org/html/search.phtml` website: *The Mathematics Genealogy Project.* Department of Mathematics, North Dakota State University, Fargo, accessed: 24 January 2006.
39. Council of Biology Editors Style Manual Committee. *Scientific Style and Format: The CBE Manual for Authors, Editors, and Publishers*, 6th edition. Cambridge University Press, 1994. (Original version: *Style Manual for Biological Journals* prepared by the Committee on Form and Style of the Conference of Biological Editors, American Institute of Biological Sciences, 1960 (2nd edition: 1964); 3rd–5th editions: *CBE Style Manual: A Guide for Authors, Editors, and Publishers in the Biological Sciences*, 1972–1983.)
40. Cruz, Joan Carroll. *Relics.* Our Sunday Visitor, Huntington, Indiana, 1984.
41. Curtis, Charles W. *Pioneers of Representation Theory: Frobenius, Burnside, Schur, and Brauer.* History of Mathematics 15, American Mathematical Society & London Mathematical Society, 1999.
42. David, F. N.; Neyman, J. Extension of the Markoff Theorem on least squares. *Statistical Research Memoirs*, 2 (1938), 105–116.
43. Davidson, Russell; MacKinnon, James G. *Estimation and Inference in Econometrics.* Oxford University Press, New York, 1993.
44. Davidson, Russell; MacKinnon, James G. *Econometric Theory and Methods.* Oxford University Press, New York, 2004.
45. Davis, Philip J. *The Education of a Mathematician.* A. K. Peters, Natick, Massachusetts, 2000.
46. Davis, Scott L. `homepages.tscnet.com/omard1/m.htm` website: *Natural Magick Books,* Silverdale, Washington, accessed 1 February 2006.
47. Delehaye, Père Hippolyte. Review of Meisen.[102] *Analecta Bollandiana*, 50, 176–181 (1932).
48. Deutsch, Karl W.; Platt, John; Senghaas, Dieter. Conditions favoring major advances in social science. *Science*, 171 (1971), 450–459.
49. Dharmananda, Subhuti. Myrrh and frankincense, 8 pp. `www.itmonline.org/arts/myrrh.htm` website: Institute for Traditional Medicine, Portland, Oregon, accessed 3 February 2006.
50. Dragomir, Sever Silvestru. *Advances in Inequalities of the Schwarz, Grüss and Bessel Type in Inner Product Spaces.* Nova Science Publishers, Hauppauge, New York, 2005.
51. Du Cange, Sieur Charles Du Fresne. Constantinopolis Christians, lib. iv. c. 6, n. 67. Codinus Orig. Conatan, p, 62. (Reference #2 from [55].)
52. Duncan, W. J. Some devices for the solution of large sets of simultaneous linear equations. (With an Appendix on the reciprocation of partitioned matrices.) *The London, Edinburgh, and Dublin Philosophical Magazine and Journal of Science, Seventh Series*, 35 (1944), 660–670.
53. Emrich, Duncan. A certain Nicholas of Patara. *American Heritage*, 12 (1960), 22–27.

54. *Encyclopædia Britannica* `www.search.eb.com/eb/article?eu=39437` website: "Gymnasium" article accessed 29 December 2003.
55. Eternal Word Television Network. `www.ewtn.com/library/MARY/STNICH.htm` website: *St Nicholas, Confessor, Archbishop of Myra – A.D. 342*, provided courtesy of Eternal Word Television Network, Irondale, Alabama, accessed 12 February 2006.
56. Farebrother, Richard William. A. C. Aitken and the consolidation of matrix theory. *Linear Algebra and its Applications*, 264 (1997), 3–12.
57. Farebrother, R. William; Styan, George P. H.; Tee, Garry J. Gottfried Wilhelm von Leibniz: 1646–1716. *Image: The Bulletin of the International Linear Algebra Society*, no. 30 (2003), pp. 13–16.
58. Fiebig, Denzil G.; Krämer, Walter; Bartels, Robert. The Frisch–Waugh theorem and generalized least squares. *Econometric Reviews*, 15 (1996), 431–443.
59. Fournier, Catherine. `www.domestic-church.com/CONTENT.DCC/19981101/SAINTS/nicholas.htm` website: *Saint Nicholas, Bishop of Myra.* Domestic Church Communications Ltd., accessed 19 January 2004.
60. Frazer, Robert A.; Duncan, W. J.; Collar, A. R. *Elementary Matrices and Some Applications to Dynamics and Differential Equations.* Cambridge University Press, 1938. (Reprinted: 1947–1963. Reprint edition: AMS Press, New York, 1982.)
61. Frei, Günter. Zur Geschichte des Crelleschen Journals. *Journal für die reine und angewandte Mathematik*, 500 (1998), 1–4.
62. Frisch, Ragnar; Waugh, Frederick V. Partial time regressions as compared with individual trends. *Econometrica*, 1 (1933), 387–401.
63. Fritzsche, Bernd; Kirstein, Bernd, eds. *Ausgewählte Arbeiten zu den Ursprüngen der Schur-Analysis: Gewidmet dem großen Mathematiker Issai Schur (1875–1941).* Edited and with a foreword and afterword by Bernd Fritzsche and Bernd Kirstein; with contributions by G. Herglotz, I. Schur, G. Pick, R. Nevanlinna, H. Weyl, W. Ledermann & W. K. Hayman. Teubner-Archiv zur Mathematik 16, B. G. Teubner Verlagsgesellschaft, Stuttgart, 1991.
64. Gohberg, I., ed. *I. Schur Methods in Operator Theory and Signal Processing.* Operator Theory: Advances and Applications OT18, Birkhäuser Verlag, Basel, 1986.
65. de Voraigne, Jacobus. *The Golden Legend: Readings on the Saints, Volume I.*, translated by William Granger Ryan. Princeton University Press, 1993. (This translation by William Granger Ryan of *Legenda aurea* by Jacobus de Voragine (*c.* 1229–1298) is "based on the only modern Latin edition of the work, produced by Dr. Th. Graesse in 1845".)
66. Gordon, Cosmo. The first English books on book-keeping. *Accounting Research*, 5 (1954), 215–218. (Reprinted in Solomons & Zeff.[147])
67. Grala, Jolanta; Markiewicz, Augustyn; Styan, George P. H. Tadeusz Banachiewicz: 1882–1954. *Image: The Bulletin of the International Linear Algebra Society*, no. 25 (2000), page 24.

68. Gross, Jürgen; Puntanen, Simo. Extensions of the Frisch–Waugh–Lovell theorem. *Discussiones Mathematicae: Probability and Statistics*, 25 (2005), 39–49.
69. Guttman, Louis. Enlargement methods for computing the inverse matrix. *The Annals of Mathematical Statistics*, 17 (1946), 336–343.
70. Harrison, R. Martin. *Mountain and Plain: From the Lycian Coast to the Phrygian Plateau in the Late Roman and Early Byzantine Period,* edited by Wendy Young. University of Michigan Press, Ann Arbor, 2001. (Posthumous work based on the author's notes which were sorted and edited by Wendy Young.)
71. Haubold, Rev Dr. A. `www.nikolaikirche-leipzig.de` website: *Nicolaikirche Leipzig: A Short Architectural History,* accessed 19 January 2004.
72. Haynsworth, E. V. On the Schur complement. *Basel Mathematical Notes*, BMN 20, 17 pp., June 1968. (*Basle Mathematical Notes* on cover page.)
73. Haynsworth, E. V. Determination of the inertia of a partitioned Hermitian matrix. *Linear Algebra and its Applications*, 1 (1968), 73–81.
74. Haynsworth, Emilie V.; Ostrowski, Alexander M. On the inertia of some classes of partitioned matrices. *Basel Mathematical Notes*, BMN 18, 25 pp., August 1967. (*Basle Mathematical Notes* on cover page. Same title as Haynsworth & Ostrowski.[75])
75. Haynsworth, Emilie V.; Ostrowski, Alexander M. On the inertia of some classes of partitioned matrices. *Linear Algebra and its Applications*, 1 (1968), 299–316. (Same title as Haynsworth & Ostrowski.[74])
76. Henn, P. *Ahn's Second German Book: Being The Second Division of Ahn's Rudiments of the German Language.* Steiger's German Series, E. Steiger, New York, 1873.
77. Hocquél-Schneider, Sabine. *Alte Nikolaischule Leipzig.* Kulturstiftung Leipzig, Edition Leipzig, 1994.
78. *InstaPLANET presents Saint Nicholas of Myra: The Man and the Russian ICON.* `www.instaplanet.com/icon.html` website. Elsner SF.
79. Intellectual Reserve, Inc. `www.familysearch.org/eng/Search/RG/guide/German_Gothic99-36316.ASP` website: *Handwriting Guide: German Gothic Resource Guide*, Family History Library, Salt Lake City, accessed 28 December 2003.
80. *The Jewish Encyclopedia.* `www.jewishgen.org/belarus/je_mogilev.htm` website: Mogilev from *The Jewish Encyclopedia*, Moghilef (Mohilev). JewishGen Belarus SIG, accessed 28 December 2003.
81. Johnson, Norman Lloyd; Kotz, Samuel, eds. *Leading Personalities in Statistical Sciences: From the Seventeenth Century to the Present.* Wiley, New York, 1997.
82. Jones, Charles W. *Saint Nicholas of Myra, Bari, and Manhattan: Biography of a Legend.* The University of Chicago Press, 1978.
83. Joseph, Anthony; Melnikov, Anna. In memoriam: Issai Schur. In Joseph, Melnikov & Rentschler.[84]
84. Joseph, Anthony; Melnikov, Anna; Rentschler, Rudolf, eds. *Studies in Memory of Issai Schur.* Papers from the Paris Midterm Workshop of the Euro-

pean Community Training and Mobility of Researchers Network held in Chevaleret, France, May 21–25, 2000, and the Schur Memoriam Workshop held in Rehovot, Israel, December 27–31, 2000. Progress in Mathematics 210, Birkhäuser, Boston, 2003.

85. Kats, P. The "Nouuelle Instruction" of Jehan Ympyn Christophle: I, II. *The Accountant*, vol. 77, no. 2750, pp. 261–269 & no. 2751, pp. 287–296, 1927.
86. Katz, Elihu. Louis Guttman, 1916–1987. *Public Opinion Quarterly*, 52 (1988), 240–242.
87. Kavalieris, Laimonis; Lam, Fred C.; Roberts, Leigh A.; Shanks, John A., eds. *Proceedings of the A. C. Aitken Centenary Conference (incorporating the 3rd Pacific Statistical Congress, the Annual Meeting of the New Zealand Statistical Association and the 1995 New Zealand Mathematics Colloquium, 28 August – 1 September 1995)*. Otago Conference Series 5, University of Otago Press, Dunedin, New Zealand, 1996.
88. Kedar, Benjamin Z. *The Changing Land between the Jordan and the Sea: Aerial Photographs from 1917 to the Present.* Yad Ben-Zvi Press, Jerusalem & MOD Publishing House, Tel Aviv, 1999.
89. Klimchalk, Joan M. *Santa Claus: A Living Legend*, Stamp Exhibit winning the NTSS-2005 Reserve Grand Award, as announced and illustrated in *Topical Time*, vol. 56, no. 5 (September–October 2005), pp. 21 & 25.
90. Klimchalk, Joan M. `www.hwcn.org/link/cpc/cpc_klimch.html` website: "Who is Santa Claus?": article winning the 1998 Ken Mackenzie Writer's Award for the best article in *Yule Log*, 8 pp., on the Christmas Philatelic Club's Home Page, accessed 31 January 2006.
91. Kotz, Samuel; Johnson, Norman L.; Read, Campbell B., eds. *Encyclopedia of Statistical Sciences, Volume 3: Faà di Bruno's Formula–Hypothesis Testing.* Wiley, New York, 1983.
92. Kunisch. Letter relieving Issai Schur from his duties at the University of Berlin, 28 September 1935. (Reprinted in Soifer,[146] courtesy of Dr. W. Schultze and the Archive of the Humboldt University of Berlin.)
93. Ledermann, W. Issai Schur and his school in Berlin. *Bulletin of the London Mathematical Society*, 15 (1983), 97–106. (Reprinted in Fritzsche & Kirstein.[63])
94. Ledermann, Walter; Neumann, Peter M. The life of Issai Schur through letters and other documents. In Joseph, Melnikov & Rentschler.[84]
95. Lee, Peter M. `www.york.ac.uk/depts/maths/histstat/people/welcome.htm` website: *Portraits of Statisticians.* Department of Mathematics, The University of York, York, England, UK, accessed 11 January 2004.
96. Lovell, Michael C. Seasonal adjustment of economic time series and multiple regression analysis. *Journal of the American Statistical Association*, 58 (1963), 993–1010.
97. Lovell, Michael C. `mlovell.web.wesleyan.edu/` website: Department of Economics, Wesleyan University, Middletown, Connecticut, accessed 25 January 2006.
98. Martin, Stuart. *Schur Algebras and Representation Theory.* Cambridge University Press, 1993.

99. Mashey, Anne B. *A Guide to Olde German Handwriting of the Mid-1800's, No. 2*, Revised edition. Anne B. Mashey, Wexford, Pennsylvania, 1982. (Original version: 1979.)
100. McCulloh, John M. Review of Jones.[82] *The American Historical Review*, 84, page 720 (1979).
101. McLachlan, Gordon W. *Germany: The Rough Guide,* 4th edition. Rough Guides, London, 1998. (Original version: 1992.)
102. Meisen, Karl. *Nikolauskult und Nikolausbrauch im Abendlande: eine kultgeographisch-volkskundliche Untersuchung*, Reprint edition. Quellen und Abhandlungen zur mittelrheinischen Kirchengeschichte 41: Forschungen zur Volkskunde Heft 9-12, Schwann, Düsseldorf, 1981. (An enlargement of the author's Habilitationsschrift, Universität Bonn, 1928. Original version: 1931.)
103. *Merriam-Webster's Collegiate Dictionary*, 11th edition with CD-ROM and online subscription. Merriam-Webster, Springfield, Massachusetts, 2003.
104. *Merriam-Webster's Geographical Dictionary*, 3rd edition. Merriam-Webster, Springfield, Massachusetts, 1997 & 2001. (Original version: *Webster's New Geographical Dictionary*, 1972 & 1984.)
105. Mietelski, Jan. `www.oa.uj.edu.pl/history/history.html` website: *Two Hundred Years' History of the Cracow Astronomical Observatory*, The Observatory in the years of T. Banachiewicz's management (1919–1954), accessed 11 January 2004.
106. Miller, Jeff. `jeff560.tripod.com/` website: *Images of Mathematicians on Postage Stamps*. Gulf High School, New Port Richey, Florida.
107. Ne'eman, Yuval. Issai Schur died here: some background comments, in memoriam. In Joseph, Melnikov & Rentschler.[84]
108. Nes, Solrunn. *The Mystical Language of Icons*, 2nd edition. Eerdmans, Grand Rapids, Michigan, 2005. (First published 2004 by Eastern Christian Publications, Fairfax, Virginia.)
109. O'Connor, J. J.; Robertson, E. F., eds. MacTutor [History of Mathematics Archive] website: `www-history.mcs.st-andrews.ac.uk/Search/historysearch.html` School of Mathematics and Statistics, University of St Andrews, St Andrews, Scotland.
110. Odell, Patrick L. Gauss–Markov theorem. In Kotz, Johnson & Read.[91]
111. Ouellette, Diane Valérie. Schur complements and statistics. *Linear Algebra and its Applications*, 36 (1981), 187–295.
112. Pinl, Max; Furtmüller, Lux. Mathematicians under Hitler. In Weltsch.[166]
113. *PlanetWare: Your Unlimited Guide to the World.* `www.planetware.com/picture/myra-kale-tr-myra1.htm` website: PlanetWare, Inc., accessed 31 January 2006.
114. *Prominent Poles: A New Polish Online Biography Project.* Tadeusz Banachiewicz, astronomer and mathematician, website, accessed 31 January 2006: `www.polishwashington.com/prominent-poles/tadeusz.banachiewicz.htm`
115. Pugsley, A. G. Robert Alexander Frazer: 1891–1959. *Biographical Memoirs of Fellows of the Royal Society*, 7 (1961), 75–84.

116. Pukkila, Tarmo; Puntanen, Simo, eds. *Proceedings of the First International Tampere Seminar on Linear Statistical Models and their Applications: University of Tampere, Tampere, Finland, August 30th to September 2nd, 1983.* Department of Mathematical Sciences/Statistics, University of Tampere, Tampere, Finland, 1985.
117. Puntanen, Simo; Styan, George P. H. Historical introduction: Issai Schur and the early development of the Schur complement. Chapter 0 and Bibliography in Zhang.[174]
118. Puntanen, Simo; Styan, George P. H. Schur complements in statistics and probability. Chapter 6 and Bibliography in Zhang.[174]
119. Puntanen, Simo; Styan, George P. H. Some comments on Issai Schur (1875–1941), the Nicolai Gymnasium in Libau (Kurland), and publications under and I. Schur and J. Schur. Report 2005-06, Department of Mathematics and Statistics, McGill University, Montréal, November 2005.
120. Puntanen, Simo; Styan, George P. H. Some comments about Issai Schur (1875–1941) and the early history of Schur complements. Report 2006-02, Department of Mathematics and Statistics, McGill University, Montréal, February 2006.
121. Puntanen, Simo; Styan, George P. H. A philatelic introduction to matrices and statistics. Invited article in preparation for presentation at the 15th International Workshop on Matrices and Statistics, Uppsala, Sweden, June 2006.
122. Puntanen, Simo; Styan, George P. H. Stochastic stamps: a philatelic introduction to chance. Invited article in preparation for publication in *Chance* Magazine in celebration of its 20th birthday in 2007.
123. Radwin, Michael J. `www.hebcal.com/converter/` website: *Hebrew Date Converter*, accessed 28 September 2005.
124. Réau, Louis. *Iconographie de l'Art Chrétien.* Presses Universitaire de France, Paris, 1955–1959.
125. Reingold, Edward M.; Dershowitz, Nachum. *Calendrical Calculations: The Millennium Edition.* Cambridge University Press, 2001.
126. Relf, Ernest F. William Jolly Duncan: 1894–1960. *Biographical Memoirs of Fellows of the Royal Society*, 7 (1961), 37–51.
127. The Royal Society. `www.royalsoc.ac.uk/DServe/DServe.exe?dsqApp=Archive&dsqCmd=Index.tcl` website: *Library and Archive Catalogues.* The Royal Society, London, accessed 25 January 2006.
128. Schoeps, Julius H.; Grözinger, Karl E.; Mattenklott, Gert, eds. *Menora: Jahrbuch für deutsch-jüdische Geschichte 1999: im Auftrag des Moses Mendelssohn-Zentrums für europäisch-jüdische Studien.* Philo Verlagsgesellschaft mbH, Berlin & Bodenheim bei Mainz, 1999.
129. Schreiber, Peter. The Cauchy-Bunyakovsky-Schwarz inequality. In Schreiber.[130]
130. Schreiber, Peter, ed. *Hermann Graßmann, Werk und Wirkung: Internationale Fachtagung anläßlich des 150. Jahrestages des ersten Erscheinens der "Linealen Ausdehnungslehre" (Lieschow/Rügen, 23.–28.5.1994)*, Ernst-Moritz-Arndt-Universität, Greifswald, 1995. [English translation of title:

Hermann Graßmann, Work and Influence: Proceedings of the International Conference on the Occasion of the One Hundred and Fiftieth Anniversary of the First Publication of the "Lineale Ausdehnungslehre" (Lieschow/Rügen, 23.-28.5.1994).]

131. Schur, Issai. Über eine Klasse von Matrizen, die sich einer gegebenen Matrix zuordnen lassen. Doctoral dissertation, Universität Berlin, 1901; reprinted in Brauer & Rohrbach.[28])
132. Schur, J. [Schur, Issai.] Über Potenzreihen, die im Innern des Einheitskreises beschränkt sind [I]. *Journal für die reine und angewandte Mathematik*, 147 (1917), 205–232; reprinted in Brauer & Rohrbach[28] and in Fritsche & Kirstein.[63] (Translated into English as Schur.[137])
133. Schur, J. [Schur, Issai.] Über Potenzreihen, die im Innern des Einheitskreises beschränkt sind [II]. *Journal für die reine und angewandte Mathematik*, 148, 122–145 (1918); reprinted in Brauer & Rohrbach[28] and in Fritsche & Kirstein.[63] (Translated into English as Schur.[137])
134. Schur, Issai. *Biographischen Mitteilungen*, welche die Kaiserliche Leopoldino-Carolina Deutsche Akademie der Naturforscher nach §7 der Statuten von ihren neueintretenden Mitgliedern zur Aufbewahrung im Archiv erbittet, 24 June 1919. (In German. English translation of title: *Biographical communications*, forming the application for membership in the Academia Leopoldino-Carolina Naturae Curiosorum = German Academy of Natural Scientists.) Reprinted in Fritzsche & Kirstein.[63]
135. Schur, J. [Schur, Issai.] *Die Algebraischen Grundlagen der Darstellungstheorie der Gruppen.* Vorlesungen über Darstellungstheorie, gehalten auf Einladung des Mathematischen Seminars der Eidg. Techn. Hochschule Zürich, bearbeitet und herausgegeben von Dr. E. Stiefel, Graph. Anstalt Gebr. Frey & Kratz, Zürich, 1936. (Mimeographed lecture notes from Schur's 1936 lectures at the Eidgenössische Technische Hochschule (ETH) Zürich, edited and issued by E. Stiefel. See also Stammbach.[150])
136. Schur, Issai. *Vorlesungen über Invariantentheorie.* Bearbeitet und herausgegeben von Helmut Grunsky. Die Grundlehren der mathematischen Wissenschaften in Einzeldarstellungen mit besonderer Berücksichtigung der Anwendungsgebiete 143. Springer-Verlag, Berlin, 1968.
137. Schur, Issai. On power series which are bounded in the interior of the unit circle: I, II. In Gohberg.[64] (English translation of Schur.[132,133])
138. Schwarz, H. A. Miscellen aus dem Gebiete der Minimalflächen. *Vierteljahrsschrift der Naturforschenden Gesellschaft in Zürich*, 19 (1874), 243–271. (Revised as Ref. 139.)
139. Schwarz, H. A. Miscellen aus dem Gebiete der Minimalflächen. *Journal für die reine und angewandte Mathematik*, 80 (1875) 280–300. (Revised version of Ref. 138. Reprinted in Schwarz.[141])
140. Schwarz, H. A. Ueber ein die Flächen kleinsten Flächeninhalts betreffendes Problem der Variationsrechnung: Festschrift zum Jubelgeburtstage des Herrn Karl Weierstrass. *Acta Societatis Scientiarum Fennicæ (Helsinki)* 15:315–362 (1888). (See pp. 343–345. Preface dated 31 October 1885. Reprinted in Schwarz;[141] see pp. 251–253.)

141. Schwarz, H. A. *Gesammelte Mathematische Abhandlungen von H. A. Schwarz*, Julius Springer, Berlin, 1890. (Available online at `www.hti.umich.edu/u/umhistmath/` website: The University of Michigan Historical Mathematics Collection, Ann Arbor.)
142. Shye, Samuel. Louis Guttman. In Johnson & Kotz.[81]
143. Silverstone, H. Alexander Craig Aitken, M.A., D.Sc., Ll.D., F.R.S.E., F.R.S. (1895–1967). *Journal of the Royal Statistical Society, Series A*, 131 (1968), 259–261.
144. Smith, Steven Bradley. *The Great Mental Calculators: The Psychology, Methods, and Lives of Calculating Prodigies, Past and Present.* With a foreword by Wim Klein and an introduction by Hans Eberstark. Columbia University Press, New York, 1983.
145. Snyder, Stephen, project coordinator. `www.jewishgen.org/yizkor/libau/libau.html` website: *A Town Named Libau (Liepaja, Latvia)*, JewishGen, accessed 27 December 2003. (Translation of the 36-page booklet: *A Town Named Libau* in English, German and Hebrew and additional material about Libau, Editor and Publisher of booklet unknown, believed to have been published in Israel, 1985. Translation of Booklet: Naomi Arond; additional material translated and donated by Harry Hurwitz.)
146. Soifer, Alexander. Issai Schur: Ramsey theory before Ramsey. *Geombinatorics*, 5 (1995), 6–23.
147. Solomons, David; Zeff, Stephen A., eds. Accounting Research, *1948–1958: Volume 1, Selected Articles on Accounting History.* Garland, New York. (Reprints of selected articles from *Accounting Research*, 1948–1958.)
148. Soot-Ryen, Tron. `www.nrk.no/underholdning/store_norske/4380950.html` website: *Ragnar Frisch* (in Norwegian, NRK, Trondheim, accessed 4 February 2006. (Article includes a photograph of Ragnar Frisch: ©NTB arkiv/SCANPIX.)
149. *St. Nicholas Center Discovering the Truth about Santa Claus* website: `www.stnicholascenter.org/Brix?pageID=37` St. Nicholas Center, Holland, Michigan, accessed 29 January 2006.
150. Stammbach, Urs. *Die Zürcher Vorlesung von Issai Schur über Darstellungstheorie.* (With forewords by Walter Ledermann and Eduard Stiefel.) Schriftenreihe A: Wissenschaftsgeschichte 5, ETH-Bibliothek, Zürich, 2004. (See also Schur.[135])
151. Steele, J. Michael. *The Cauchy–Schwarz Master Class: An Introduction to the Art of Mathematical Inequalities.* Cambridge University Press, 2004.
152. Stigler, Stephen M. *Statistics on the Table: The History of Statistical Concepts and Methods.* Harvard University Press, 1999.
153. Storrer, Norman J.; Jensen, Larry O. *A Genealogical and Demographic Handbook of German Handwriting: 17th–19th Centuries, V.1: Births & Baptisms.* Norman J. Storrer, Pleasant Grove, Utah, 1977.
154. Strzałkowski, Adam. *Tadeusz Banachiewicz: Mistrz i Nauczyciel* (in Polish), website: `www.zwoje-scrolls.com/zwoje41/TB_01.jpg` accessed 24 January 2006. [English translation of title: *Tadeusz Banachiewicz: Champion and Scientist.*]

155. Styan, George P. H. Schur complements and linear statistical models. In Pukkila & Puntanen.[116]
156. Styan, George P. H., ed. Louis Guttman: 1916–1987. *The IMS Bulletin*, 17, page 284 (1988).
157. Styan, George P. H., ed. *The Collected Papers of T. W. Anderson: Volume 1, Volume 2.* Wiley, New York, 1990.
158. Tinbergen, J. Professor Ragnar Frisch. *Journal of the Royal Statistical Society Series A*, vol. 136 (1973), page 483.
159. Tucker, Suzetta. `http://ww2.netnitco.net/users/legend01/myrrh.htm` website: *ChristStory Christmas Symbols: Myrrh* (1999), ChristStory Christian Bestiary, accessed 4 February 2006.
160. *United States Government Printing Office Style Manual*, Revised edition. United States Government Printing Office, Washington, D. C., 1959. (Original version: 1894.)
161. URANOS Group. `www.uranos.eu.org/biogr/banache.html` website: *Tadeusz Banachiewicz (1882–1954)*, accessed 11 January 2004.
162. Vogt, Annette. Issai Schur: als Wissenschaftler vertrieben. In Schoeps, Grözinger & Mattenklott.[128]
163. Wales, Jimmy, founder. `en.wikipedia.org/wiki/Main_Page` website: *Wikipedia: The Free Encyclopedia.* Wikimedia Foundation, accessed 21 August 2005.
164. Watson, Geoffrey S.; Alpargu, Gülhan; Styan, George P. H. Some comments on six inequalities associated with the inefficiency of ordinary least squares with one regressor. *Linear Algebra and its Applications*, 264 (1997), 13–53.
165. Frederick Vail Waugh (1898–1974): unsigned obituary. *American Journal of Agricultural Economics*, 56 (1974), 680–681.
166. Weltsch, Robert, ed. *Year Book XVIII.* Publications of the Leo Baeck Institute, Secker & Warburg, London, 1973.
167. Whittaker, J. M.; Bartlett, M. S. Alexander Craig Aitken: 1895–1967. *Biographical Memoirs of Fellows of the Royal Society*, 14 (1968), 1–14.
168. Williams, Nicola; Herrmann, Debra; Kemp, Cathryn. *Lonely Planet Guide; Estonia, Latvia & Lithuania*, 3rd edition. Lonely Planet Publications, Melbourne, 2003. (Original version by John Noble, Nicola Williams & Robin Gauldie: 1997; 2nd edition by Nicola Williams, Kate Galbraith & Steve Kokker, 2000.)
169. Wilson, Robin J. *Stamping Through Mathematics.* Springer-Verlag, New York, 2001.
170. Wimp, Jet. Reviews: *To catch the spirit: the memoir of A. C. Aitken* [Univ. Otago Press, Dunedin, 1995], *Determinants and matrices* [Oliver and Boyd, Edinburgh, 1939], *The case against decimalisation* [ibid., 1962], *Gallipoli to the Somme: recollections of a New Zealand infantryman* [Oxford Univ. Press, Oxford, 1963] by A. C. Aitken. *The Mathematical Intelligencer*, 20 (1998), no. 2, 62–79.
171. Yamey, Basil S.; Kojima, Osamu, eds. *A Notable and Very Excellente Woorke by Jan Ympyn Christoffels.* Daigakudo Books, Kyoto, 1975.

172. Ympyn Christoffels, Jan. *A Notable and Very Excellente Woorke: expressyng and declaryng the maner and forme how to kepe a boke of acco[m]ptes or reconynges ...* translated with great diligence out of the Italian toung into Dutche, and out of Dutche into French, and now out of French (*Nouuelle instruction*, 1543) into Englishe. Goldsmiths'-Kress Library of Economic Literature 40.3, Microfilm reel, Research Publications, New Haven, Connecticut, 1974. (Original version published by Sir Richard Grafton, London, 1547.)
173. Zawada, Anna Karolina. *Observo ergo sum: Tadeusz Banachiewicz 1882–1954.* Muzeum Uniwerstytetu Jagiellońskiego Collegium Maius, Karków, 2004.
174. Zhang, Fuzhen, ed. *The Schur Complement and Its Applications.* Numerical Methods and Algorithms 4, Springer, New York, 2005.

PART B

Applications of Statistics

Estimating the Number of SARS Cases in Mainland China in 2002-3

Joe Gani

CMA, Mathematical Sciences Institute, Australian National University, Canberra ACT 0200, Australia
E-mail: gani@maths.anu.edu.au

This note is concerned with estimating the number of SARS cases in the 2002–3 epidemic in mainland China. Some recorded data from Hong Kong, and partial data from the mainland are used, while SARS incidence is modelled by a Beta function in order to derive likely outcomes. The conclusion is that there were many more cases of SARS in mainland China than were officially recorded.

Keywords: SARS; incidence; epidemic; recorded data.

AMS Classification: 92C60.

In their paper on the SARS epidemic in Hong Kong, Riley *et al.* (2003) provided graphs of the daily and weekly incidence of infectives beween the end of February and April 2003. The daily incidence graph looked very much like a Beta function, and suggested that the partial data available from the Chinese mainland, discussed by Lai (2005), and displayed in his Figure 1 reproduced below, might also be fitted by a similar Beta function.

While the SARS epidemic in mainland China lasted between the end of November 2002 and the end of May 2003 (say, 183 days from 30 November 2002 to 31 May 2003), data on SARS incidence was recorded for only the 41 days from 21 April to 31 May 2003, with a maximum incidence of 200 on 28 April 2003.

We first propose a Beta function with parameters $p = 2, q = 3$, for the partial 41-day data of the SARS epidemic in mainland China, this being the simplest function with the appropriate shape. We then extend our scope to the 183 days of the epidemic, and derive some tentative estimates of the total SARS cases during that period. Let T be the duration of the epidemic,

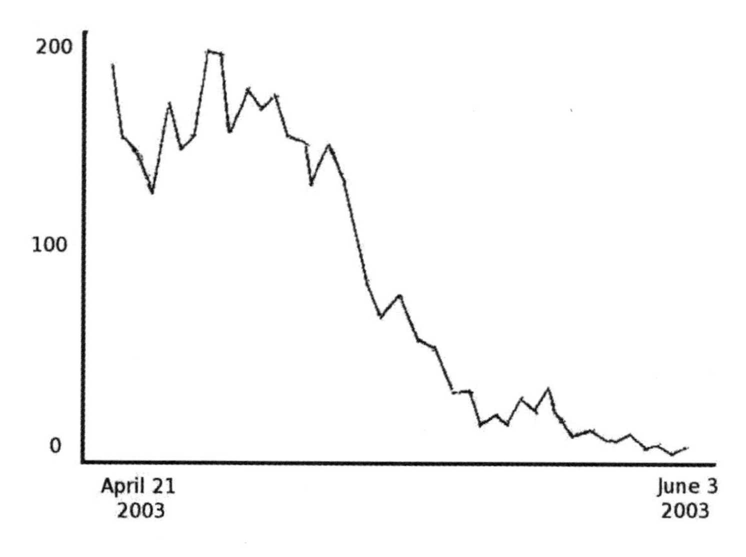

Fig. 1. Daily new SARS cases in mainland China from April 21 to June 3 2003. Reproduced from Figure 1 of Lai (2005).

and let the incidence of SARS at time $0 \leq t \leq T$ be given by

$$\begin{aligned} N(t) &= K\left(\frac{t}{T}\right)^{p-1}\left[\frac{T-t}{T}\right]^{q-1} \\ &= \frac{Kt^{p-1}(T-t)^{q-1}}{T^{p+q-2}} \end{aligned} \tag{1}$$

where K is a constant to be estimated later.

The time M at which the mode of $N(t)$ is reached occurs when

$$\begin{aligned} \frac{dN(t)}{dt} &= \frac{K[(p-1)t^{p-2}(T-t)^{q-1} - (q-1)t^{p-1}(T-t)^{q-2}]}{T^{p+q-2}} \\ &= 0, \end{aligned} \tag{2}$$

so that the maximum incidence is at time

$$M = \frac{(p-1)T}{(p+q-2)}, \tag{3}$$

or $T/3$ when $p = 2, q = 3$. This maximum incidence is then given by

$$\begin{aligned} N(M) &= \frac{KM^{p-1}(T-M)^{q-1}}{T^{p+q-2}} \\ &= \frac{K\,(T/3)\,(2T/3)^2}{T^3} \\ &= \frac{4K}{27}. \end{aligned} \tag{4}$$

Now, for the 41-day data, the mode was 200, a value which is possibly applicable for the longer 183-day series, although the lower peak of 160 reached on 8 May 2003 is perhaps more representative. If we adopt this as the value of $N(M)$, then

$$160 = \frac{4K}{27}, \quad \text{or} \quad K = 1\,080.$$

From this, we can approximate the total number of SARS cases during the 41-day period as the integral

$$\begin{aligned} \int_0^{41} N(t)dt &= \frac{1}{41^3}\int_0^{41} Kt(41-t)^2 dt \\ &= \left.\frac{K\left(41^2t^2/2 - 82t^3/3 + t^4/4\right)}{41^3}\right|_0^{41} \\ &= 3\,690. \end{aligned} \tag{5}$$

The observed number was 3 669, a figure close to our approximation, which tends to support our empirical approach.

Using similar methods over the total 183-day period of the epidemic, and now assuming 200 as the mode $N(M)$, we have that

$$200 = \frac{4K}{27}, \quad \text{or} \quad K = 1\,350.$$

Thus, the estimated total of expected SARS cases in the 183-day period would be

$$\begin{aligned} \int_0^{183} N(t)dt &= \frac{1}{183^3}\int_0^{183} Kt(183-t)^2 dt \\ &= \left.\frac{K\left(183^2t^2/2 - 366t^3/3 + t^4/4\right)}{183^3}\right|_0^{183} \\ &= 20\,587.5 \end{aligned} \tag{6}$$

This is probably closer to the true number of SARS cases in mainland China in 2002–3 than the officially recorded figure of 3 669. An alternative rough estimate may be derived as follows: the average number of daily recorded cases in the 41-day period was $3\,669/41 = 89.49$, so that if the same number of daily cases occurred over the 183-day epidemic, then the total number of cases would be

$$89.49 \times 183 = 16\,376,$$

a figure slightly lower than the estimated 20 588 above.

WHO reported the total number of SARS cases worldwide as 8 098 (WHO 2003), 3 669 of them in mainland China and 4 429 elsewhere. Even assuming the lower figure of 16 376 as an estimate of the total number of SARS cases in mainland China, we see that the worldwide total of cases was likely to be closer to

$$4\,429 + 16\,376 = 20\,805. \tag{7}$$

It should be pointed out in all caution that these are speculative estimates rather than established facts; the only certain conclusion is that mainland China suffered a heavier SARS toll in 2002–3 than has been reported.

Acknowledgment

I should like to thank Dr Linda Stals of the Department of Mathematics, Australian National University, for her help with the reproduction of Figure 1.

References

1. Lai, D. (2005) Monitoring the SARS epidemic in China: a time series analysis. *J.Data Sci.* 3, 279–294.
2. Riley, S., Fraser, C., Donnelly, C.A., Ghani, A.C., Abu-Raddad, L.J., Hedley, A.J., Leung, G.M., Ho, L-M., Lam, T-H., Thach, T.Q., Chau, P., Chan, K-P., Lo, S-U., Leung, P-Y., Tsang, T., Ho, W., Lee, K-H., Lau, E.M.C., Ferguson, N.M. and Anderson, R.M. (2003) Transmission dynamics of the etiological agent of SARS in Hong Kong: Impact of Public Health interventions. *Science* 300, 1961–1966.
3. WHO (2003) Cumulative number of reported probable cases of SARS. World Health Organization, Geneva.

Using Statistics to Determine the Effectiveness of Prescribed Burning

Karen J. King

Bushfire Cooperative Research Centre, School of Resources, Environment and Society, Australian National University, Canberra ACT 0200, Australia
E-mail: karen.king@anu.edu.au

Joanne Chapman

School of Physical, Environmental and Mathematical Sciences, UNSW@ADFA, Canberra ACT 2600, Australia
E-mail: j.chapman@adfa.edu.au

Prescribed burning is an important tool for managing landscapes for both ecological values and people and property protection. Although widely used, the long-term effectiveness of prescribed burning in meeting management objectives is often difficult to determine. Recent work is using a computer simulation model, FIRESCAPE-SWTAS, to investigate the effectiveness of a range of prescribed burning treatments in reducing risks posed by unplanned fires using likelihood methods. Initial studies were conducted in the World Heritage Area in south-west Tasmania. This talk discusses the linking of ecological modelling and statistical methods in determining the effectiveness of a range of prescribed burning scenarios in meeting management objectives in this landscape, and the implications for future work.

Keywords: simulation modelling; FIRESCAPE; unplanned fire; prescribed burning; likelihood methods.

1. Introduction

Fire, originating from both natural and anthropogenic sources, globally, is the principal natural disturbance event in terrestrial ecosystems (Gardner *et al.*, 1999). Fossil and palynological evidence suggests the occurrence of fires in the Australian landscape as far back as the Devonian (350 Ma BP) (Martin, 1996), with anthropogenic fires contributing to the historical fire regime for at least the last thirty to forty thousand years (Gill, 1981; Kershaw, 1986; Bowman, 1998; Perry, 1998; McKenzie, 2002).

It is implicit that successful fire management of Australian landscapes requires an accurate understanding of the temporal and spatial components of fire regimes in a range of ecosystems. Factors contributing to the nature of resultant fire regimes include ignition potentials, meteorological conditions, and the natural heterogeneity present in any landscape as a consequence of topography, the modern vegetation mosaic, and the historical fire regime (e.g. Rothermel, 1983; Pinol *et al.*, 2005). Knowledge of the interplay of these factors assists in determining the potential incidences and areas burnt by unplanned fires at different locations in landscapes. Fire management objectives are usually directed at reducing the risk bushfires pose to identified values, including the protection of people, property, biodiversity and water and air quality.

A prominent management option is to manipulate vegetation, and therefore fuel loads, through the implementation of an appropriate prescribed burning program (Gill and McCarthy, 1998; Pinol *et al.*, 2005). Consequently, in many landscapes prescribed burning is undertaken in the expectation that it is effective in reducing fuel loads and fuel connectivity, and therefore fire spread rates, intensities and resultant sizes. Reductions in spread rates and intensities, by definition, improve the facilitation of fire suppression efforts (Byram 1959; Gill *et al.*, 1987; Gill and McCarthy, 1998). While the logical basis of these premises is simple, they have rarely been validated with field and computer simulation data obtained over adequate time periods, and encompassing natural variability in meteorological and environmental conditions (Fernandes and Botelho, 2003).

The present study investigated the relationship between prescribed burning effort and the resultant unplanned fire sizes in south-west Tasmania, Australia. This study is part of a larger study investigating management options for reducing bushfire risk to a range of values, and in a diversity of Australian landscapes. South-west Tasmania consists of a World Heritage Area and Conservation Area, containing two large man-made hydro-electric scheme lakes (study area approximately 1.7 million hectares) (Fig. 1). With human occupation remaining low, both since European settlement in the early 1800s (Marsden-Smedley, 1998) and by Aborigines during the Holocene (Kiernan, 1983; Ryan, 1996), the wilderness and ecological values of this region remain high. This region is currently managed primarily for its conservation and biodiversity values. In this landscape there are a diversity of vegetation communities, including highly flammable buttongrass moorland (*Gymnoschoenus sphaerocephalus*) communities ($\sim$ 23%), less fire-prone wet scrub and wet sclerophyll forests, in addition to remnant patches of rainforest and alpine vegetation (Reid *et al.*, 1999).

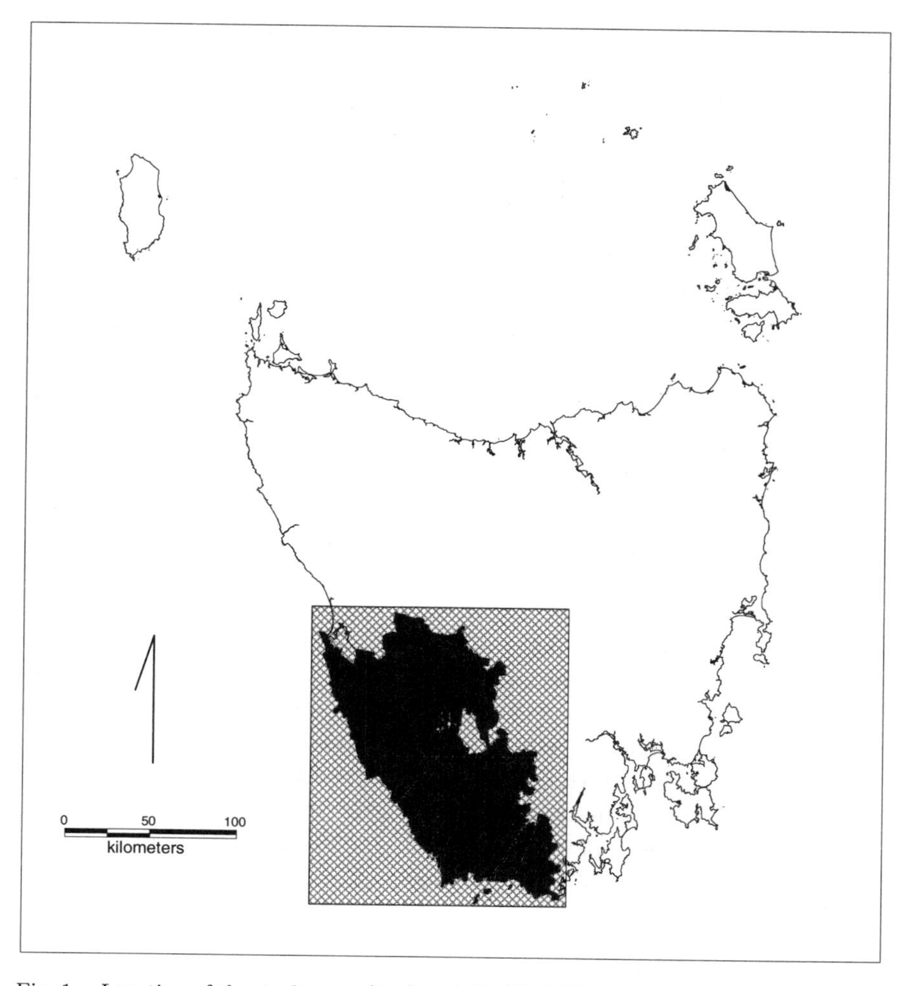

Fig. 1. Location of the study area (box) and the World Heritage Area (black) in southwest Tasmania, Australia.

As there is limited fire history and vegetation data available for southwest Tasmania, informed insights into the effects of long-term prescribed burning strategies is likely only through computer simulation modelling. FIRESCAPE-SWTAS (King, 2004), the computer simulation model used in this study, provides a method for predicting the long-term effects of different management treatments on the unplanned fire regime in the topographically complex landscape of south-west Tasmania. In this study, statistical analyses of both unplanned fire sizes and mean annual areas burnt resulting from different prescribed burning treatments were performed. Further, a statistical model was fitted for predicting the mean annual area burnt by unplanned fires under different prescribed burning treatments.

2. Methods

For this study the process-based computer simulation model designed for use in this region, FIRESCAPE-SWTAS (King, 2004), was used over a landscape including approximately 1.7 million pixels, each one hectare in size. This model incorporates a landscape fire regime simulator and a dynamic vegetation model. In this investigation one proposed and various hypothetical management options were explored to identify their impact on unplanned fire sizes and mean annual areas burnt by unplanned fires.

In FIRESCAPE-SWTAS a stochastic weather generator based on the work of Richardson (1981) was used to simulate daily meteorological conditions. Known monthly correlations of weather variables both within and between days were used in the development of this generator. In south-west Tasmania approximate mean winter and summer minimum and maximum temperatures are $3°C$ and $10°C$, and $10°C$ and $20°C$ respectively. Precipitation exhibits a winter peak, a strong east-west gradient, and an annual average of between 1500 and 3000 mm. Wind is predominantly from the north-west, and of greater velocity in the warmer months.

Fire behaviour algorithms in FIRESCAPE-SWTAS pertain to existing vegetation communities at each location (buttongrass moorlands – Marsden-Smedley and Catchpole, 1995a, 1995b, 2001; Marsden-Smedley *et al.*, 1999, 2001; heathlands – Catchpole *et al.*, 1998; Marsden-Smedley, 2002; forest – McArthur, 1967; Noble *et al.*, 1980). Vegetation responses to particular fire regimes used in the model follow the concept of 'ecological drift' proposed by Jackson (1968).

In south-west Tasmania, management burning is presently restricted to buttongrass moorland communities. To remain consistent with these practices, prescribed burning was simulated only in this vegetation community. During all simulations, fuel remaining after prescribed burns was determined based on meteorological conditions for the day, and therefore the expected fire behaviour during the prescribed burn. Consequently, fuel levels were reduced following prescribed burning, but not totally eliminated. In this study a series of hypothetical prescribed burning treatments, and one proposed treatment, were simulated. The proposed treatment was that outlined in the draft Tasmanian Wilderness World Heritage Area Tactical Management Plan 2004/05 (Tasmanian Parks and Wildlife Service, 2004). Prescribed burning treatment units in this management plan include approximately 22% of all the buttongrass moorland in the study area, with prescribed burns in these patches resulting in a mean inter-fire interval of 7 years (range 5 – 15 years).

For the hypothetical treatments, the landscape was divided into 89 large, approximately equal blocks. In different simulations various proportions of buttongrass moorland vegetation (0%, 2%, 5%, 10%, 20%, 33% and 50%) were burnt annually, with the assumption that all buttongrass in selected blocks was burnt. Two alternate spatial selection patterns (deterministic and random blocks) were simulated. The deterministic pattern burned blocks in a defined sequence across the landscape, such that patches exhibited a one year difference in their time since fire (Fig. 2). Consequently, in these simulations a consistent maximum inter-fire interval exists in all buttongrass moorland communities, with approximately equal proportions of buttongrass in each annual age class extending to this maximum age class. The alternate method for selecting prescribed burning blocks involved the random selection of a pre-defined proportion of blocks each year, again assuming all buttongrass was burnt in selected blocks (Fig. 2). Due to the random selection process, some blocks were burnt more frequently than others, resulting in a diversity of inter-fire intervals in buttongrass moorland communities.

Each simulated prescribed burning treatment was performed over 250 years, and with twenty replicates. In all simulations, unplanned fires originating from lightning occurred at approximately their present frequency and distribution, as determined from historical fire records (King, 2004). Lightning fires were ignited in simulations on days with similar meteorological conditions as they occurred historically. Analyses excluded the first fifty years of simulated data, so that all outputs reflected the response to the simulated prescribed burning treatments. Initially log-likelihood methods were used to fit the log-Normal distribution to each replicate of the fire size distribution data, and a likelihood ratio test performed to determine that all replicates of the same simulated prescribed burning treatment came from the same distribution. Therefore data from replicates were pooled for comparisons of fire size distributions and mean annual areas burnt by unplanned fires between simulated treatments.

Cochran's tests were performed prior to analyses of variance to test

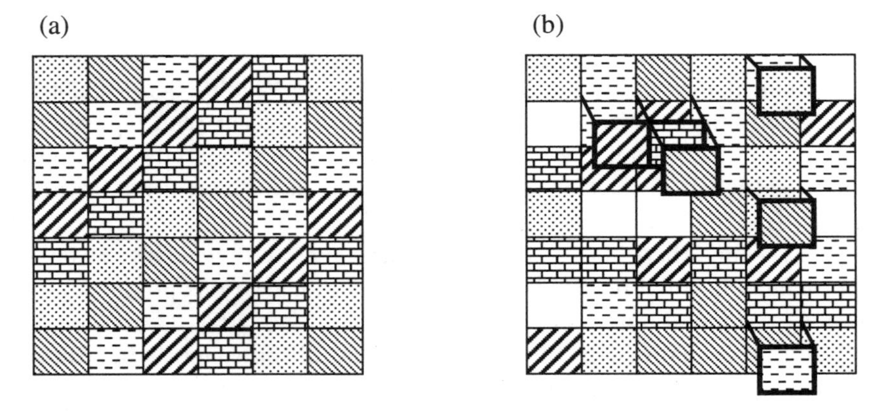

Fig. 2. Schematic diagram of (a) deterministic and (b) random selection strategies for simulations depicting hypothetical prescribed burning treatments.

for heterogeneity of variances. Two-way analyses of variance tests were then used to test the effects of simulated treatments (annual prescribed burning levels and spatial pattern) on the mean annual area burnt by unplanned fires. Multiple comparisons were then made between treatments using Tukey tests. Paired comparisons were made between the proposed treatment and all treatment levels using t-tests.

3. Results

3.1. *Comparison of replicate simulations*

Initially comparisons were made between replicates of the same simulated prescribed burning treatments to determine whether each set of fire size data came from the same distribution. For each treatment and strategy combination, a likelihood ratio test showed the replicates came from the same log-Normal distribution, assuming a log-Normal distribution described the data well. However the Kolmogorov-Smirnov goodness-of-fit test and probability plots showed the log-Normal distribution to be unsatisfactory for the data. Plots of the data revealed a very large number of one hectare fires in all data sets. As many of these fires occurred in conditions estimated to be marginally suitable for ignition, and as assumptions were made during model construction as to the conditions for initial ignition of fires, there is potential for significant inaccuracies in the number of one hectare fires in simulations. By marginally increasing the threshold of the data sets to a minimum of two hectares, these potentially unreliable data points were removed. Hutchinson (1995) similarly truncated sets of rainfall data that contained significantly large numbers of very small and potentially inaccurate values that were shown to be distorting the final distributions. Fire size distributions for all replicates of each treatment were then shown to come from the same log-Normal distribution (Table 1) and goodness of fit tests now showed the truncated data to be well approximated by this log-Normal distribution. The parameter estimates for each data set are shown in Table 2. All subsequent analyses were performed on these truncated data sets. A log-logistic distribution was also fitted, and although acceptable, did not provide such a good description of the data.

Table 1. P-values for the test that fire size data sets for replications of the same simulated prescribed burning treatment are from the same distribution.

	Percentage of buttongrass treated annually							
	0	**2**	**5**	**10**	**20**	**33**	**50**	**Proposed**
Deterministic	0.98	0.72	0.96	0.30	0.22	0.29	0.76	0.83
Random	0.98	0.98	0.89	0.44	0.22	0.76	0.74	0.83

Table 2. Parameter estimates on fitting the Log-Normal distribution to pooled data.

Percentage treated	Deterministic		Random	
	μ	σ	μ	σ
0%	5.553	2.491	5.553	2.491
2%	5.319	2.460	5.362	2.454
5%	4.985	2.348	5.172	2.426
10%	4.498	2.200	4.725	2.315
20%	3.766	1.897	4.210	2.185
33%	3.309	1.696	3.825	2.002
50%	3.024	1.637	3.415	1.904
Proposed	5.330	2.461	5.330	2.461

3.2. *Size distributions of unplanned fires*

As all replicates for each simulated prescribed burning treatment were shown to come from the same distribution (Table 1), data for each treatment were pooled for comparisons of fire size distributions between simulated treatments. Probability distributions of fire sizes (areas burnt) were compared between all simulated treatments, using logarithmic cumulative frequency plots (Fig. 3). These illustrate the proportion of fires (ordinate) that are as large as or larger than a given area in hectares (abscissa). All fire size distributions in this study demonstrated a majority of small fires, with a small number of large fires contributing to most of the total area burnt.

Likelihood ratio tests were performed to test for significant differences between fire size distributions for different levels of prescribed burning. Within each strategy, a significant difference was found between the fire size distributions for each consecutive level of hypothetical prescribed burning treatment (Table 3). This finding is supported by Fig. 4, which illustrates the parameter estimates with associated 95% confidence intervals against proportion of prescribed burning for each strategy. For both selection strategies there is no overlap between the 95% confidence intervals for μ. However there is a slight overlap in the intervals for σ at the lowest levels of burning.

Table 3. Likelihood ratio statistic, D, and p-values for the likelihood ratio test from a chi-squared distribution on two degrees of freedom for comparisons between simulated prescribed burning treatments.

Data	**Deterministic**		**Random**	
Comparisons	**D**	**p**	**D**	**p**
0% and 2%	60	$P < 0.001$	35.6	$P < 0.001$
2% and 5%	127.5	$P < 0.001$	34.8	$P < 0.001$
5% and 10%	274	$P < 0.001$	208.6	$P < 0.001$
10% and 20%	772	$P < 0.001$	277	$P < 0.001$
20% and 33%	345.5	$P < 0.001$	201.4	$P < 0.001$
33% and 50%	98	$P < 0.001$	167	$P < 0.001$
0% and Proposed	46.2	$P < 0.001$	46.2	$P < 0.001$
2% and Proposed	0.102	$P = 0.9504$	1.0	$P = 0.607$
5% and Proposed	133.5	$P < 0.001$	25	$P < 0.001$

The fire size distribution for the proposed treatment was compared to the results for the hypothetical data, and found not to be significantly different from the fire size distributions when 2% of buttongrass was burnt annually under either patch selection strategy. The proposed treatment was significantly different when compared to all other hypothetical treatments (Table 3).

Additionally, comparisons were made between spatial strategies (deterministic and random) for identical levels of prescribed burning. These identified significant differences in the resultant fire size distributions of unplanned fires for the two spatial selection processes when the proportion of buttongrass burnt annually was equal to, or greater than, 5%, with larger

(a)

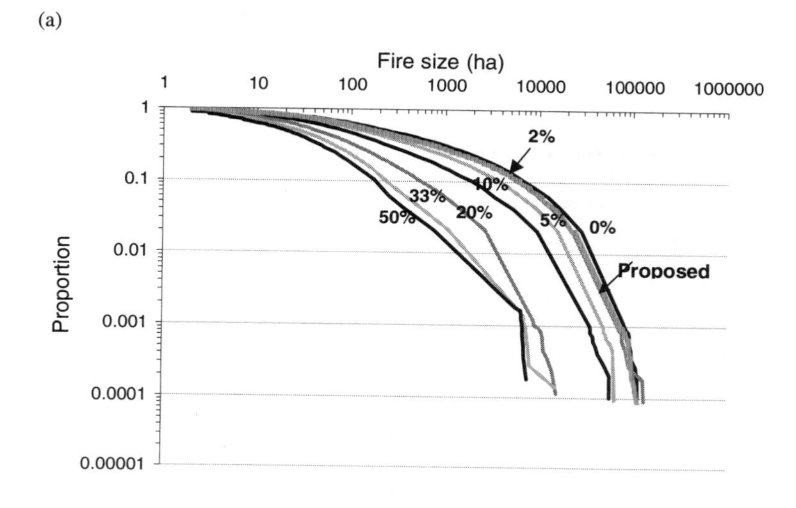

(b)

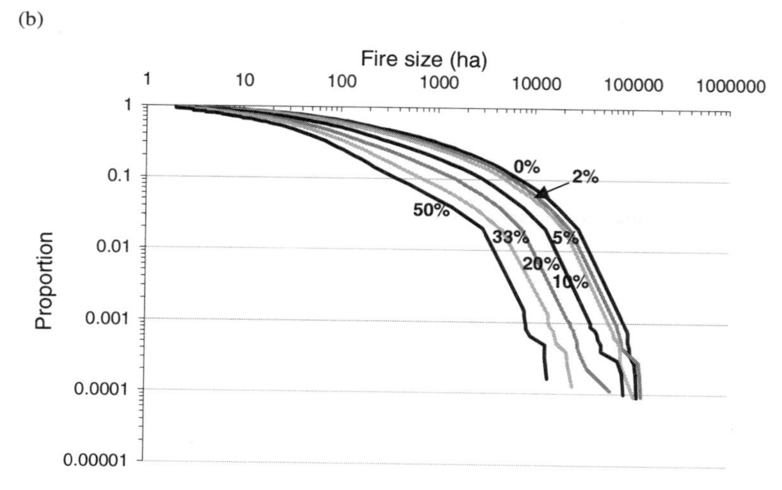

Fig. 3. The proportion of all unplanned fires as large as or larger than a given size for simulations with different buttongrass moorlands prescribed burn treatments, and blocks selected in either (a) a deterministic or (b) a random spatial strategy.

fire sizes observed for the random selection strategy (Table 4). These results are also supported by Fig. 4 and the non-overlap of the parameter intervals after the 5% level.

The ordered nature of the data, and the apparent inverse relationship

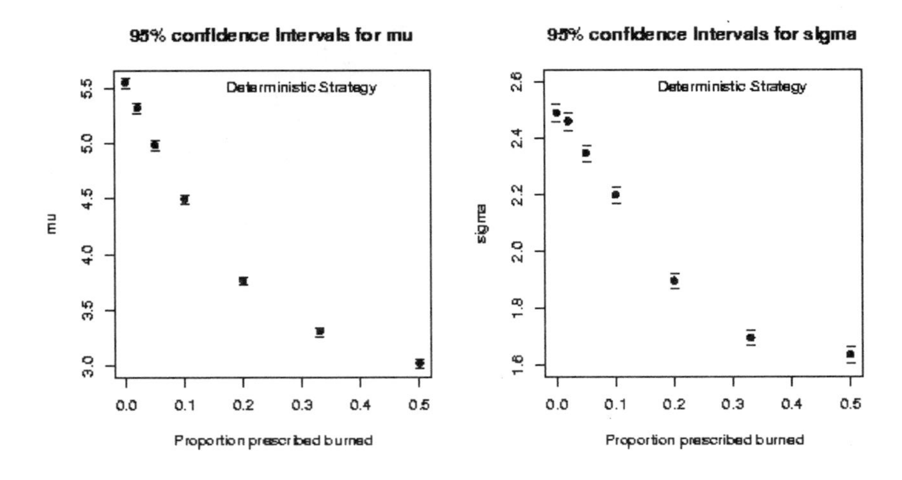

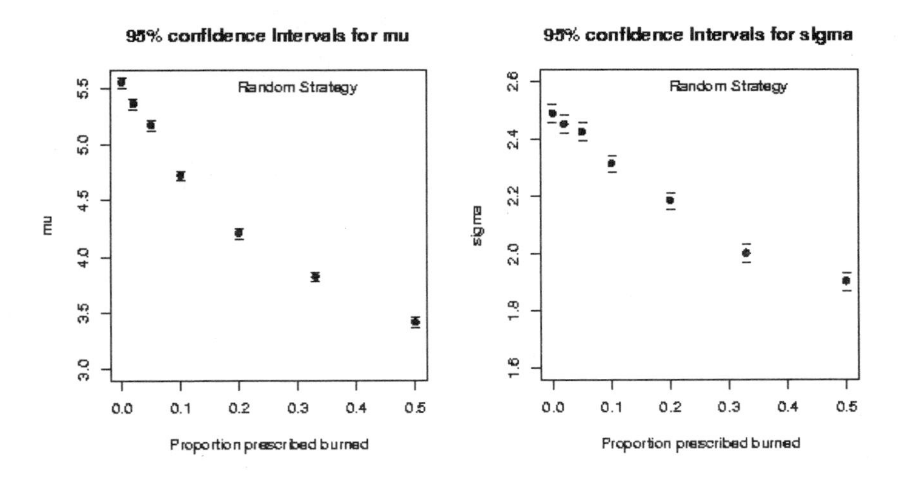

Fig. 4. Parameter estimates and associated 95% confidence intervals under both spatial strategies at each level of prescribed burning.

Table 4. Comparison of fire size distributions between patch selection strategies at each level of prescribed burning.

Percentage treated	D	p
2%	1.6	0.45
5%	44.2	2.52E-10
10%	76.4	0
20%	381.8	0
33%	485.5	0
50%	270.6	0

between the level of prescribed burning and the parameter estimates, suggested an alternative approach. Instead of fitting separate distributions to the data at different levels of prescribed burning, a single log-Normal distribution was fitted to all data within each spatial selection strategy, with parameters dependent on the proportion of buttongrass annually prescribed burnt. The results from this model showed that the proportion of buttongrass prescribed burnt has a significant effect on the parameter estimates, with smaller parameter estimates occurring for higher levels of annual prescribed burning (Table 5).

Table 5. Parameter estimates on fitting the Log-Normal distribution with parameters dependent on x, the proportion of prescribed burning.

	Deterministic	**Random**
μ	$\exp(1.676 - 1.348x)$	$\exp(1.689 - 1.028x)$
σ	$\exp(0.896 - 0.927x)$	$\exp(0.907 - 0.569x)$

3.3. *Mean annual areas burnt*

The mean annual area burnt by unplanned fires during simulations shows an approximate negative exponential distribution with respect to the proportion of buttongrass moorlands included in annual prescribed burning (Fig. 4). This implies that for linear increases in prescribed burning treatment there is a diminishing effect in reducing the mean annual area burnt by unplanned fires. This reflects the important contribution of the largest unplanned fires to the mean annual area burnt, and the observed decline in the size of the largest unplanned fires with increased prescribed burning treatment (Fig. 5).

A Cochran's test indicated that heterogeneity of variances was not significant following square root transformation of data for mean annual area

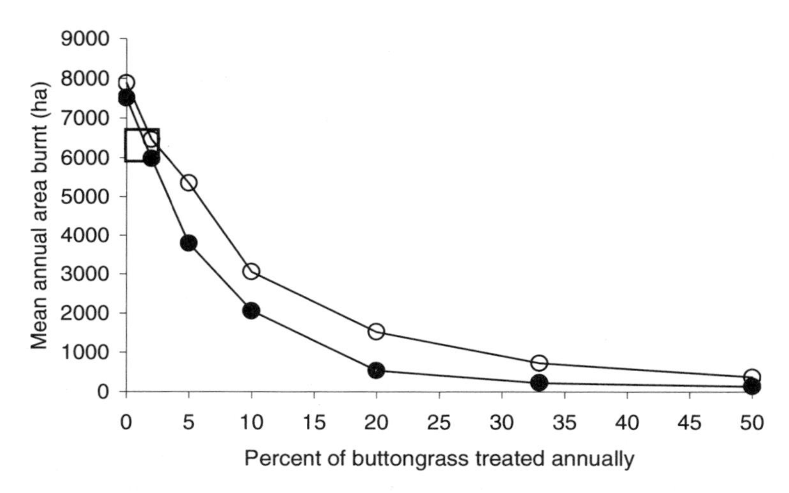

Fig. 5. Mean annual area burnt (ha) by unplanned fires for all simulated prescribe burning treatments (Deterministic — filled circles; Random — open circles; Proposed — squares).

burnt (Cochran's = 0.133; C_{crit} = 0.153). A two-way analysis of variance on these transformed data demonstrated that there was a significant interaction between effects of prescribed burning level and spatial strategy on mean annual area burnt (d.f. = 6; interaction $F = 20.1$; $P < 0.01$; levels $F = 2283$; $P < 0.05$). Multiple comparisons using Tukey tests demonstrated that means were significantly different between all treatment levels, with mean area burnt declining with increasing level of treatment. There was a significantly greater mean annual area burnt under a random, rather than deterministic, patch selection strategy for 5%, 10% and 20% treatment levels. There was no significant effect of spatial strategy at the other levels of treatment. These observations are consistent with the earlier observation that there was a significant difference in fire size distributions between spatial selection strategies when at least 5% of buttongrass moorlands was treated annually (Table 3).

T-tests between the proposed treatment and all hypothetical treatments identified non-significant differences in the mean annual area burnt only between the proposed and both 2% hypothetical treatments. This is in agreement with the results from Table 3 that indicate that the fire size distribution for the proposed treatment is not significantly different from that under either strategy at the 2% prescribed burning level.

4. Discussion

In this study, increasing the prescribed burning treatment level resulted in reductions in both the unplanned fire sizes and the mean annual area burnt by unplanned fires. Statistical analyses demonstrated significant differences in the fire size distributions between all treatment levels, with significantly greater areas burnt by unplanned fires for the random, rather than deterministic, patch selection strategy, at treatment levels of 5% or greater. These observations are consistent with a suite of studies examined by Fernandes and Botelho (2003) from North America, Europe and Australia, and a recent study by Pinol *et al.* (2005), all of which observed consistent declines in both fire sizes and mean annual areas burnt with increasing prescribed burning treatments.

In our study there was an observed approximate negative exponential relationship between the level of prescribed burning and the mean annual area burnt by unplanned fires. This implies that for a linear increase in prescribed burning effort, maximal reductions in the mean annual area burnt by unplanned fires occur when there is initially no, or minimal, initial prescribed burning. The significance of spatial strategy was observed to be the greatest between 5% and 20% treatment levels, with enhanced reductions in the mean annual area burnt with deterministic patch selection. This relationship was confirmed by the trend in the parameters of the fitted distributions.

Observed differences between the fire size distributions and mean annual areas burnt between patch selection strategies could be attributed to the fuel arrays resulting from these two spatial strategies. Fires propagate more readily where there is a continuous fuel bed with sufficient fuel loads. Random spatial strategies result in the potential for higher fuel loads to be maintained in areas not treated for extended periods, a factor conducive to the spread of large fires across the landscape. In contrast, deterministic spatial selection results in a maximum fuel age reflecting the proportion of buttongrass annually treated (maximum fuel age = $(\text{proportion treated annually})^{-1}$). Therefore it is possible that there was a greater potential for larger fires under the random spatial selection strategy, as some fires could propagate through areas not burnt for extended periods.

Observing the distribution of fire sizes for each different treatment combination provides more useful information than looking at the means and standard deviations alone. The truncated data were shown to be well modelled by a log-Normal distribution, which is heavily skewed to the right.

This shape is further emphasised with the inclusion of the one hectare fires. Large, high intensity, unplanned fires are those that pose the greatest threat to identified values in landscapes, as both they are responsible for the majority of the burnt area, and it is impossible to implement suppression efforts under the extreme fire weather conditions under which they occur. Consequently, management practices that reduce the incidence and size of the largest unplanned fires are those most effective in reducing the risk of the adverse effects from unplanned fires. The fitted distributions in this study provide a method for determining the probability of observing these large fires in the landscape under different treatment scenarios.

The design of the simulated proposed treatment was based on current knowledge of fire behaviour, and the location of identified values. Patch sizes, shapes, locations, and burning frequencies were selected in the anticipation of meeting all identified management objectives. Our simulations indicated that this treatment had no enhanced impact on reducing unplanned fire sizes or mean annual areas burnt. However, investigations beyond this study did indicate that this strategy had an enhanced effect in meeting other management objectives (King, pers. observ.).

It is possible to adopt the methodology used in this study to similar studies investigating alternate prescribed burning treatment options in a range of landscapes and ecosystems. Further studies can be performed to determine optimal patch locations, sizes, shapes and burning frequencies for meeting defined fire management objectives. Additionally investigations can be undertaken into the implications of climate and global change on management strategies. Principles attained through statistical analyses of simulation work can be applied to designing, modifying and prioritising the burning of treatment units.

References

1. Bowman, D.M.J.S. (1998) Tansley Review No. 101. The impact of Aboriginal landscape burning on the Australian biota. *New Phytologist.* 140, 385–410.
2. Byram, G.M. (1959) Combustion of forest fuels. In *Forest Fire: control and use.* (Ed K.P. Davis) pp 61-89. (McGraw-Hill: New York)
3. Catchpole, W.R., Bradstock, R.A., Choate, J., Fogarty, L.G., Gellie, N., McCarthy, G.J., McCaw, W.L., Marsden-Smedley, J.B. and Pearce, G. (1998) Co-operative development of equations for heathland fire behaviour. In *Proceedings of the 3rd International Conference of Forest Fire Research and 14th Conference of Fire and Forest Meteorology.* (Ed D.X. Viegas) pp 631–645. (University of Coimbra: Portugal)
4. Fernandes, P.M. and Botelho, H.S. (2003) A review of prescribed burning

effectiveness in fire hazard reduction. *International Journal of Wildland Fire* 12, 117–128.

5. Gardner, R.H., Romme, W.H. and Turner, M.G. (1999) Predicting forest fire effects at landscape scales. In Mlandenoff, D.J. and Baker, W.L. (eds) *Spatial Modeling of Forest Landscape Change.* Cambridge University Press, pp 186–209.
6. Gill, A.M. (1981) Adaptive responses of Australian vascular plant species to fire. In Gill, A.M., Groves, R.H. and Noble, I.R. (eds) *Fire and the Australian biota.* Australian Academy of Science, Canberra. pp 243–272.
7. Gill, A.M. and McCarthy, M.A. (1998) Intervals between prescribed fires in Australia: what intrinsic variations should apply? *Biological Conservation* 85, 161–169.
8. Gill, A.M., Christian, K.R., Moore, P.H.R., and Forrester, R.I. (1987) Bushfire incidence, fire hazard and fuel reduction burning. *Australian Journal of Ecology* 12 (3), 299–306.
9. Hutchinson, M.F. (1995) Stochastic space-time weather models from ground-based data. *Agricultural and Forest Meteorology* 73, 237–264.
10. Jackson, W.D. (1968) Fire, air, water and earth — An elemental ecology of Tasmania. *Proceedings of the Ecological Society of Australia* 3, 9–16.
11. Kershaw, A.P. (1986) The last two glacial/interglacial cycles from north-eastern Australia: implications for climate change and Aboriginal burning. *Nature.* 322, 47–49.
12. Kiernan, K.B. (1983) Relationship of cave fills to glaciation in the Nelson River Valley, central western Tasmania. *Australian Geographer* 15, 367–375.
13. King, K.J. (2004) Simulating the effects of anthropogenic burning on patterns of biodiversity. *PhD Thesis. Australian National University, Canberra.*
14. Marsden-Smedley, J.B. (1998) Changes in the south western Tasmanian fire regime since the early 1800s. *Papers and Proceedings of the Royal Society of Tasmania* 132, 15–29.
15. Marsden-Smedley, J.B. (2002) Version 1 Scrub Fire Danger Rating: outline of a prediction system. (Unpublished data) Tasmanian Department of Primary Industries, Water and Environment. Hobart. 10 pp.
16. Marsden-Smedley, J.B. and Catchpole, W.R. (1995a) Fire behaviour modelling in Tasmanian buttongrass moorlands. I. Fuel characteristics. *International Journal of Wildland Fire* 5, 203–214.
17. Marsden-Smedley, J.B. and Catchpole, W.R. (1995b) Fire behaviour modelling in Tasmanian buttongrass moorlands. II. Fire behaviour. *International Journal of Wildland Fire* 5, 215–228.
18. Marsden-Smedley, J.B. and Catchpole, W.R. (2001) Fire modelling in Tasmanian buttongrass moorlands. III Dead fuel moisture. *International Journal of Wildland Fire* 10, 241–253.
19. Marsden-Smedley, J.B., Catchpole, W.R. and Pyrke, A. (2001) Fire modelling in Tasmanian buttongrass moorlands. IV Sustaining versus non-sustaining fires. *International Journal of Wildland Fire* 10, 255–262.
20. Marsden-Smedley, J.B., Rudman, T., Pyrke, A. and Catchpole, W.R. (1999) Buttongrass moorland fire-behaviour prediction and management. *Tasforests*

11, 87–107.
21. Martin, H.A. (1996) Wildfires in past ages. *Proceedings of the Linnean Society of New South Wales.* 116, 3–17.
22. McArthur, A.G. (1967) Fire behaviour in eucalypt forests. Commonwealth of Australia Forest and Timber Bureau Leaflet No. 107.
23. McKenzie, G.M. (2002) The late Quaternary vegetation history of the south-central highlands of Victoria, Australia. II. Sites below 900m. *Australian Ecology*. 27, 32–54.
24. Noble, I.R., Bary, G.A.V., and Gill, A.M. (1980) McArthur's fire-danger meters expressed as equations. *Australian Journal of Ecology* 5, 201–203.
25. Perry, G.L.W. (1998) Current approaches to modelling the spread of wildland fire: a review. *Progress in Physical Geography.* 22 (2), 222–245.
26. Pinol, J., Beven, K. and Viegas, D.X. (2005) Modelling the effect of fire-exclusion and prescribed fire on wildfire size in Mediterranean ecosystems. *Ecological Modelling* 183. 397–409.
27. Reid, J.B., Hill, R.S., Brown, M.J. and Hovenden, M.J. (1999) (Eds) *Vegetation of Tasmania.* Environment Australia, Hobart.
28. Richardson, C.W. (1981) Stochastic simulation of daily precipitation, temperature, and solar radiation. *Water Resources Research* 17(1), 182–190.
29. Rothermel, R.C. (1983) *How to predict the spread and intensity of forest and range fires.* USDA Forest Service General Technical Report INT-143. Ogden, UT. 161 pp.
30. Ryan, L. (1996) *The Aboriginal Tasmanians.* Allen and Unwin, Singapore.
31. Tasmanian Parks and Wildlife Service, (2004 draft) *Tasmanian Wilderness World Heritage Area Tactical Management Plan. Version 4. 2004/5.* Fire Management Section. Parks and Wildlife Service, Department of Tourism, Parks, Heritage and The Arts, Hobart.

A Fair Tennis Scoring System for Doubles in the Presence of Sun and Wind Effects – An Application of Probability

Graham Pollard

School of Information Sciences and Engineering, University of Canberra, Canberra ACT 2601, Australia
E-mail: graham@foulsham.com.au

The present tennis scoring system is unfair in doubles for the case in which the two players in the team (or pair) are not equally effective on service. This unfairness can be resolved by making relatively small changes to the scoring system. It is even possible, by making additional changes, to produce a scoring system for doubles which is fair in the presence of wind and sun effects.

Keywords: unfairness in tennis doubles; sun and wind effects in tennis.

1. Introduction

A fair scoring system for a set of tennis between two doubles pairs A and B has the characteristic that Prob(A wins the set) = Prob(B wins the set) = 0.5 when the pairs are equal. The unfairness of the present tennis scoring system when used in doubles, for the case in which the two players in each doubles pairing are not equally effective on service, is demonstrated in Section 3. The methodology for converting the present scoring system into a fair one for doubles is also outlined in Section 3, and a solution to this unfairness in doubles is given in Section 4. This methodology outlined in Section 3 is then used to produce a doubles scoring system which is fair in the presence of sun and wind effects.

2. Singles

Two numerical examples are given in this section in order to give some insights that are useful for the remainder of the paper.

Example 2.1. This example appears in the paper by Newton and Pollard [1]. It considers a set of singles between two players A and B. Player A has a constant probability $PA = 0.7$ of winning each of his service games, and player B has a constant probability $PB = 0.6$ of winning each of his service games. Games are assumed to be statistically independent.

(a) Using "conventional" serving sequence, $ABABABABAB$...

$$
\begin{aligned}
P(A \text{ wins the set } 6\text{–}0) &= 0.021952 \\
P(A \text{ wins the set } 6\text{–}1) &= 0.0889056 \\
P(A \text{ wins the set } 6\text{–}2) &= 0.09567936 \\
P(A \text{ wins the set } 6\text{–}3) &= 0.187463808 \\
P(A \text{ wins the set } 6\text{–}4) &= 0.109593388 \\
\text{SUM} &= 0.503594156
\end{aligned}
$$

(b) Using the "extreme" serving sequence, $AAAAABBBBB/AB$...

$$
\begin{aligned}
P(A \text{ wins the set } 6\text{–}0) &= 0.067228 \\
P(A \text{ wins the set } 6\text{–}1) &= 0.0979608 \\
P(A \text{ wins the set } 6\text{–}2) &= 0.11310768 \\
P(A \text{ wins the set } 6\text{–}3) &= 0.115704288 \\
P(A \text{ wins the set } 6\text{–}4) &= 0.109593388 \\
\text{SUM} &= 0.503594156
\end{aligned}
$$

Although the "extreme" sequence is of no practical relevance, it can be seen that the probability that player A wins the set with a score of 6–0, 6–1, 6–2, 6–3 or 6–4 is the same for both serving sequences. It can be shown that, in general, the probability that player A wins the set is unaffected by using any service order whatsoever for the first 10 games, provided each player serves (or is planned to serve) 5 games each.

Example 2.2. This example is an extension of Example 2.1 to the case in which PA is not constant. We assume PA has values 0.5, 0.55, 0.6, 0.65 and 0.7 for Player A's service game 1 through to game 5 and that PA remains at 0.7 for service games 6 onwards. This might represent a case in which player A takes some time to reach full effect on his service. We assume player B is an identical and equal player (i.e. $PB = 0.5, 0.55, 0.6, 0.65, 0.7, \ldots$). It can be shown that for both the "conventional" and "extreme" sequences, we have $P(A \text{ wins the set}) = P(B \text{ wins the set}) = 0.5$. That is, the set scoring system is fair. Indeed, fairness is maintained under any arrangement of the five game probabilities (0.5, 0.55, 0.6, 0.65, 0.7) for either player.

3. The Present Doubles Scoring System

The doubles pair or team A has two players $A1$ and $A2$ and the pair or team B has two players $B1$ and $B2$. If we assume team A serves in the first game of the set, there are 4 service-game orders for the set:

(a) $A1, B1, A2, B2, \ldots$
(b) $A1, B2, A2, B1, \ldots$
(c) $A2, B1, A1, B2, \ldots$
(d) $A2, B2, A1, B1, \ldots$

Pollard [4] noted that the "games-structure" within the tiebreak set scoring system is "first to 6 games leading by at least 2 games; if 5 games each is reached, first to 7 games leading by 2 games; and if 6 games each is reached, play the tiebreak game", which he denoted by

$$F(6,2); \quad \text{if 5–5, } F(7,2); \quad \text{if 6–6, play } TB.$$

He also noted that this games-structure can be seen as "best of 10 games; if 5 games each is reached, best of 12 games; and if 6 games each is reached, play the tiebreak game", which he denoted by

$$B(10); \quad \text{if 5–5, } B(12); \quad \text{if 6–6, play } TB.$$

We now consider the "points-structure" within the tiebreak game. The order of the types of service points, by the rules of tennis, must be the same as the order of the types of service games. Thus, the tiebreak point-order for order (a) above is

$$A1, B1, B1, A2, A2, B2, B2, A1, A1, B1, B1, A2, A2, B2, B2, A1, \ldots.$$

Pollard [4] noted that the tiebreak stopping rule is "first to 7 points leading by at least 2 points; if 6 points each is reached first to 8 points leading by 2 points; if 7 points each is reached, first to 9 points leading by 2 points;..." which he denoted by

$$TB(7,2); \quad \text{if 6–6, } TB(8,2); \quad \text{if 7–7, } TB(9,2); \quad \ldots.$$

He also noted that this tiebreak stopping rule can be seen as

$$B(12); \quad \text{if 6–6, } B(14); \quad \text{if 7–7, } B(16); \quad \ldots.$$

The following numerical example appears in the paper by Pollard and Noble [2]. Player $A1$ is assumed to have a constant probability $pa1$ of winning a point on service whilst player $A2$ is assumed to have a constant probability $pa2$ of winning a point on service. It is not uncommon in practice, particularly in mixed doubles, for one of these probabilities to be greater than the other. Player $A1$ is assumed to be the more effective on service, with $pa1 = 0.7$ and $pa2 = 0.5$. The B pair is assumed to be identical (or equal) to the A pair. Thus $pb1 = 0.7$ and $pb2 = 0.5$. It is clear that when we have two identical and equal teams we can assume without loss of generality that team A serves first, and hence only the 4 service-orders (a), (b), (c) and (d) above need to be considered.

In Table 1 the unfairness of the "games-structure" within the tiebreak set scoring system can be seen for service orders (b) and (c). For order (b) the more effective team A player, $A1$, serves 3 service games within the $B(10)$-games (sub-) structure whereas the more effective team B player, $B1$, serves just 2 service games. Not surprisingly, for order (b), $P(A$ wins the set$) > P(B$ wins the set$)$, demonstrating the unfairness of the $B(10)$-games (sub-) structure. Analogously, it can be seen that $P(A$ wins the set$) < P(B$ wins the set$)$ for service order (c). It is noted that the games-structure for orders (a) and (d) is fair.

Table 1. The probability team $(A1, A2)$ wins the tiebreak set when $pa1 = pb1 = 0.7$ and $pa2 = pb2 = 0.5$, for the four orders of service (correct to 4 decimal places).

Order	(a) $A1, B1, A2, B2$	(b) $A1, B2, A2, B1$	(c) $A2, B1, A1, B2$	(d) $A2, B2, A1, B1$
$P(A$ wins set before tiebreak$)$	0.4209	0.4858	0.3712	0.3825
$P(B$ wins set before tiebreak$)$	0.4209	0.3712	0.4858	0.3825
$P($Tiebreak is played$)$	0.1582	0.1430	0.1430	0.2349
$P(A$ wins if tiebreak is played$)$	0.4618	0.5287	0.4719	0.5356
$P(A$ wins in a tiebreak$)$	0.0731	0.0756	0.0675	0.1258
$P(A$ wins the set$)$	0.4940	0.5614	0.4387	0.5083

The unfairness of the tiebreak points-structure within the tiebreak game stopping rule can be seen in Table 1 for all service point orders. In order (a) for example, player $B1$ serves on 4 occasions within the $B(12)$-points (sub-) structure, whereas player $B2$ serves on just 2 occasions, whilst players $A1$ and $A2$ serve on 3 occasions each. Similarly, the $B(14)$-points (sub-) structure within the tiebreak game stopping rule can be seen to be unfair, whilst the $B(16)$-points (sub-) structure is indeed fair as each player serves on 4 occasions each. Thus it can be seen that the tiebreak game is unfair for all orders of serving, and hence the tiebreak set itself must be unfair for all orders of serving.

The (longer) tiebreak game, "$TB(10,2)$; if 9–9, $TB(11,2)$; if 10–10, $TB(12,2)$, ...", an optional scoring system within the rules of tennis and the one used for the third set in the Australian Open Mixed Doubles Championship, is also unfair for similar reasons.

4. Solution to the Unfairness of the Present Scoring System for Doubles

Pollard [4] showed that the "games-structure" within the set could be made fair by using only $B(n)$-games (sub-) structures in which n is always a multiple of 4 (the number of players or servers). He also showed that the tiebreak game structure within the set could be made fair by using only $B(n)$-points (sub-) structures in which n is a multiple of 8 (4 players each serving 2 points at a time). Thus, he noted the following fair structures.

Games Structure :

$B(8)$; if 4–4, $B(12)$; if 6–6, $B(16)$; ... (advantage set)

(i.e. $F(5,2)$; if 4–4, $F(7,2)$; if 6–6, $F(9,2)$; ... (advantage set))

or

$B(8)$; if 4–4, play modified TB (tiebreak set)

(i.e. $F(5,2)$; if 4–4, play modified TB (tiebreak set))

Modified Tiebreak Game Structure :

$B(16)$; if 8–8, $B(24)$; if 12–12, $B(32)$; ...

(i.e. $TB(9,2)$; if 8–8, $TB(13,2)$; if 12–12, $TB(17,2)$; ...)

5. A Fair Scoring System for a Set of Doubles in the Presence of Sun and Wind Effects

The probability of winning a point can differ from one end of the court to another, for example in the presence of a wind effect. This is a playing-from-a-particular-end effect, E. Also, the sun can affect serves from one end of the court. This is a sun or serving-from-a-particular-end effect, S. Fair stopping rules can be devised for the case in which these effects exist, and the following is an extension of earlier work by Pollard and Noble [3].

It is assumed that players $A1$ and $B1$'s game-probabilities when serving from end 1 become $PA1 + E1 + S1$ and $PB1 + E1 + S1$ respectively, where $E1$ is the effect both players $A1$ and $B1$ have from being at end 1 (whether serving or receiving) and $S1$ is the effect both players $A1$ and $B1$ have when serving (but not receiving) from end 1. Correspondingly it is assumed players $A2$ and $B2$'s game-probabilities when serving from end 1 are $PA2 + E2 + S2$ and $PB2 + E2 + S2$. It is assumed that team A's probability of winning a game from end 1 is $QB1 + E3$ when $B1$ is serving from end 2, and $QB2 + E4$ when $B2$ is serving from end 2, where $QB1 = 1-PB1$ and $QB2 = 1-PB2$. Also, team B's probability of winning a game from end 1 is assumed to be $QA1 + E3$ when $A1$ is serving from end 2, and $QA2+E4$ when $A2$ is serving from end 2. We assume the teams change ends after 2, 6, 10,... games are played.

Assuming the service order is A1, B1, A2, B2 with team A commencing service at end 1, team A's game-probabilities for games 1 through to 8 can be seen to be $PA1 + E1 + S1$, $QB1 + E3$, $PA2 - E4$, $QB2 - E2 - S2$, $PA1 - E3$, $QB1 - E1 - S1$, $PA2 + E2 + S2$ and $QB2 + E4$ respectively. Correspondingly, team B's game-probabilities for games 1 through to 8 are $QA1 - E1 - S1$, $PB1 - E3$, $QA2 + E4$, $PB2 + E2 + S2$, $QA1 + E3$, $PB1 + E1 + S1$, $QA2 - E2 - S2$ and $PB2 - E4$. It can be seen that when $PA1 = PB1$ and $PA2 = PB2$, the 8 probabilities for team A are equal to the 8 probabilities for team B although the orderings are different. It follows that stopping rules making use of 8 games at a time are fair. Thus, in the presence of wind and sun effects, and changing ends after 2, 6, 10, ... games, the following stopping rules are fair.

Games Structure:

$B(8)$; if 4–4, $B(16)$; if 8–8, $B(24)$; ... (advantage set)

(i.e. $F(5,2)$; if 4–4, $F(9,2)$; if 8–8, $F(13,2)$; ... (advantage set))

or

$B(8)$; if 4–4, play "extended" modified TB (tiebreak set)

(i.e. $F(5,2)$; if 4–4, play "extended" modified TB (tiebreak set))

where the "extended" modified TB game is described in the following paragraph.

We now design an "extended" modified TB game for the case in which there are end and sun effects. It is assumed players $A1$ and $B1$'s point-probabilities when serving from end 1 are $pa1 + e1 + s1$ and $pb1 + e1 + s1$, whilst players $A2$ and $B2$'s point-probabilities when serving from end 1 are $pa2+e2+s2$ and $pb2+e2+s2$ respectively. It is assumed team A's probability of winning a point when receiving from end 1 is $qb1+e3$ when $B1$ is serving from end 2, and $qb2+e4$ when $B2$ is serving from end 2 ($qa1 = 1-pa1, \ldots$). It is also assumed that team B's probability of winning a point from end 1 is $qa1 + e3$ when $A1$ is serving from end 2, and $qa2 + e4$ when $A2$ is serving from end 2. It is assumed that the teams change ends after 8, 24, 40, ... points.

Assuming the point service order is (just for a change!) $A1, B2, A2, B1$ with player $A1$ starting service from end 1, it can be seen that team A's point-probabilities for points 1 through to 16 are $pa1 + e1 + s1$, $qb2 + e4$, $qb2+e4$, $pa2+e2+s2$, $pa2+e2+s2$, $qb1+e3$, $qb1+e3$, $pa1+e1+s1$, $pa1-e3$, $qb2 - e2 - s2$, $qb2 - e2 - s2$, $pa2 - e4$, $pa2 - e4$, $qb1 - e1 - s1$, $qb1 - e1 - s1$ and $pa1 - e3$. It can be seen that, when $pa1 = pb1$ and $pa2 = pb2$, the above 16 probabilities for team A are equal to the corresponding 16 probabilities for team B, although the orderings are different. It follows that stopping rules making use of 16 points at a time are fair. Thus, in the presence of the wind and sun effects, and changing ends after 8, 24, 40, ... points, the following stopping rule is fair.

"Extended" modified tiebreak structure:

$B(16)$; if 8–8, $B(32)$; if 16–16, $B(48)$; ...

or

$TB(9,2)$; if 8–8, $TB(17,2)$; if 16–16, $TB(25,2)$; ...

It can be seen that there is a price paid for including wind and sun effects, in that there are now fewer "stopping situations". For example, within the games-structure for the advantage set, winning games' scores of 7–4 and 7–5 are no longer fair in the presence of end and sun effects. Correspondingly, within the "extended" modified tiebreak game, winning points' scores of 13–8, 13–9, 13–10 and 13–11 are also no longer fair.

It can be shown that the "extended" fair set model presented in this section remains fair in the presence of (serving to the) forehand- and backhand-court effects. For example, $pa1$ becomes $pa1 + f1$ when serving to the forehand-court and it becomes $pa1 + b1$ when serving to the backhand-court ($f1 + b1 = 0$). Correspondingly $pa2$ becomes $pa2 + f2$ and $pa2 + b2$ ($f2 + b2 = 0$), and we make associated changes to $pb1$ and $pb2$. The reader is referred to Pollard and Noble [3] for an example using forehand-court and backhand-court effects.

The methods in this paper can also be used to produce fair tiebreak scoring systems for the win by at least 4 (and win by at least 6, ...) points structures within the tiebreak game. It can be seen that win by at least 3 (and win by at least 5, ...) points structures are unfair.

References

1. P. Newton and G. Pollard (2004) Service neutral scoring strategies in tennis. *Proceedings of the Seventh Australian Conference on Mathematics and Computers in Sport*, edited by R. Hugh Morton and S. Ganesalingam, Massey University, NZ (pp 221–225).
2. G. Pollard and K. Noble (2003) A solution to the unfairness of tiebreak tennis doubles. *Tennis Science and Technology 2*, edited by S. Miller (International Tennis Federation, Roehampton, London) (pp 325–332).
3. G. Pollard and K. Noble (2004) Some attractive properties of the 16-point tiebreak game in tennis. *Proceedings of the Seventh Australasian Conference on Mathematics and Computers in Sport*, edited by R. Hugh Morton and S. Ganesalingam, Massey University, NZ (pp 257–261).
4. G. Pollard (2005) On solving an aspect of unfairness in the tennis doubles scoring system. *Chance*, Vol 18, No. 4 (pp 17–19).

PART C

Theoretical Issues in Probability and Statistics

Perturbed Markov Chains

Jeffrey J. Hunter
Institute of Information and Mathematical Sciences, Massey University, Auckland 1330, New Zealand
E-mail: j.hunter@massey.ac.nz

For finite irreducible discrete time Markov chains, whose transition probabilities are subjected to a perturbation, the "mean first passage times" and the "mixing times" play an important role in determining, respectively, the relative and absolute differences between the stationary probabilities in the perturbed and unperturbed situations. Bounds for these differences are given and are illustrated by means of an example.

Keywords: Markov chain; generalised inverse; stationary distribution; mean first passage time; mixing time; perturbation theory.

1. Introduction

Let $P = [p_{ij}]$ be the transition matrix of a finite irreducible m-state Markov chain. Let $\tilde{P} = [\tilde{p}_{ij}] = P + \mathbf{E}$ be the transition matrix of the perturbed Markov chain where $\mathbf{E} = [\epsilon_{ij}]$ is the matrix of perturbations. We assume that the perturbed Markov chain is also irreducible with the same state space $S = \{1, 2, \ldots, m\}$. Let $\boldsymbol{\pi}^T = (\pi_1, \pi_2, \ldots, \pi_m)$ and $\tilde{\boldsymbol{\pi}}^T = (\tilde{\pi}_1, \tilde{\pi}_2, \ldots, \tilde{\pi}_m)$ be the stationary probability vectors for the Markov chains with transition matrices P and $\tilde{P}$, respectively. Let $M = [m_{ij}]$ be the "mean first passage time" matrix and let $N = [n_{ij}] = [(1 - \delta_{ij})m_{ij}\pi_j]$ be the "matrix of mixing times" (Hunter [8]) of the Markov chain with transition matrix P. Let $\mathbf{e}^T = (1, 1, \ldots, 1)$.

In Section 2 we develop a general relationship between $\boldsymbol{\pi}^T$ and $\tilde{\boldsymbol{\pi}}^T$ in terms of generalised inverses (g-inverses) of $I - P$. This leads to a series of special cases, some of which are new. Alternative relationships in terms of M and N are also derived.

In Section 3 these structural results are used to derive absolute component-wise and relative error bounds between the stationary probabilities of the two Markov chains.

In Section 4 the special case involving perturbations in a single row are considered and some new interrelationships derived.

In Section 5 an example illustrating these results is provided.

2. Updating Stationary Probability Vectors Under General Perturbations

We show that for any general perturbation $\mathbf{E}$ the difference $\tilde{\boldsymbol{\pi}}^T - \boldsymbol{\pi}^T$ can always be expressed in the form

$$\tilde{\boldsymbol{\pi}}^T - \boldsymbol{\pi}^T = \tilde{\boldsymbol{\pi}}^T \mathbf{E} H. \tag{1}$$

The most general expression for H can be given in terms of g-inverses of $I - P$. We make use of the following relevant results that appear in Hunter [5], [6].

- A g-inverse A^- of a matrix A has the property that $AA^-A = A$.
- For finite irreducible Markov chains, $I - P + \mathbf{t}\mathbf{u}^T$ is non-singular if and only if $\mathbf{t}$ and $\mathbf{u}$ are vectors such that $\boldsymbol{\pi}^T\mathbf{t} \neq 0$ and $\mathbf{u}^T\mathbf{e} \neq 0$.
- Further, if $\boldsymbol{\pi}^T\mathbf{t} \neq 0$ and $\mathbf{u}^T\mathbf{e} \neq 0$, then $[I - P + \mathbf{t}\mathbf{u}^T]^{-1}$ is a g-inverse of $I - P$.
- All g-inverses of $I - P$ are of the form $[I - P + \mathbf{t}\mathbf{u}^T]^{-1} + \mathbf{e}\mathbf{f}^T + \mathbf{g}\boldsymbol{\pi}^T$ for arbitrary $\mathbf{f}$ and $\mathbf{g}$.

Theorem 2.1. *If G is any g-inverse of $I - P$, then*

$$\tilde{\boldsymbol{\pi}}^T - \boldsymbol{\pi}^T = \tilde{\boldsymbol{\pi}}^T \mathbf{E} G(I - \Pi) \tag{2}$$

where $\Pi = \mathbf{e}\boldsymbol{\pi}^T$, so that in (1) H can be taken as $G(I - \Pi)$.

Proof. Since $\boldsymbol{\pi}^T(I - P) = \mathbf{0}^T$ and $\tilde{\boldsymbol{\pi}}^T(I - \tilde{P}) = \tilde{\boldsymbol{\pi}}^T(I - P - \mathbf{E}) = \mathbf{0}^T$,

$$(\tilde{\boldsymbol{\pi}}^T - \boldsymbol{\pi}^T)(I - P) = \tilde{\boldsymbol{\pi}}^T \mathbf{E}. \tag{3}$$

Thus post-multiplying (3) by $G(I - \Pi)$ yields

$$(\tilde{\boldsymbol{\pi}}^T - \boldsymbol{\pi}^T)(I - P)G(I - \Pi) = \tilde{\boldsymbol{\pi}}^T \mathbf{E} G(I - \Pi). \tag{4}$$

By taking G as the general form $[I - P + \mathbf{t}\mathbf{u}^T]^{-1} + \mathbf{e}\mathbf{f}^T + \mathbf{g}\boldsymbol{\pi}^T$ and using the observations that $\Pi = \mathbf{e}\boldsymbol{\pi}^T$, $\boldsymbol{\pi}^T\mathbf{e} = 1$, $P\mathbf{e} = \mathbf{e}$, and the results (see [5]) that

$$[I - P + \mathbf{t}\mathbf{u}^T]^{-1}\mathbf{t} = \frac{\mathbf{e}}{\mathbf{u}^T\mathbf{e}}\,, \tag{5}$$

$$(I - P)[I - P + \mathbf{t}\mathbf{u}^T]^{-1} = I - \frac{\mathbf{t}\boldsymbol{\pi}^T}{\boldsymbol{\pi}^T\mathbf{t}}\,, \tag{6}$$

it can be seen that, for any g-inverse G of $I-P$,

$$(I-P)G(I-\Pi) = I-\Pi. \tag{7}$$

Thus, from (4) and (7),

$$(\tilde{\boldsymbol{\pi}}^T - \boldsymbol{\pi}^T)(I-\Pi) = \tilde{\boldsymbol{\pi}}^T \mathbf{E} G(I-\Pi).$$

Further,

$$(\tilde{\boldsymbol{\pi}}^T - \boldsymbol{\pi}^T)(I-\Pi) = (\tilde{\boldsymbol{\pi}}^T - \boldsymbol{\pi}^T)(I-\mathbf{e}\boldsymbol{\pi}^T) = (\tilde{\boldsymbol{\pi}}^T - \boldsymbol{\pi}^T)$$

and (2) follows. □

Theorem 2.1 was first given by Hunter ([7]). All known published results for the difference $\tilde{\boldsymbol{\pi}}^T - \boldsymbol{\pi}^T$ can be obtained from this result. In particular we have the following special cases.

Theorem 2.2.

(i) If $\boldsymbol{\pi}^T\mathbf{t} \neq 0$ *and* $\mathbf{u}^T\mathbf{e} \neq 0$*, then*

$$\tilde{\boldsymbol{\pi}}^T - \boldsymbol{\pi}^T = \tilde{\boldsymbol{\pi}}^T \mathbf{E}[I-P+\mathbf{t}\mathbf{u}^T]^{-1}(I-\Pi). \tag{8}$$

(ii) If $\mathbf{u}^T\mathbf{e} \neq 0$*, then*

$$\tilde{\boldsymbol{\pi}}^T - \boldsymbol{\pi}^T = \tilde{\boldsymbol{\pi}}^T \mathbf{E}[I-P+\mathbf{e}\mathbf{u}^T]^{-1}. \tag{9}$$

(iii) If $G = [I-P+\mathbf{e}\mathbf{u}^T]^{-1} + \mathbf{e}\mathbf{f}^T$ *with* $\mathbf{u}^T\mathbf{e} \neq 0$ *and* $\mathbf{f}^T$ *an arbitrary vector, then*

$$\tilde{\boldsymbol{\pi}}^T - \boldsymbol{\pi}^T = \tilde{\boldsymbol{\pi}}^T \mathbf{E} G. \tag{10}$$

(iv) If $Z \equiv [I-P+\Pi]^{-1}$ *is Kemeny and Snell's 'fundamental matrix' ([10]), then*

$$\tilde{\boldsymbol{\pi}}^T - \boldsymbol{\pi}^T = \tilde{\boldsymbol{\pi}}^T \mathbf{E} Z. \tag{11}$$

(v) If $A^{\#} \equiv Z - \Pi$ *is the 'group inverse' ([12]) of* $I-P$*, then*

$$\tilde{\boldsymbol{\pi}}^T - \boldsymbol{\pi}^T = \tilde{\boldsymbol{\pi}}^T \mathbf{E} A^{\#}. \tag{12}$$

Proof.

(i) For (8), substitution of G into (2) leads to the $\mathbf{e}\mathbf{f}^T$ term vanishing since $\mathbf{E}\mathbf{e} = \mathbf{0}$. Similarly the $\mathbf{g}\boldsymbol{\pi}^T$ term cancels since $\boldsymbol{\pi}^T(I-\mathbf{e}\boldsymbol{\pi}^T) = \mathbf{0}^T$.

(ii) Equation (9) follows from (8) upon substitution of $\mathbf{t} = \mathbf{e}$ since, from (5), $[I - P + \mathbf{e}\mathbf{u}^T]^{-1}\mathbf{e} = \mathbf{e}/(\mathbf{u}^T\mathbf{e})$ and $\mathbf{E}\mathbf{e} = \mathbf{0}$.

(iii) Substitution of the form of G into (2) leads to the $\tilde{\boldsymbol{\pi}}^T\mathbf{E}G\Pi$ term vanishing since
$\mathbf{E}G\Pi = \mathbf{E}([I - P + \mathbf{e}\mathbf{u}^T]^{-1} + \mathbf{e}\mathbf{f}^T)\mathbf{e}\boldsymbol{\pi}^T = \mathbf{E}\mathbf{e}\{(1/(\mathbf{u}^T\mathbf{e})) + (\mathbf{f}^T\mathbf{e})\}\boldsymbol{\pi}^T = O$, where $O = [\ 0\ ]_{m\times m}$.

(iv) Equation (11) follows from (9) or (10) with $\mathbf{u}^T = \boldsymbol{\pi}^T$.

(v) Equation (12) follows from (10) with $\mathbf{u}^T = \boldsymbol{\pi}^T$ and $\mathbf{f}^T = \boldsymbol{\pi}^T$. □

The general result (8) was first given by Hunter [7]. The other results, or special cases of them, appear in the literature but with *ad hoc* derivations. Results (9) and (10) appear in Seneta [16], while result (10) appears in Seneta [18]. Result (11) was initially given by Schweitzer [15]. Result (12) is due to Meyer [13].

Note that Z and $A^{\#}$ are both g-inverses of $I - P$ (see e.g. Hunter [5], [6]; Meyer [12]).

If G is any g-inverse of $I - P$, then (Hunter [5], [6])

$$M = [G\Pi - E(G\Pi)_d + I - G + EG_d]D, \tag{13}$$

where $E = \mathbf{e}\mathbf{e}^T = [\ 1\ ]_{m\times m}$ and $D = M_d = (\Pi_d)^{-1}$.

Theorem 2.3. *If M is the mean first passage time matrix of the finite irreducible Markov chain with transition matrix P, then for any general perturbation* $\mathbf{E}$ *of P,*

$$\tilde{\boldsymbol{\pi}}^T - \boldsymbol{\pi}^T = -\tilde{\boldsymbol{\pi}}^T\mathbf{E}(M - M_d)(M_d)^{-1}. \tag{14}$$

Proof. Let G be any g-inverse of $I - P$ and let $H = G(I - \Pi)$. From (13) observe that,

$$(M - M_d)(M_d)^{-1} = G\Pi - E(G\Pi)_d - G + EG_d = EH_d - H.$$

Thus

$$H = EH_d - (M - M_d)(M_d)^{-1}.$$

$\mathbf{E}G(I - \Pi) = \mathbf{E}H = \mathbf{E}EH_d - \mathbf{E}(M - M_d)(M_d)^{-1} = -\mathbf{E}(M - M_d)(M_d)^{-1}$, and since $\mathbf{E}\mathbf{e} = \mathbf{0}$ implies $\mathbf{E}EH_d = \mathbf{E}\mathbf{e}\mathbf{e}^T H_d = 0$, (14) follows from (6). □

Result (14) appears in Hunter [7]. It can be simplified, using the "mixing time matrix", $N = [n_{ij}]$ (see also Hunter [8]).

Theorem 2.4. *Let* $N = [n_{ij}] = [(1-\delta_{ij})m_{ij}/m_{jj}] = [(1-\delta_{ij})m_{ij}\pi_j]$. *Then for any general perturbation* $\mathbf{E}$,

$$\tilde{\boldsymbol{\pi}}^T - \boldsymbol{\pi}^T = -\tilde{\boldsymbol{\pi}}^T \mathbf{E} N. \tag{15}$$

Proof. (15) follows directly from (14) since $N = (M - M_d)(M_d)^{-1}$. □

Thus every expression for $\tilde{\boldsymbol{\pi}}^T - \boldsymbol{\pi}^T$ (e.g. (8), (9), (10), (11), (12), (14) and (15)) is of the form given by (1) where H involves a g-inverse, is a g-inverse, or involves M or N.

3. Bounds on Stationary Distributions Under General Perturbations

We wish to explore absolute component-wise bounds of the form

$$|\pi_j - \tilde{\pi}_j| \leq k_l \left\|\mathbf{E}\right\|_\infty$$

and relative error bounds of the form

$$\left|\frac{\pi_j - \tilde{\pi}_j}{\pi_j}\right| \leq k_l \left\|\mathbf{E}\right\|_\infty .$$

For a summary of previous key results see Cho and Meyer [1]. They summarise and compare results due to Schweitzer [15], Meyer [13], Haviv and Van der Heyden [4], Kirkland *et al.* [11], Funderlic and Meyer [3], Meyer [14], Seneta [17], [18] and [19], Ipsen and Meyer [9], and Cho and Meyer [2].

Result (15) shows that elemental expressions for $\tilde{\pi}_j - \pi_j$ can be expressed in terms of $n_{ij} = m_{ij}/m_{jj} = m_{ij}\pi_j$ $(i \neq j)$ with $n_{jj} = 0$. In particular for each $j = 1, 2, \ldots, m$,

$$\pi_j - \tilde{\pi}_j = \sum_{l \neq j} \alpha_l n_{lj} \quad \text{where } \alpha_l = \sum_{k=1}^{m} \tilde{\pi}_k \epsilon_{kl}. \tag{16}$$

Theorem 3.1. *For a general perturbation* $\boldsymbol{E} = [\epsilon_{ij}]$, *for each fixed index* j $(1 \leq j \leq m)$,

$$|\pi_j - \tilde{\pi}_j| \leq \frac{\|\boldsymbol{E}\|_\infty}{2} \max_{i \neq j} \{n_{ij}\}, \tag{17}$$

where $\|\boldsymbol{E}\|_\infty = \max_{1 \leq k \leq m} \sum_{l=1}^{m} |\epsilon_{kl}|$.

Proof. The proof is based upon the result (see [4]) that for any vectors $\mathbf{c}$ and $\mathbf{d}$ such that $\mathbf{c}^T\mathbf{e} = \sum_{i=1}^m c_i = 0$, then

$$\left|\mathbf{c}^T\mathbf{d}\right| = \left|\sum_{i=1}^m c_i d_i\right| \leq \left(\sum_{i=1}^m |c_i|\right) \frac{\max_{r,s}|d_r - d_s|}{2}. \tag{18}$$

From (16), for each j, since $n_{jj} = 0$, $\pi_j - \tilde{\pi}_j$ can be expressed as $\sum_{l=1}^m \alpha_l n_{lj}$ where

$$\sum_{l=1}^m \alpha_l = \sum_{l=1}^m \sum_{k=1}^m \tilde{\pi}_k \epsilon_{kl} = \sum_{k=1}^m \tilde{\pi}_k \sum_{l=1}^m \epsilon_{kl} = 0.$$

Applying (18) yields

$$|\pi_j - \tilde{\pi}_j| \leq \sum_{l=1}^m |\alpha_l| \left(\frac{\max_{r,s}|n_{rj} - n_{sj}|}{2}\right). \tag{19}$$

Now

$$\sum_{l=1}^m |\alpha_l| \leq \sum_{l=1}^m \sum_{k=1}^m \tilde{\pi}_k |\epsilon_{kl}| = \sum_{k=1}^m \tilde{\pi}_k \sum_{l=1}^m |\epsilon_{kl}| \leq \sum_{k=1}^m \tilde{\pi}_k \left\{\max_{1\leq k\leq m} \sum_{l=1}^m |\epsilon_{kl}|\right\}$$

$= \|\mathbf{E}\|_\infty$. Since $n_{jj} = 0$,

$$\max_{r,s}|n_{rj} - n_{sj}| = \max\left\{\max_{r\neq j, s\neq j}(|n_{rj} - n_{sj}|), \max_{s\neq j} n_{sj}\right\}.$$

Further, since $n_{rj} \geq 0$ and $\min\{n_{rj}\} = 0$, for all r, s we have

$$|n_{rj} - n_{sj}| \leq \max\{n_{rj}\} - \min\{n_{rj}\} = \max\{n_{rj}\} \tag{20}$$

leading to $\max_{r,s}|n_{rj} - n_{sj}| = \max_{i\neq j} n_{ij}$. □

The bounds (17) were also derived by Cho and Meyer [2]. Their different proof uses the results of equation (12), the properties of $A^{\#}$ and the inequality (18).

The following corollary, giving bounds for relative differences, follows from (20) since, for $i \neq j$, $n_{ij} = \pi_j m_{ij}$ (see Cho and Meyer [2]).

Corollary 3.1. *For a general perturbation* $\mathbf{E} = [\epsilon_{ij}]$, *for each fixed index* j $(1 \leq j \leq m)$,

$$\left|\frac{\pi_j - \tilde{\pi}_j}{\pi_j}\right| \leq \frac{\|\mathbf{E}\|_\infty}{2} \max_{i\neq j}\{m_{ij}\}. \tag{21}$$

Note that the bound for absolute differences depends on the magnitudes of the n_{ij} whereas the bound for relative differences depends on the magnitudes of the m_{ij}.

4. Stationary Distributions Under Row Perturbations

We consider the effect of perturbations made in a single row of the transition matrix, say the r-th row. Let $\mathbf{p}_r^T = \mathbf{e}_r^T P$ denote the r-th row of the transition matrix P. Now let $\mathbf{E} = \mathbf{e}_r \boldsymbol{\epsilon}_r^T$ where $\boldsymbol{\epsilon}_r^T = \tilde{\mathbf{p}}_r^T - \mathbf{p}_r^T$. The perturbation results from changing the r-th row of the transition matrix P to the r-th row of the transition matrix $\tilde{P}$. Suppose that $\boldsymbol{\epsilon}_r^T = (\epsilon_1, \epsilon_2, \ldots, \epsilon_m)$ where $\boldsymbol{\epsilon}_r^T \mathbf{e} = 0$. Substitution in equation (15) yields

$$\boldsymbol{\pi}^T - \tilde{\boldsymbol{\pi}}^T = \tilde{\boldsymbol{\pi}}^T \mathbf{e}_r \boldsymbol{\epsilon}_r^T N = \tilde{\pi}_r \boldsymbol{\epsilon}_r^T N,$$

so that in elemental form, for $j = 1, 2, \ldots, m$,

$$\pi_j - \tilde{\pi}_j = \tilde{\pi}_r \sum_{i \neq j} \epsilon_i n_{ij} = \pi_j \tilde{\pi}_r \sum_{i \neq j} \epsilon_i m_{ij}. \tag{22}$$

Following the arguments used to develop Theorem 3.1 and Corollary 3.1 we can find general bounds for $\pi_j - \tilde{\pi}_j$ as follows.

Theorem 4.1. *For a perturbation $\boldsymbol{\epsilon}_r^T = \tilde{\mathbf{p}}_r^T - \mathbf{p}_r^T = (\epsilon_1, \epsilon_2, \ldots, \epsilon_m)$ involving only the elements of the r-th row of the transition matrix, for each fixed index j $(1 \leq j \leq m)$,*

$$|\pi_j - \tilde{\pi}_j| \leq \tilde{\pi}_r \frac{\|\mathbf{E}\|_\infty}{2} \max_{i \neq j}\{n_{ij}\} = \pi_j \tilde{\pi}_r \frac{\|\mathbf{E}\|_\infty}{2} \max_{i \neq j}\{m_{ij}\} \tag{23}$$

where $\|\mathbf{E}\|_\infty = \sum_{i=1}^m |\epsilon_i|$.

The simplest perturbation arises by considering a single row, say the r-th, by decreasing the (r, a)-th element of P by an amount ϵ and increasing the (r, b)-th element of P by the same amount to obtain the new transition matrix $\tilde{P}$. Thus $\tilde{p}_{ra} = p_{ra} - \epsilon < p_{ra}$ and $\tilde{p}_{rb} = p_{rb} + \epsilon > p_{rb}$ ($\epsilon_a = -\epsilon$, $\epsilon_b = \epsilon$). We assume that the stochastic and irreducible nature of both P and $\tilde{P}$ is preserved. This requires $\epsilon < p_{ra} \leq 1$, and $0 \leq p_{rb} < 1 - \epsilon$. For this special case we obtain the following results.

Theorem 4.2. *If the transition probability p_{ra} in an irreducible finite Markov chain is decreased by an amount ϵ while p_{rb} is increased by an amount ϵ, then, with the irreducibility preserved,*

$$\pi_j - \tilde{\pi}_j = \begin{cases} \epsilon \tilde{\pi}_r n_{ba} = \epsilon \pi_a \tilde{\pi}_r m_{ba}, & j = a, \\ -\epsilon \tilde{\pi}_r n_{ab} = -\epsilon \pi_b \tilde{\pi}_r m_{ab}, & j = b, \\ \epsilon \tilde{\pi}_r (n_{bj} - n_{aj}) = \epsilon \pi_j \tilde{\pi}_r (m_{bj} - m_{aj}), & j \neq a, b. \end{cases} \tag{24}$$

We now make use of the following easily proved relationship between the mean first passage times m_{ij} between states i and j in a finite irreducible Markov chain. For all i, j and k,

$$m_{ij} \leq m_{ik} + m_{kj}. \tag{25}$$

A consequence of (25) is that $m_{aj} \leq m_{ab} + m_{bj}$ and that $m_{bj} \leq m_{ba} + m_{aj}$, so that, for $j \neq a, b$,

$$-m_{ab} \leq m_{bj} - m_{aj} \leq m_{ba}. \tag{26}$$

This leads to the following corollary to Theorem 4.2.

Corollary 4.1. *Under the conditions of Theorem 4.2, the maximum relative change between the stationary probabilities π_j and $\tilde{\pi}_j$ is given by the following bound. For all j, $1 \leq j \leq m$,*

$$\left|\frac{\pi_j - \tilde{\pi}_j}{\pi_j}\right| \leq \epsilon\tilde{\pi}_r \max\{m_{ab}, m_{ba}\} = \max\left\{\left|\frac{\pi_a - \tilde{\pi}_a}{\pi_a}\right|, \left|\frac{\pi_b - \tilde{\pi}_b}{\pi_b}\right|\right\}. \tag{27}$$

Result (27) provides a new bound that cannot be improved, as it is achieved at one of the states $j = a$ or b. This is a significant improvement over the bound given by (21) since $\|\mathbf{E}\| = 2\epsilon$ and thus the relative differences for the stationary probabilities for state j are bounded as

$$\left|\frac{\pi_j - \tilde{\pi}_j}{\pi_j}\right| \leq \epsilon \max_{i \neq j}\{m_{ij}\}, \tag{28}$$

i.e. the bound on the *relative changes* in the stationary probabilities at any state j depends only on the mean first passage times m_{ab} and m_{ba} and not on the m_{ij} for $i \neq j$.

Noting that $|n_{bj} - n_{aj}| = n_{ba}$ or n_{ab} according as $j = a$ or b, respectively, we deduce from (24) the following corollary:

Corollary 4.2. *Under the conditions of Theorem 4.2, the maximum absolute change between the stationary probabilities π_j and $\tilde{\pi}_j$ is given by the following bound. For all j, $1 \leq j \leq m$,*

$$|\pi_j - \tilde{\pi}_j| = \epsilon\tilde{\pi}_r\,|n_{bj} - n_{aj}| \leq \epsilon\tilde{\pi}_r \max_s\{|n_{bs} - n_{as}|\}, \tag{29}$$

i.e. the bound on the *absolute changes* in the stationary probabilities depends on the maximum difference of mixing times from states a and b to any state and not the maximum over all the mixing times from a general state to any other state. In general, from (23), since $n_{jj} = 0$,

$$|\pi_j - \tilde{\pi}_j| \leq \epsilon\tilde{\pi}_r \max_i\{n_{ij}\} \leq \epsilon\tilde{\pi}_r \max_{s,t}\{n_{st}\}. \tag{30}$$

5. Example

Funderlic and Meyer [3] provide an example involving the analysis of radio-phosphorus kinetics in an aquarium system. This leads to a Markov chain with eight states and state space S = {1, 2, 3, 4, 5, 6, 7, 8}, transition matrix

$$P = \begin{bmatrix} 0.740 & 0.110 & 0 & 0 & 0 & 0 & 0 & 0.150 \\ 0 & 0.689 & 0 & 0 & 0.011 & 0 & 0 & 0.300 \\ 0 & 0 & 0 & 0.400 & 0 & 0 & 0 & 0.600 \\ 0 & 0 & 0 & 0.669 & 0.011 & 0 & 0 & 0.320 \\ 0 & 0 & 0 & 0 & 0.912 & 0 & 0 & 0.088 \\ 0 & 0 & 0 & 0 & 0 & 0.740 & 0 & 0.260 \\ 0 & 0 & 0 & 0 & 0 & 0 & 0.870 & 0.130 \\ 0.150 & 0 & 0.047 & 0 & 0 & 0.055 & 0.270 & 0.478 \end{bmatrix},$$

stationary probability vector
$\boldsymbol{\pi}^T \approx (0.137, 0.049, 0.011, 0.014, 0.008, 0.050, 0.494, 0.238)$,
mean first passage time matrix

$$M = \begin{bmatrix} 7.29 & 39.99 & 92.50 & 225.69 & 1437.84 & 78.00 & 13.26 & 5.38 \\ 26.28 & 20.61 & 90.74 & 223.93 & 1406.17 & 76.24 & 11.50 & 3.62 \\ 25.02 & 65.02 & 89.49 & 133.19 & 1437.27 & 74.98 & 10.24 & 2.36 \\ 26.06 & 66.06 & 90.53 & 74.05 & 1409.09 & 76.02 & 11.28 & 3.40 \\ 34.03 & 74.02 & 98,49 & 231.68 & 128.99 & 83.99 & 19.25 & 11.36 \\ 26.51 & 66.50 & 90.97 & 224.16 & 1458.24 & 19.88 & 11.73 & 3.85 \\ 30.35 & 70.35 & 94.82 & 228.01 & 1462.09 & 80.32 & 2.03 & 7.69 \\ 22.66 & 62.66 & 87.13 & 220.32 & 1454.40 & 72.62 & 7.88 & 4.21 \end{bmatrix}$$

and matrix of mixing times,

$$N = \begin{bmatrix} 0 & 1.940 & 1.034 & 3.048 & 11.147 & 3.923 & 6.549 & 1.278 \\ 3.605 & 0 & 1.014 & 3.024 & 10.902 & 3.835 & 5.680 & 0.860 \\ 3.432 & 3.154 & 0 & 1.798 & 11.143 & 3.771 & 5.059 & 0.561 \\ 3.575 & 3.205 & 1.012 & 0 & 10.924 & 3.824 & 5.572 & 0.808 \\ 4.668 & 3.591 & 1.101 & 3.129 & 0 & 4.224 & 9.505 & 2.702 \\ 3.636 & 3.227 & 1.017 & 3.027 & 11.305 & 0 & 5.793 & 0.914 \\ 4.164 & 3.413 & 1.060 & 3.079 & 11.335 & 4.040 & 0 & 1.830 \\ 3.109 & 3.040 & 0.974 & 2.975 & 11.276 & 3.653 & 3.894 & 0 \end{bmatrix}.$$

Let us suppose that we carry out a two-element perturbation in some row of P. Let the changes be made at positions (a, b) in the r-th row.

Tables 1 and 2 look at the relevant bounds given by (27) and (29) for the maximum relative and absolute changes to the stationary probabilities when changes are made at all possible sites (a, b). These results are compared with the universal bounds given by (28) and (30).

Since the maximum value of m_{ij} for $i \neq j$ occurs at state $j = 5$ (viz. 1462.09) we expect that the maximal relative change will occur at state 5. This is substantiated by observing that the first seven row entries in Table 1, ranked according to the magnitude of the bound, all involve state 5.

The maximal bound is achieved under a two-element perturbation involving states 5 and 7. This is possible only by decreasing p_{25}, p_{45} or p_{55} and, respectively, increasing p_{27}, p_{47} or p_{57} when perturbing in rows 2, 4 or 5 of the transition matrix.

Note that the smallest relative changes to any stationary probabilities will take place under a two-element perturbation involving states 7 and 8 as indicated in the last row of Table 1. Note further that the overall changes here are much smaller than those predicted by any of the bounds for the relative changes at any state as given by (28). Note that typically one would increase p_{r7} and decrease p_{r8}. This is possible only for $r = 1, 2, 3, 4, 5$ or 6.

The entries in Table 1 are partitioned into two regions to designate where the bounds based on (28) are an improvement for the bounds for particular relative differences involving state j.

Using bounds for the absolute differences in the stationary probabilities based upon the group inverse, as in (12), Funderlic and Meyer comment that "π_3 and π_8 must be very insensitive to perturbations while π_5 may be slightly more sensitive."

From the above table we note that for the first seven pairs, all involving making a perturbation to the transition probabilities involving a transition into state 5, we obtain the largest possible change to the stationary probability of being in state 5, which is much larger than the changes at any other state.

The smallest possible changes overall typically arise when a change is made to the transition probabilities into states 3 and 8 (typically decreasing p_{r8} and increasing p_{r3}).

Overall we see that the smallest changes to the stationary probabilities occur at state 3 and the largest changes at state 5. Note however, by considering the (1,3) entries, if one makes a two-element perturbation to the transition probabilities into states 1 and 3 (which by considering the transition probabilities is possible only from states 1 or 8) we make only a minimal change to the stationary probability of being in state 5.

Note that, even though the smallest relative changes to the stationary

Table 1. Bounds for relative differences based on magnitudes of m_{ij}.

a,b	m_{ab}	m_{ba}	max	$\max_{i \neq j} m_{ij}$							
				j = 8	j = 7	j=1	j= 2	j= 6	j=3	j= 4	j = 5
5,7	19.25	1462.09	1462.09	11.36	19.25	34.03	74.02	83.99	98.49	231.68	1462.09
5,6	83.99	1458.24	1458.24	11.36	19.25	34.03	74.02	83.99	98.49	231.68	1462.09
5,8	11.36	1454.40	1454.40	11.36	19.25	34.03	74.02	83.99	98.49	231.68	1462.09
1,5	1437.84	34.03	1437.84	11.36	19.25	34.03	74.02	83.99	98.49	231.68	1462.09
3,5	1437.27	98.49	1437.27	11.36	19.25	34.03	74.02	83.99	98.49	231.68	1462.09
4,5	1409.09	231.68	1409.09	11.36	19.25	34.03	74.02	83.99	98.49	231.68	1462.09
2,5	1406.17	74.02	1406.17	11.36	19.25	34.03	74.02	83.99	98.49	231.68	1462.09
4,7	11.28	228.01	228.01	11.36	19.25	34.03	74.02	83.99	98.49	231.68	1462.09
1,4	225.69	26.06	225.69	11.36	19.25	34.03	74.02	83.99	98.49	231.68	1462.09
4,6	76.02	224.16	224.16	11.36	19.25	34.03	74.02	83.99	98.49	231.68	1462.09
2,4	223.93	66.06	223.93	11.36	19.25	34.03	74.02	83.99	98.49	231.68	1462.09
4,8	3.40	220.32	220.32	11.36	19.25	34.03	74.02	83.99	98.49	231.68	1462.09
3,4	133.19	90.53	133.19	11.36	19.25	34.03	74.02	83.99	98.49	231.68	1462.09
3,7	10.24	94.82	94.82	11.36	19.25	34.03	74.02	83.99	98.49	231.68	1462.09
1,3	92.50	25.02	92.50	11.36	19.25	34.03	74.02	83.99	98.49	231.68	1462.09
3,6	74.98	90.97	90.97	11.36	19.25	34.03	74.02	83.99	98.49	231.68	1462.09
2,3	90.74	65.02	90.74	11.36	19.25	34.03	74.02	83.99	98.49	231.68	1462.09
3,8	2.36	87.13	87.13	11.36	19.25	34.03	74.02	83.99	98.49	231.68	1462.09
6,7	11.73	80.32	80.32	11.36	19.25	34.03	74.02	83.99	98.49	231.68	1462.09
1,6	78.00	26.51	78.00	11.36	19.25	34.03	74.02	83.99	98.49	231.68	1462.09
2,6	76.24	66.50	76.24	11.36	19.25	34.03	74.02	83.99	98.49	231.68	1462.09
6,8	3.85	72.62	72.62	11.36	19.25	34.03	74.02	83.99	98.49	231.68	1462.09
2,7	11.50	70.35	70.35	11.36	19.25	34.03	74.02	83.99	98.49	231.68	1462.09
2,8	3.62	62.66	62.66	11.36	19.25	34.03	74.02	83.99	98.49	231.68	1462.09
1,2	39.99	26.28	39.99	11.36	19.25	34.03	74.02	83.99	98.49	231.68	1462.09
1,7	13.26	30.36	30.36	11.36	19.25	34.03	74.02	83.99	98.49	231.68	1462.09
1,8	5.38	22.66	22.66	11.36	19.25	34.03	74.02	83.99	98.49	231.68	1462.09
7,8	7.69	7.89	7.89	11.36	19.25	34.03	74.02	83.99	98.49	231.68	1462.09

Table 2. Bounds for absolute differences based on magnitudes of n_{ij}.

	$\max n_{i3}$ 1.101	$\max n_{i8}$ 2.702	$\max n_{i4}$ 3.129	$\max n_{i2}$ 3.591	$\max n_{i6}$ 4.224	$\max n_{i1}$ 4.667	$\max n_{i7}$ 9.505	$\max n_{i5}$ 11.335	max
a,b	$n_{a3}-n_{b3}$	$n_{a8}-n_{b8}$	$n_{a4}-n_{b4}$	$n_{a2}-n_{b2}$	$n_{a6}-n_{b6}$	$n_{a1}-n_{b1}$	$n_{a7}-n_{b7}$	$n_{a5}-n_{b5}$	$\|n_{aj}-n_{bj}\|$
5,7	0.041	0.873	0.050	0.178	0.185	0.504	9.505	−11.335	11.335
5,6	0.084	1.787	0.102	0.365	4.224	1.031	3.712	−11.305	11.305
5,8	0.127	2.702	0.153	0.551	0.572	1.559	5.612	−11.276	11.276
1,5	−0.067	−1.424	−0.081	−1.651	−0.301	−4.667	−2.957	11.147	11.147
3,5	−1.101	−2.141	−1.330	−0.437	−0.453	−1.235	−4.446	11.143	11.143
4,5	−0.089	−1.894	−3.129	−0.386	−0.401	−1.093	−3.933	10.924	10.924
2,5	−0.087	−1.842	−0.105	−3.591	−0.390	−1.063	−3.825	10.902	10.902
1,7	−0.026	−0.551	−0.031	−1.473	−0.116	−4.164	6.549	−0.188	6.549
6,7	−0.043	−0.914	−0.052	−0.187	−4.040	−0.528	5.793	−0.030	5.793
2,7	−0.046	−0.969	−0.055	−3.413	−0.205	−0.559	5.680	−0.434	5.680
4,7	−0.048	−1.021	−3.079	−0.208	−0.216	−0.589	5.572	−0.411	5.572
3,7	−1.060	−1.268	−1.280	−0.259	−0.268	−0.732	5.059	−0.192	5.059
1,6	0.017	0.364	0.021	−1.286	3.923	−3.636	0.756	−0.158	3.923
7,8	0.086	1.829	0.104	0.373	0.387	1.055	−3.894	0.060	3.894
2,6	−0.003	−0.054	−0.003	−3.227	3.835	−0.031	−0.113	−0.404	3.835
4,6	−0.005	−0.106	−3.027	−0.022	3.824	−0.061	−0.221	−0.381	3.824
3,6	−1.017	−0.353	−1.229	−0.072	3.771	−0.204	−0.734	−0.163	3.771
6,8	0.043	0.914	0.052	0.187	−3.653	0.528	1.899	0.030	3.653
1,2	0.020	0.418	0.024	1.940	0.088	−3.605	0.869	0.246	3.605
1,4	0.022	0.470	3.048	−1.264	0.099	−3.575	0.977	0.223	3.575
1,3	1.034	0.717	1.249	−1.214	0.152	−3.432	1.490	0.004	3.432
2,4	0.002	0.052	3.024	−3.205	0.011	0.030	0.108	−0.023	3.205
2,3	1.014	0.299	1.225	−3.154	0.063	0.173	0.621	−0.241	3.154
1,8	0.060	1.278	0.073	−1.100	0.270	−3.109	2.655	−0.128	3.109
2,8	0.040	0.860	0.049	−3.040	0.182	0.496	1.786	−0.374	3.040
4,8	0.038	0.808	−2.975	0.165	0.171	0.466	1.678	−0.351	2.975
3,4	−1.012	−0.247	1.799	−0.050	−0.052	−0.143	−0.513	0.219	1.799
3,8	−0.974	0.561	−1.177	0.114	0.119	0.324	1.165	−0.133	1.177

probabilities occur in the two-element perturbation at the (7,8) case, this does not effect the smallest absolute change to the stationary probabilities. Similarly the converse deduction applies for the (3,8) case. Thus, there is no guarantee of reciprocity of minimal (or maximal) changes holding simultaneously for both the absolute and relative difference cases.

The main results (Sections 1 to 4) of this paper appear in Hunter [7]. The example of Section 5 has been provided to give an appreciation of the difficulties involved in establishing universal results that predict how Markov chains behave under perturbations.

References

1. Cho, G.E. and Meyer, C.D. (2001) Comparison of perturbation bounds for a stationary distribution of a Markov chain, *Linear Algebra Appl.*, 335, 137–150.
2. Cho, G.E. and Meyer, C.D. (2000) Markov chain sensitivity measured by mean first passage time, *Linear Algebra Appl.*, 316, 21–28.
3. Funderlic, R. and Meyer, C.D. (1986) Sensitivity of the stationary distribution vector for an ergodic Markov chain, *Linear Algebra Appl.*, 76, 1–17.
4. Haviv, M. and Van der Heyden, L. (1984) Perturbation bounds for the stationary probabilities of a finite Markov chain, *Adv. Appl. Probab.*, 16, 804–818.
5. Hunter, J.J. (1982) Generalized inverses and their application to applied probability problems, *Linear Algebra Appl.*, 45, 157–198.
6. Hunter, J.J. (1983) *Mathematical Techniques of Applied Probability, Volume 2, Discrete Time Models: Techniques and Applications*, Academic, New York.
7. Hunter, J.J. (2005) Stationary distributions and mean first passage times of perturbed Markov chains, *Linear Algebra Appl.*, 410, 217–243.
8. Hunter, J.J. (2006) Mixing times with applications to perturbed Markov chains. (To appear in *Linear Algebra Appl.*)
9. Ipsen, I.C.F. and Meyer, C.D. (1994) Uniform stability of Markov chains, *SIAM J. Matrix Anal. Appl.* 4, 1061–1074.
10. Kemeny, J.G. and Snell, J.L. (1960) *Finite Markov Chains*, Van Nostrand, New York.
11. Kirkland, S.J., Neumann, M. and Shader, B.L. (1998) Applications of Paz's inequality to perturbation bounds for Markov chains, *Linear Algebra Appl.*, 268, 183–196.
12. Meyer, C.D. (1975) The role of the group generalized inverse in the theory of finite Markov chains, *SIAM Rev.*, 17, 443–464.
13. Meyer, C.D. (1980) The condition of a finite Markov chain and perturbation bounds for the limiting probabilities, *SIAM J. Algebraic Discrete Methods*, 1, 273–283.
14. Meyer, C.D. (1994) Sensitivity of the stationary distribution of a Markov chain, *SIAM J. Matrix Appl.*, 15, 715–728.

15. Schweitzer, P. (1968) Perturbation theory and finite Markov chains, *J. Appl. Probab.*, 5, 410–413.
16. Seneta, E. (1988) Sensitivity to perturbation of the stationary distribution: Some refinements, *Linear Algebra Appl.*, 108, 121–126.
17. Seneta, E. (1988) Perturbation of the stationary distribution measured by the ergodicity coefficients, *Adv. Appl. Probab.*, 20, 228–230.
18. Seneta, E. (1991) Sensitivity analysis, ergodicity coefficients, and rank-one updates for finite Markov chains, in: W.J. Stewart (Ed.), *Numerical Solution of Markov Chains*, Marcel-Dekker, New York, pp. 121–129.
19. Seneta, E. (1993) Sensitivity of finite Markov chains under perturbation, *Statist. Probab. Lett.*, 17, 163–168.

Matrix Tricks for Linear Statistical Models: A Short Review of our Personal Top Fourteen

Jarkko Isotalo and Simo Puntanen

Department of Mathematics, Statistics and Philosophy
FI-33014 University of Tampere, Finland
E-mail (1): jarkko.isotalo@uta.fi
E-mail (2): sjp@uta.fi

George P. H. Styan

Department of Mathematics and Statistics, McGill University
Burnside Hall Room 1005
805 rue Sherbrooke Street West
Montréal (Québec), Canada H3A 2K6
E-mail: styan@math.mcgill.ca

In teaching a course in linear statistical models to first year graduate students or to final year undergraduate students, say, there is no way to proceed smoothly without matrices and related concepts of linear algebra; their use is really essential. Our experience is that making some particular matrix tricks familiar to students can increase their insight into linear statistical models (and also multivariate statistical analysis). In matrix algebra, there are handy, sometimes even very simple "tricks" to simplify and clarify the problem treatment — both for the student and for the researcher. Of course, the concept of *trick* is not uniquely defined: by trick we simply mean here a central important handy result. In this paper we collect together our Top Fourteen favourite matrix tricks for linear statistical models. We merely state our tricks with some references; a more comprehensive report including proofs, examples and full references is in progress [see Isotalo, Puntanen & Styan (2005)].

Keywords: best linear unbiased estimation; column space; generalised inverse; linear model; linear regression; Löwner ordering; matrix inequalities; ordinary least squares; projector; Schur complement; singular value decomposition.

AMS Classification: 15-01, 15-02, 15A09, 15A42, 15A99, 62H12, 62J05.

Notation and linear algebraic preliminaries

The symbols $\mathbf{A}'$, $\mathbf{A}^-$, $\mathbf{A}^+$, $\mathscr{C}(\mathbf{A})$, $\mathscr{C}(\mathbf{A})^\perp$, $\mathscr{N}(\mathbf{A})$, and $\mathrm{r}(\mathbf{A})$ will stand for the transpose, a generalised inverse, the Moore–Penrose inverse, the column space, the orthogonal complement of the column space, the null space, and the rank, respectively, of $\mathbf{A}$. The set of $n \times m$ matrices is denoted as $\mathbb{R}^{n\times m}$. Occasionally we may denote $\mathbf{A}_{n\times m}$ indicating that $\mathbf{A}$ is an $n \times m$ matrix. We consider only matrices with real elements.

We recall that matrix $\mathbf{G}$ is a generalised inverse of $\mathbf{A}$ if it satisfies the equation $\mathbf{AGA} = \mathbf{A}$; we denote $\mathbf{A}^- = \mathbf{G}$. An equivalent characterisation is that $\mathbf{G}$ is a generalised inverse of $\mathbf{A}$ if (and only if) $\mathbf{x} = \mathbf{Gy}$ is a solution of $\mathbf{Ax} = \mathbf{y}$ for any $\mathbf{y}$ which makes the equation consistent (solvable). The matrix $\mathbf{A}^+$, the Moore–Penrose inverse of $\mathbf{A}$, is defined as the unique solution to the four equations

$$\begin{aligned} &(\mathrm{mp1})\ \mathbf{AA}^+\mathbf{A} = \mathbf{A}, && (\mathrm{mp2})\ \mathbf{A}^+\mathbf{AA}^+ = \mathbf{A}^+, \\ &(\mathrm{mp3})\ \mathbf{AA}^+ = (\mathbf{AA}^+)', && (\mathrm{mp4})\ \mathbf{A}^+\mathbf{A} = (\mathbf{A}^+\mathbf{A})'. \end{aligned} \tag{1}$$

The set of all generalised inverses of $\mathbf{A}_{n\times m}$ is denoted as

$$\{\mathbf{A}^-\} = \{\, \mathbf{G}_{m\times n} : \mathbf{AGA} = \mathbf{A} \,\}. \tag{2}$$

The column space $\mathscr{C}(\mathbf{A})$ of an $n \times m$ matrix $\mathbf{A}$ is a subspace of $\mathbb{R}^n$ spanned by the columns of $\mathbf{A}$:

$$\mathscr{C}(\mathbf{A}) = \{\, \mathbf{y} \in \mathbb{R}^n : \exists\, \mathbf{x} \in \mathbb{R}^m \text{ such that } \mathbf{y} = \mathbf{Ax} \,\}, \tag{3}$$

and, correspondingly, the null space is $\mathscr{N}(\mathbf{A}) = \{\, \mathbf{x} \in \mathbb{R}^m : \mathbf{Ax} = \mathbf{0} \,\}$. Notation $(\mathbf{A} : \mathbf{B})$ stands for the partitioned matrix with $\mathbf{A}_{n\times m}$ and $\mathbf{B}_{n\times p}$ as submatrices.

We may also recall that for any idempotent $n \times n$ matrix $\mathbf{P}$ the following direct sum decomposition holds:

$$\mathbb{R}^n = \mathscr{C}(\mathbf{P}) \oplus \mathscr{C}(\mathbf{I}_n - \mathbf{P}) = \mathscr{C}(\mathbf{P}) \oplus \mathscr{N}(\mathbf{P}). \tag{4}$$

Hence an idempotent matrix $\mathbf{P}$ is an *oblique projector* (or shortly projector) onto $\mathscr{C}(\mathbf{P})$ along $\mathscr{C}(\mathbf{I}_n - \mathbf{P})$. The column space $\mathscr{C}(\mathbf{I}_n - \mathbf{P})$, which (for any idempotent $\mathbf{P}$) has the property $\mathscr{C}(\mathbf{I}_n - \mathbf{P}) = \mathscr{N}(\mathbf{P})$, determines the direction in which the projection is done; if this direction is orthogonal with respect to a given inner product, then $\mathbf{P}$ is an *orthogonal projector* w.r.t. this inner product.

Unless otherwise stated, we assume that the inner product between two vectors $\mathbf{x}, \mathbf{y} \in \mathbb{R}^n$ is defined as $\mathbf{x}'\mathbf{Iy} = \mathbf{x}'\mathbf{y}$. Then, the matrix $\mathbf{P}$ is an orthogonal projector with respect to the standard inner product (shortly w.r.t. $\mathbf{I}$) if it is idempotent and symmetric, that is,

$$\mathbf{P} \text{ is orthogonal projector (w.r.t. } \mathbf{I}) \iff \mathbf{P}^2 = \mathbf{P} \text{ and } \mathbf{P}' = \mathbf{P}. \tag{5}$$

If an idempotent symmetric $\mathbf{P}$ has the property $\mathscr{C}(\mathbf{P}) = \mathscr{C}(\mathbf{A})$, then $\mathbf{P}$ is the orthogonal projector onto $\mathscr{C}(\mathbf{A})$, denoted as $\mathbf{P_A}$. Clearly we have

$$\mathbf{P_A} = \mathbf{AA}^+ = \mathbf{A}(\mathbf{A}'\mathbf{A})^-\mathbf{A}', \tag{6}$$

which is invariant for any choice of $(\mathbf{A}'\mathbf{A})^-$.

By $\mathbf{A}^\perp$ we denote any matrix satisfying

$$\mathscr{C}(\mathbf{A}^\perp) = \mathscr{N}(\mathbf{A}') = \mathscr{C}(\mathbf{A})^\perp; \tag{7}$$

this is the orthocomplement (w.r.t. $\mathbf{I}$) of $\mathscr{C}(\mathbf{A})$.

If a symmetric matrix $\mathbf{A}$ is nonnegative definite, we denote $\mathbf{A} \geq \mathbf{0}$, which means that there exists a matrix $\mathbf{L}$ such that $\mathbf{A} = \mathbf{LL}'$. Notation $\mathbf{A} > \mathbf{0}$ indicates that $\mathbf{A}$ is positive definite, i.e., $\mathbf{A}$ is a nonsingular nonnegative definite matrix. Similarly, notation $\mathbf{A} \geq \mathbf{B}$ means that the difference $\mathbf{A} - \mathbf{B}$ can be expressed as $\mathbf{A} - \mathbf{B} = \mathbf{KK}'$ for some matrix $\mathbf{K}$. We say that matrix $\mathbf{B}$ is below $\mathbf{A}$ with respect to the Löwner partial ordering.

When the inner product is defined as $\mathbf{x}'\mathbf{V}\mathbf{y}$, where $\mathbf{V}$ is a positive definite symmetric matrix, then the orthogonal projector is characterised as follows:

$$\mathbf{P} \text{ is orthogonal projector (w.r.t. } \mathbf{V}) \iff \mathbf{P}^2 = \mathbf{P} \ \&\ (\mathbf{VP})' = \mathbf{VP}. \tag{8}$$

If, in addition, $\mathscr{C}(\mathbf{P}) = \mathscr{C}(\mathbf{A})$, then $\mathbf{P}$ is the orthogonal projector onto $\mathscr{C}(\mathbf{A})$, denoted as $\mathbf{P}_{\mathbf{A};\mathbf{V}}$, and its explicit (unique) representation is

$$\mathbf{P}_{\mathbf{A};\mathbf{V}} = \mathbf{A}(\mathbf{A}'\mathbf{V}\mathbf{A})^-\mathbf{A}'\mathbf{V}. \tag{9}$$

As regards the matrix factorisations needed, the most important in this paper is the *eigenvalue decomposition* (spectral decomposition): every symmetric $n \times n$ matrix $\mathbf{A}$ can be expressed as

$$\mathbf{A} = \mathbf{U}\boldsymbol{\Gamma}\mathbf{U}', \tag{10}$$

where $\mathbf{U}$ is orthogonal, $\boldsymbol{\Gamma} = \operatorname{diag}(\gamma_1, \dots, \gamma_n)$, and $\gamma_1 \geq \gamma_2 \geq \cdots \geq \gamma_n$ are ordered eigenvalues of $\mathbf{A}$. The columns of $\mathbf{U}$ are the orthonormal eigenvectors of $\mathbf{A}$. In particular we will need the eigenvalue decomposition of a nonnegative definite $n \times n$ matrix $\mathbf{V}$:

$$\mathbf{V} = \mathbf{T}\boldsymbol{\Lambda}\mathbf{T}', \quad \boldsymbol{\Lambda} = \operatorname{diag}(\lambda_1, \dots, \lambda_n), \quad \mathbf{T}'\mathbf{T} = \mathbf{I}_n. \tag{11}$$

Here $\lambda_1 \geq \lambda_2 \geq \cdots \geq \lambda_n \geq 0$, and $\mathbf{t}_i$ (the i-th column of $\mathbf{T}$) is the eigenvector corresponding to λ_i. In particular, denoting $\mathbf{T}_1 = (\mathbf{t}_1 : \dots : \mathbf{t}_v)$ and $\boldsymbol{\Lambda}_1 = \operatorname{diag}(\lambda_1, \dots, \lambda_v)$, where $v = \operatorname{r}(\mathbf{V})$, we get the symmetric nonnegative definite square root of $\mathbf{V}$:

$$\mathbf{V}^{1/2} = \mathbf{T}_1\boldsymbol{\Lambda}_1^{1/2}\mathbf{T}_1'. \tag{12}$$

With the general linear statistical model (often called the Gauss–Markov model) we mean the equation

$$\mathbf{y} = \mathbf{X}\boldsymbol{\beta} + \boldsymbol{\varepsilon}, \tag{13}$$

or in other notation,

$$\mathscr{M} = \{\mathbf{y},\, \mathbf{X}\boldsymbol{\beta},\, \sigma^2\mathbf{V}\}, \tag{14}$$

where

$$\mathrm{E}(\mathbf{y}) = \mathbf{X}\boldsymbol{\beta}, \quad \mathrm{E}(\boldsymbol{\varepsilon}) = \mathbf{0}, \quad \mathrm{cov}(\mathbf{y}) = \mathrm{cov}(\boldsymbol{\varepsilon}) = \sigma^2\mathbf{V}. \tag{15}$$

Vector $\mathbf{y}$ is an $n \times 1$ observable random vector, $\boldsymbol{\varepsilon}$ is an $n \times 1$ random error vector, $\mathbf{X}$ is a known $n \times p$ model (design) matrix, $\boldsymbol{\beta}$ is a $p \times 1$ vector of unknown parameters, $\mathbf{V}$ is a known $n \times n$ nonnegative definite matrix, and σ^2 is an unknown nonzero constant. By $\mathrm{E}(\cdot)$ and $\mathrm{cov}(\cdot)$ we denote expectation vector and covariance matrix of a random vector argument.

1. Rank of the Product

Theorem 1.1. *The rank of a partitioned matrix* $(\mathbf{A} : \mathbf{B})$ *can be expressed (for any choice of generalised inverse* $\mathbf{A}^-$*) as*

$$\begin{aligned} \mathrm{r}(\mathbf{A} : \mathbf{B}) &= \mathrm{r}(\mathbf{A}) + \mathrm{r}[(\mathbf{I} - \mathbf{A}\mathbf{A}^-)\mathbf{B}] && \text{(16a)} \\ &= \mathrm{r}(\mathbf{A}) + \mathrm{r}[(\mathbf{I} - \mathbf{A}\mathbf{A}^+)\mathbf{B}] && \text{(16b)} \\ &= \mathrm{r}(\mathbf{A}) + \mathrm{r}[(\mathbf{I} - \mathbf{P}_\mathbf{A})\mathbf{B}], && \text{(16c)} \end{aligned}$$

while the rank of the matrix product $\mathbf{AB}$ *is*

$$\begin{aligned} \mathrm{r}(\mathbf{AB}) &= \mathrm{r}(\mathbf{A}) - \dim \mathscr{C}(\mathbf{A}') \cap \mathscr{C}(\mathbf{B}^\perp) && \text{(17a)} \\ &= \mathrm{r}(\mathbf{A}) - \dim \mathscr{C}(\mathbf{A}') \cap \mathscr{N}(\mathbf{B}') && \text{(17b)} \\ &= \mathrm{r}(\mathbf{A}) - \dim \mathscr{C}(\mathbf{A}') \cap \mathscr{C}(\mathbf{I} - \mathbf{P}_\mathbf{B}). && \text{(17c)} \end{aligned}$$

References

(16): Marsaglia & Styan (1974, Theorem 4);

(17): Marsaglia & Styan (1974, Corollary 6.2), C. R. Rao (1973a, p. 28; First Edition 1965, p. 27), A. R. Rao & Bhimasankaram (2000, Thm 3.5.11), C. R. Rao & M. B. Rao (1998, p. 426).

2. Rank Cancellation Rule

Theorem 2.1. *For any conformable matrices involved,*

$$\mathbf{LAY} = \mathbf{MAY} \ \text{ and } \ \mathrm{r}(\mathbf{AY}) = \mathrm{r}(\mathbf{A}) \implies \mathbf{LA} = \mathbf{MA}, \tag{18}$$

$$\mathbf{DAM} = \mathbf{DAN} \ \text{ and } \ \mathrm{r}(\mathbf{DA}) = \mathrm{r}(\mathbf{A}) \implies \mathbf{AM} = \mathbf{AN}. \tag{19}$$

Furthermore,

$$\mathrm{r}(\mathbf{AY}) = \mathrm{r}(\mathbf{A}) \implies \mathrm{r}(\mathbf{KAY}) = \mathrm{r}(\mathbf{KA}) \text{ for every possible } \mathbf{K}. \tag{20}$$

References

Marsaglia & Styan (1974, Theorem 2), Rao & Bhimasankaram (2000, Theorem 3.5.7).

3. Sum of Orthogonal Projectors

Theorem 3.1. *Let $\mathbf{P_A}$ and $\mathbf{P_B}$ be orthogonal projectors (with respect to the standard inner product) onto $\mathscr{C}(\mathbf{A})$ and $\mathscr{C}(\mathbf{B})$, respectively. Then*

$$\mathbf{P_A} + \mathbf{P_B} \ \text{ is orthogonal projector} \iff \mathbf{A}'\mathbf{B} = \mathbf{0}, \tag{21}$$

in which case

$$\mathbf{P_A} + \mathbf{P_B} = \mathbf{P}_{(\mathbf{A}\,:\,\mathbf{B})}. \tag{22}$$

4. Decomposition of Orthogonal Projector

Theorem 4.1. *The orthogonal projector (with respect to the standard inner product) onto $\mathscr{C}(\mathbf{A} : \mathbf{B})$ can be decomposed as*

$$\mathbf{P}_{(\mathbf{A}\,:\,\mathbf{B})} = \mathbf{P_A} + \mathbf{P}_{(\mathbf{I}-\mathbf{P_A})\mathbf{B}}. \tag{23}$$

References to projectors

Baksalary (1987), Ben-Israel & Greville (2003), Groß & Trenkler (1998), Halmos (1951, Sections 26–29), Rao & Mitra (1971, Section 5.1), Rao & Yanai (1979), Seber (1980), Seber & Lee (2003, Appendix B), Takane & Yanai (1999), Trenkler (1994).

5. OLSE vs BLUE

We collect together some necessary and sufficient conditions for the equality between OLSE (ordinary least squares estimator) and BLUE (best linear unbiased estimator) of $\mathbf{X}\boldsymbol{\beta}$ under a general linear model $\{\mathbf{y},\, \mathbf{X}\boldsymbol{\beta},\, \sigma^2\mathbf{V}\}$.

Consider the eigenvalue decomposition of the covariance matrix $\mathbf{V}$:

$$\mathbf{V} = \mathbf{T}\boldsymbol{\Lambda}\mathbf{T}', \tag{24}$$

where $\mathbf{\Lambda}$ is an $n \times n$ diagonal matrix of the n eigenvalues λ_i of $\mathbf{V}$: $\lambda_1 \geq \lambda_2 \geq \cdots \geq \lambda_n \geq 0$. Matrix $\mathbf{T}$ is an $n \times n$ matrix comprising the corresponding orthonormal eigenvectors $\mathbf{t}_1, \mathbf{t}_2, \ldots, \mathbf{t}_n$. We also let $\lambda_{\{1\}} > \lambda_{\{2\}} > \cdots > \lambda_{\{t\}} \geq 0$ denote the t *distinct* eigenvalues of $\mathbf{V}$ with multiplicities $m_1, m_2, \ldots, m_t$, and let $\mathbf{T}_{\{1\}}, \mathbf{T}_{\{2\}}, \ldots, \mathbf{T}_{\{t\}}$ be matrices consisting of (sets of) corresponding orthonormal eigenvectors so that

$$\mathbf{T} = (\mathbf{T}_{\{1\}} : \mathbf{T}_{\{2\}} : \ldots : \mathbf{T}_{\{t\}}), \qquad \mathbf{T}_i'\mathbf{T}_i = \mathbf{I}_{m_i}, \quad i = 1, 2, \ldots, t, \tag{25}$$

$$\mathbf{V} = \lambda_{\{1\}}\mathbf{T}_{\{1\}}\mathbf{T}_{\{1\}}' + \cdots + \lambda_{\{t\}}\mathbf{T}_{\{t\}}\mathbf{T}_{\{t\}}', \qquad m_1 + m_2 + \cdots + m_t = n. \tag{26}$$

Theorem 5.1. *Consider the general linear model* $\{\mathbf{y}, \mathbf{X}\boldsymbol{\beta}, \mathbf{V}\}$, *where* $\mathbf{X}$ *and* $\mathbf{V}$ *need not be of full rank. Then*

$$\text{(a) } \mathbf{Gy} = \mathit{BLUE}(\mathbf{X}\boldsymbol{\beta}) \iff \text{(b) } \mathbf{G}(\mathbf{X} : \mathbf{VM}) = (\mathbf{X} : \mathbf{0}), \tag{27}$$

where $\mathbf{M} = \mathbf{I} - \mathbf{H} = \mathbf{I} - \mathbf{P}_{\mathbf{X}}$. *The corresponding condition for* $\mathbf{Ay}$ *to be the BLUE of an estimable parametric function* $\mathbf{K}\boldsymbol{\beta}$ *is*

$$\mathbf{A}(\mathbf{X} : \mathbf{VM}) = (\mathbf{K} : \mathbf{0}). \tag{28}$$

Moreover, the following statements are equivalent:

(i) $\mathbf{HV} = \mathbf{VH}$,
(ii) $\mathbf{HV} = \mathbf{HVH}$,
(iii) $\mathbf{HVM} = \mathbf{0}$,
(iv) $\mathbf{X}'\mathbf{VZ} = \mathbf{0}$, *where* $\mathscr{C}(\mathbf{Z}) = \mathscr{C}(\mathbf{M})$,
(v) $\mathscr{C}(\mathbf{VX}) \subset \mathscr{C}(\mathbf{X})$,
(vi) $\mathscr{C}(\mathbf{VX}) = \mathscr{C}(\mathbf{X}) \cap \mathscr{C}(\mathbf{V})$,
(vii) $\mathbf{HVH} \leq \mathbf{V}$, *i.e.,* $\mathbf{V} - \mathbf{HVH}$ *is nonnegative definite,*
(viii) $\mathrm{r}(\mathbf{V} - \mathbf{HVH}) = \mathrm{r}(\mathbf{V}) - \mathrm{r}(\mathbf{HVH})$,
(ix) $\mathscr{C}(\mathbf{X})$ *has a basis consisting of* r *eigenvectors of* $\mathbf{V}$, *where* $r = \mathrm{r}(\mathbf{X})$,
(x) $\mathrm{r}(\mathbf{T}_{\{1\}}'\mathbf{X}) + \cdots + \mathrm{r}(\mathbf{T}_{\{t\}}'\mathbf{X}) = \mathrm{r}(\mathbf{X})$, *where* $\mathbf{T}_{\{i\}}$ *is a matrix consisting of the orthogonal eigenvectors corresponding to the* i*-th largest eigenvalue* $\lambda_{\{i\}}$ *of* $\mathbf{V}$; $\lambda_{\{1\}} > \lambda_{\{2\}} > \cdots > \lambda_{\{t\}}$,
(xi) $\mathbf{T}_{\{i\}}'\mathbf{HT}_{\{i\}} = (\mathbf{T}_{\{i\}}'\mathbf{HT}_{\{i\}})^2$ *for all* $i = 1, 2, \ldots, t$,
(xii) $\mathbf{T}_{\{i\}}'\mathbf{HT}_{\{j\}} = \mathbf{0}$ *for all* $i, j = 1, 2, \ldots, t$, $i \neq j$,
(xiii) $\mathbf{V}$ *can be expressed as* $\mathbf{V} = \alpha\mathbf{I} + \mathbf{XAX}' + \mathbf{ZBZ}'$, *where* $\alpha \in \mathbb{R}$, *and* $\mathbf{A}$ *and* $\mathbf{B}$ *are symmetric, such that* $\mathbf{V}$ *is nonnegative definite, i.e.,*

$$\mathbf{V} \in \mathcal{V}_1 = \{\, \mathbf{V} \geq \mathbf{0} : \mathbf{V} = \alpha\mathbf{I} + \mathbf{XAX}' + \mathbf{ZBZ}',\ \mathbf{A} = \mathbf{A}',\ \mathbf{B} = \mathbf{B}' \,\},$$

(xiv) $\mathbf{V}$ *can be expressed as* $\mathbf{V} = \mathbf{XCX}' + \mathbf{ZDZ}'$, *where* $\mathbf{C}$ *and* $\mathbf{D}$ *are symmetric, such that* $\mathbf{V}$ *is nonnegative definite, i.e.,*

$$\mathbf{V} \in \mathcal{V}_2 = \{\, \mathbf{V} \geq \mathbf{0} : \mathbf{V} = \mathbf{XCX}' + \mathbf{ZDZ}',\ \mathbf{C} = \mathbf{C}',\ \mathbf{D} = \mathbf{D}' \,\},$$

(xv) $\mathbf{V}$ *can be expressed as* $\mathbf{V} = \mathbf{HEH} + \mathbf{MFM}$, *where* $\mathbf{E}$ *and* $\mathbf{F}$ *are symmetric nonnegative definite, i.e.,*

$$\mathbf{V} \in \mathcal{V}_3 = \{\, \mathbf{V} \geq \mathbf{0} : \mathbf{V} = \mathbf{HEH} + \mathbf{MFM},\ \mathbf{E} \geq \mathbf{0},\ \mathbf{F} \geq \mathbf{0} \,\}.$$

References

Anderson (1948), Baksalary (2004), Bartmann & Bloomfield (1981), Bloomfield & Watson (1975), Groß (2004), Puntanen & Styan (1989), Puntanen, Styan & Werner (2000), Rao (1967), Rao (1968), Rao (1971), Rao & Mitra (1971), Seshadri & Styan (1980), Watson (1955), Zyskind (1967), Zyskind (1969).

6. General Solution to AXB = C

Theorem 6.1. *A necessary and sufficient condition for the equation* $\mathbf{AXB} = \mathbf{C}$ *to have a solution (for* $\mathbf{X}$*) is that*

$$\mathbf{AA}^{-}\mathbf{CB}^{-}\mathbf{B} = \mathbf{C}, \tag{29}$$

in which case the general solution is

$$\mathbf{X} = \mathbf{A}^{-}\mathbf{CB}^{-} + \mathbf{Z} - \mathbf{A}^{-}\mathbf{AZBB}^{-}, \tag{30}$$

where $\mathbf{Z}$ *is an arbitrary matrix, and* $\mathbf{A}^{-}$ *and* $\mathbf{B}^{-}$ *are fixed (but arbitrary) g-inverses. In particular, the general solution to* $\mathbf{AX} = \mathbf{C}$ *is [if* $\mathbf{AA}^{-}\mathbf{C} = \mathbf{C}$*, i.e.,* $\mathscr{C}(\mathbf{C}) \subset \mathscr{C}(\mathbf{A})$*]*

$$\mathbf{X} = \mathbf{A}^{-}\mathbf{C} + (\mathbf{I} - \mathbf{A}^{-}\mathbf{A})\mathbf{Z}. \tag{31}$$

Similarly, the general solution to $\mathbf{XB} = \mathbf{C}$ *is [if* $\mathscr{C}(\mathbf{C}') \subset \mathscr{C}(\mathbf{B}')$*]*

$$\mathbf{X} = \mathbf{CB}^{-} + \mathbf{Z}(\mathbf{I} - \mathbf{BB}^{-}). \tag{32}$$

References

Rao (1968), Rao (1971), Rao (1973a), Rao (1973b), Rao & Mitra (1971) .

7. Invariance w.r.t. the Choice of Generalised Inverse

Theorem 7.1. *Let* $\mathbf{A} \neq \mathbf{0}$ *and* $\mathbf{C} \neq \mathbf{0}$*. Then the following two statements are equivalent:*

(a) $\mathbf{AB}^{-}\mathbf{C}$ *is invariant with respect to the choice of* $\mathbf{B}^{-}$,

(b) $\mathscr{C}(\mathbf{C}) \subset \mathscr{C}(\mathbf{B})$ *and* $\mathscr{C}(\mathbf{A}') \subset \mathscr{C}(\mathbf{B}')$.

In particular, $\mathbf{BB}^{-}\mathbf{C}$ *is invariant with respect to the choice of* $\mathbf{B}^{-}$ *if and only if* $\mathscr{C}(\mathbf{C}) \subset \mathscr{C}(\mathbf{B})$*; in other words,*

$$\mathbf{BB}^{-}\mathbf{C} = \mathbf{C} \ \text{ for all } \mathbf{B}^{-} \iff \mathscr{C}(\mathbf{C}) \subset \mathscr{C}(\mathbf{B}). \tag{33}$$

REFERENCES

Rao & Mitra (1971, Lemma 2.2.4, and the Complement 14, p. 43), Rao, Mitra & Bhimasankaram (1972).

8. Block Diagonalisation and the Schur Complement

Theorem 8.1. *Let* $\mathbf{A}$ *be a symmetric nonnegative definite matrix partitioned as*

$$\mathbf{A} = \begin{pmatrix} \mathbf{A}_{11} & \mathbf{A}_{12} \\ \mathbf{A}_{21} & \mathbf{A}_{22} \end{pmatrix}, \tag{34}$$

where $\mathbf{A}_{11}$ *is a square matrix. Then the following decomposition holds:*

$$\begin{aligned} \mathbf{A} &= \begin{pmatrix} \mathbf{I} & \mathbf{0} \\ \mathbf{A}_{21}\mathbf{A}_{11}^{=} & \mathbf{I} \end{pmatrix} \begin{pmatrix} \mathbf{A}_{11} & \mathbf{0} \\ \mathbf{0} & \mathbf{A}_{22} - \mathbf{A}_{21}\mathbf{A}_{11}^{-}\mathbf{A}_{12} \end{pmatrix} \begin{pmatrix} \mathbf{I} & \mathbf{A}_{11}^{\sim}\mathbf{A}_{12} \\ \mathbf{0} & \mathbf{I} \end{pmatrix} \\ &:= \mathbf{BCD}, \end{aligned} \tag{35}$$

where $\mathbf{A}_{11}^{=}$, $\mathbf{A}_{11}^{-}$, *and* $\mathbf{A}_{11}^{\sim}$ *are arbitrary generalised inverses of* $\mathbf{A}_{11}$, *and*

$$\begin{aligned} \mathbf{A}_{22\cdot 1} = (\mathbf{A}/\mathbf{A}_{11}) &= \mathbf{A}_{22} - \mathbf{A}_{21}\mathbf{A}_{11}^{-}\mathbf{A}_{12} \\ &= \text{the Schur complement of } \mathbf{A}_{11} \text{ in } \mathbf{A}. \end{aligned} \tag{36}$$

Moreover,

$$\begin{aligned} \begin{pmatrix} \mathbf{I} & \mathbf{0} \\ -\mathbf{A}_{21}\mathbf{A}_{11}^{=} & \mathbf{I} \end{pmatrix} \cdot \mathbf{A} \cdot \begin{pmatrix} \mathbf{I} & -\mathbf{A}_{11}^{\sim}\mathbf{A}_{12} \\ \mathbf{0} & \mathbf{I} \end{pmatrix} &= \begin{pmatrix} \mathbf{A}_{11} & \mathbf{0} \\ \mathbf{0} & \mathbf{A}_{22} - \mathbf{A}_{21}\mathbf{A}_{11}^{-}\mathbf{A}_{12} \end{pmatrix} \\ &= \mathbf{B}^{-1}\mathbf{A}\mathbf{D}^{-1}. \end{aligned} \tag{37}$$

REFERENCES

Carlson (1986), Cottle (1974), Henderson & Searle (1981), Ouellette (1981), Puntanen & Styan (2005a, 2005b), Styan (1985), Zhang (2005).

9. Nonnegative Definiteness of a Partitioned Matrix

Theorem 9.1. *Let* $\mathbf{A}$ *be a symmetric matrix partitioned as*

$$\mathbf{A} = \begin{pmatrix} \mathbf{A}_{11} & \mathbf{A}_{12} \\ \mathbf{A}_{21} & \mathbf{A}_{22} \end{pmatrix} \tag{38}$$

where $\mathbf{A}_{11}$ *is a square matrix. Then the following three statements are equivalent:*

(a) $\mathbf{A} \geq \mathbf{0}$,
(b) $\mathbf{A}_{11} \geq \mathbf{0}$, $\mathscr{C}(\mathbf{A}_{12}) \subset \mathscr{C}(\mathbf{A}_{11})$ *and* $\mathbf{A}_{22} - \mathbf{A}_{21}\mathbf{A}_{11}^{-}\mathbf{A}_{12} \geq \mathbf{0}$,
(c) $\mathbf{A}_{22} \geq \mathbf{0}$, $\mathscr{C}(\mathbf{A}_{21}) \subset \mathscr{C}(\mathbf{A}_{22})$ *and* $\mathbf{A}_{11} - \mathbf{A}_{12}\mathbf{A}_{22}^{-}\mathbf{A}_{21} \geq \mathbf{0}$.

REFERENCES

Albert (1969), Dey, Hande & Tiku (1994).

10. The Matrix $\dot{\mathbf{M}}$

In this section we consider the properties of matrix

$$\dot{\mathbf{M}} = \mathbf{M}(\mathbf{MVM})^{-}\mathbf{M}. \tag{39}$$

We observe that $\dot{\mathbf{M}}$ is unique if and only if $\mathscr{C}(\mathbf{M}) \subset \mathscr{C}(\mathbf{MV})$, which is further equivalent to

$$\mathbb{R}^n = \mathscr{C}(\mathbf{X} : \mathbf{V}). \tag{40}$$

Even though $\dot{\mathbf{M}}$ is not necessarily unique, the matrix product

$$\mathbf{P_V}\dot{\mathbf{M}}\mathbf{P_V} = \mathbf{P_V}\mathbf{M}(\mathbf{MVM})^{-}\mathbf{M}\mathbf{P_V} := \ddot{\mathbf{M}} \tag{41}$$

is, however, clearly invariant for any choice of $(\mathbf{MVM})^{-}$, i.e.,

$$\mathbf{P_V}\dot{\mathbf{M}}\mathbf{P_V} = \mathbf{P_V}\mathbf{M}(\mathbf{MVM})^{-}\mathbf{M}\mathbf{P_V} = \mathbf{P_V}\mathbf{M}(\mathbf{MVM})^{+}\mathbf{M}\mathbf{P_V}. \tag{42}$$

The matrices $\dot{\mathbf{M}}$ and $\ddot{\mathbf{M}}$ appear to be very handy in many considerations related to the linear model $\mathscr{M} = \{\mathbf{y}, \mathbf{X}\boldsymbol{\beta}, \sigma^2\mathbf{V}\}$. Therefore, we will state a theorem collecting together some of their properties.

Theorem 10.1. *Consider the linear model* $\mathscr{M} = \{\mathbf{y}, \mathbf{X}\boldsymbol{\beta}, \sigma^2\mathbf{V}\}$, *where* $\mathbf{X}$ *and* $\mathbf{V}$ *may not have full column ranks. Let the matrices* $\dot{\mathbf{M}}$ *and* $\ddot{\mathbf{M}}$ *be defined as*

$$\dot{\mathbf{M}} = \mathbf{M}(\mathbf{MVM})^{-}\mathbf{M}, \qquad \ddot{\mathbf{M}} = \mathbf{P_V}\dot{\mathbf{M}}\mathbf{P_V}. \tag{43}$$

Assume that the condition

$$\mathbf{H}\mathbf{P_V}\mathbf{M} = \mathbf{0} \tag{44}$$

holds. Then

(a) $\ddot{\mathbf{M}} = \mathbf{P_V M}(\mathbf{MVM})^{-}\mathbf{MP_V} = \mathbf{V}^{+} - \mathbf{V}^{+}\mathbf{X}(\mathbf{X}'\mathbf{V}^{+}\mathbf{X})^{-}\mathbf{X}'\mathbf{V}^{+}$,

(b) $\ddot{\mathbf{M}} = \mathbf{MV}^{+}\mathbf{M} - \mathbf{MV}^{+}\mathbf{X}(\mathbf{X}'\mathbf{V}^{+}\mathbf{X})^{-}\mathbf{X}'\mathbf{V}^{+}\mathbf{M}$,

(c) $\ddot{\mathbf{M}} = \mathbf{M}\ddot{\mathbf{M}} = \ddot{\mathbf{M}}\mathbf{M} = \mathbf{M}\ddot{\mathbf{M}}\mathbf{M}$,

(d) $\mathbf{P_V}\ddot{\mathbf{M}}\mathbf{P_V} = \mathbf{P_V}(\mathbf{MVM})^{+}\mathbf{P_V}$,

(e) $\ddot{\mathbf{M}}\mathbf{V}\ddot{\mathbf{M}} = \ddot{\mathbf{M}}$, *i.e.*, $\mathbf{V} \in \{(\ddot{\mathbf{M}})^{-}\}$,

(f) $\mathrm{r}(\ddot{\mathbf{M}}) = \mathrm{r}(\mathbf{VM}) = \mathrm{r}(\mathbf{V}) - \dim \mathscr{C}(\mathbf{X}) \cap \mathscr{C}(\mathbf{V}) = \mathrm{r}(\mathbf{X} : \mathbf{V}) - \mathrm{r}(\mathbf{X})$
$= \mathrm{r}[\mathbf{M}(\mathbf{MVM})^{+}\mathbf{M}]$,

(g) *If* $\mathbf{Z}$ *is a matrix with property* $\mathscr{C}(\mathbf{Z}) = \mathscr{C}(\mathbf{M})$, *then*

$$\ddot{\mathbf{M}} = \mathbf{P_V Z}(\mathbf{Z}'\mathbf{VZ})^{-}\mathbf{Z}'\mathbf{P_V}, \qquad \mathbf{V}\dot{\mathbf{M}}\mathbf{V} = \mathbf{VZ}(\mathbf{Z}'\mathbf{VZ})^{-}\mathbf{Z}'\mathbf{V}. \tag{45}$$

(h) *Let* $(\mathbf{X} : \mathbf{Z})$ *be orthogonal. Then always*

$$[(\mathbf{X} : \mathbf{Z})'\mathbf{V}(\mathbf{X} : \mathbf{Z})]^{+} = (\mathbf{X} : \mathbf{Z})'\mathbf{V}^{+}(\mathbf{X} : \mathbf{Z}). \tag{46}$$

If $\mathbf{HP_V M} = \mathbf{0}$ *or equivalently* $\mathscr{C}(\mathbf{VH}) \cap \mathscr{C}(\mathbf{VM}) = \{\mathbf{0}\}$, *then*

$$\begin{aligned}
&[(\mathbf{X} : \mathbf{Z})'\mathbf{V}(\mathbf{X} : \mathbf{Z})]^{+} \\
&= \begin{pmatrix} [\mathbf{X}'\mathbf{VX} - \mathbf{X}'\mathbf{VZ}(\mathbf{Z}'\mathbf{VZ})^{-}\mathbf{Z}'\mathbf{VX}]^{+} & \cdot \\ \cdot & [\mathbf{Z}'\mathbf{VZ} - \mathbf{Z}'\mathbf{VX}(\mathbf{X}'\mathbf{VX})^{-}\mathbf{X}'\mathbf{VZ}]^{+} \end{pmatrix} \\
&= \begin{pmatrix} \mathbf{X}'\mathbf{V}^{+}\mathbf{X} & \cdot \\ \cdot & \mathbf{Z}'\mathbf{V}^{+}\mathbf{Z} \end{pmatrix}.
\end{aligned} \tag{47}$$

(i) *If* $\mathbf{V}$ *is positive definite, and* $\mathbf{Z}$ *as in* (g) *then*

$$\begin{aligned}
\dot{\mathbf{M}} = \ddot{\mathbf{M}} &= \mathbf{M}(\mathbf{MVM})^{-}\mathbf{M} = (\mathbf{MVM})^{+} \\
&= \mathbf{Z}(\mathbf{Z}'\mathbf{VZ})^{-1}\mathbf{Z}' \\
&= \mathbf{V}^{-1} - \mathbf{V}^{-1}\mathbf{X}\big(\mathbf{X}'\mathbf{V}^{-1}\mathbf{X}\big)^{-}\mathbf{X}'\mathbf{V}^{-1} \\
&= \mathbf{V}^{-1}\big(\mathbf{I} - \mathbf{P}_{\mathbf{X};\mathbf{V}^{-1}}\big),
\end{aligned} \tag{48}$$

$$\mathbf{X}\big(\mathbf{X}'\mathbf{V}^{-1}\mathbf{X}\big)^{-}\mathbf{X}' = \mathbf{V} - \mathbf{VZ}(\mathbf{Z}'\mathbf{VZ})^{-1}\mathbf{Z}'\mathbf{V}, \tag{49}$$

$$\begin{aligned}
\mathbf{X}\big(\mathbf{X}'\mathbf{V}^{-1}\mathbf{X}\big)^{-}\mathbf{X}'\mathbf{V}^{-1} &= \mathbf{I} - \mathbf{VZ}(\mathbf{Z}'\mathbf{VZ})^{-1}\mathbf{Z}' \\
&= \mathbf{I} - \mathbf{P}'_{\mathbf{Z};\mathbf{V}},
\end{aligned} \tag{50}$$

(j) *If* $\mathbf{V}$ *is positive definite, and the columns of* $\mathbf{X}$ *and* $\mathbf{Z}$ *are orthonormal, then*

$$\big(\mathbf{X}'\mathbf{V}^{-1}\mathbf{X}\big)^{-1} = \mathbf{X}'\mathbf{VX} - \mathbf{X}'\mathbf{VZ}(\mathbf{Z}'\mathbf{VZ})^{-1}\mathbf{Z}'\mathbf{VX}. \tag{51}$$

REFERENCES

Bhimasankaram & Saha Ray (1997), Bhimasankaram & Sengupta (1996), Bhimasankaram, Shah & Saha Ray (1998), Chipman (1998), Groß & Puntanen (2000), Nurhonen & Puntanen (1992), Puntanen (1996, 1997), Werner & Yapar (1995).

11. Disjointness of Column Spaces

Theorem 11.1. *For conformable matrices, the following statements are equivalent:*

(a) $\mathscr{C}(\mathbf{A}) \cap \mathscr{C}(\mathbf{B}) = \{\mathbf{0}\}$,

(b) $\begin{pmatrix} \mathbf{A}' \\ \mathbf{B}' \end{pmatrix} (\mathbf{AA}' + \mathbf{BB}')^{-}(\mathbf{AA}' : \mathbf{BB}') = \begin{pmatrix} \mathbf{A}' & \mathbf{0} \\ \mathbf{0} & \mathbf{B}' \end{pmatrix}$,

(c) $\mathbf{A}'(\mathbf{AA}' + \mathbf{BB}')^{-}\mathbf{AA}' = \mathbf{A}'$,

(d) $\mathbf{A}'(\mathbf{AA}' + \mathbf{BB}')^{-}\mathbf{B} = \mathbf{0}$,

(e) $(\mathbf{AA}' + \mathbf{BB}')^{-}$ *is a generalised inverse of* $\mathbf{AA}'$,

(f) $\mathbf{A}'(\mathbf{AA}' + \mathbf{BB}')^{-}\mathbf{A} = \mathbf{P}_{\mathbf{A}'}$,

(g) $\mathscr{C}\begin{pmatrix} \mathbf{0} \\ \mathbf{B}' \end{pmatrix} \subset \mathscr{C}\begin{pmatrix} \mathbf{A}' \\ \mathbf{B}' \end{pmatrix}$,

(h) $\mathbf{Y}(\mathbf{A} : \mathbf{B}) = (\mathbf{0} : \mathbf{B})$ *has a solution for* $\mathbf{Y}$.

References

Magnus & Neudecker (1999).

12. Full Rank Decomposition

Theorem 12.1. *Let* $\mathbf{A}$ *be an* $n \times m$ *matrix with rank* $r > 0$. *Then* $\mathbf{A}$ *can be written as a product*

$$\mathbf{A} = \mathbf{UV}', \tag{52}$$

where $\mathrm{r}(\mathbf{U}_{n\times r}) = \mathrm{r}(\mathbf{V}_{m\times r}) = r$, *i.e.,* $\mathbf{U}$ *and* $\mathbf{V}$ *have full column ranks.*

References

Marsaglia & Styan (1972), Marsaglia & Styan (1974), Rao & Bhimasankaram (2000, p. 132).

13. Singular Value Decomposition

Theorem 13.1. *Let* $\mathbf{A}$ *be an* $n \times m$ $(m \leq n)$ *matrix with rank* $r > 0$. *Then* $\mathbf{A}$ *can be written as a product*

$$\begin{aligned} \mathbf{A} &= (\mathbf{U}_1 : \mathbf{U}_0) \begin{pmatrix} \mathbf{\Delta}_{1(r\times r)} & \mathbf{0}_{r\times(m-r)} \\ \mathbf{0}_{(n-r)\times r} & \mathbf{0}_{(n-r)\times(m-r)} \end{pmatrix} \begin{pmatrix} \mathbf{V}_1' \\ \mathbf{V}_0' \end{pmatrix} \\ &= \mathbf{U\Delta V}' = \mathbf{U}_1\mathbf{\Delta}_1\mathbf{V}_1' = \mathbf{U}_*\mathbf{\Delta}_*\mathbf{V}' \\ &= \delta_1\mathbf{u}_1\mathbf{v}_1' + \cdots + \delta_r\mathbf{u}_r\mathbf{v}_r' = \mathbf{U}\begin{pmatrix} \mathbf{\Delta}_{*(m\times m)} \\ \mathbf{0}_{(n-m)\times m} \end{pmatrix}\mathbf{V}', \end{aligned} \tag{53}$$

where

$$
\begin{aligned}
\delta_i &= \mathrm{sg}_i(\mathbf{A}) = \sqrt{\mathrm{ch}_i(\mathbf{A}'\mathbf{A})} = \text{i-}th\ singular\ value\ of\ \mathbf{A}, \quad i = 1, \dots, m, \\
\boldsymbol{\Delta}_1 &= \mathrm{diag}(\delta_1, \dots, \delta_r), \quad \delta_1 \geq \dots \geq \delta_r > 0, \quad \boldsymbol{\Delta} \in \mathbb{R}^{n\times m}, \\
\boldsymbol{\Delta} &= \begin{pmatrix} \boldsymbol{\Delta}_1 & \mathbf{0} \\ \mathbf{0} & \mathbf{0} \end{pmatrix} = \begin{pmatrix} \boldsymbol{\Delta}_* \\ \mathbf{0} \end{pmatrix} \in \mathbb{R}^{n\times m}, \quad \boldsymbol{\Delta}_1 \in \mathbb{R}^{r\times r}, \quad \boldsymbol{\Delta}_* \in \mathbb{R}^{m\times m}, \\
\delta_{r+1} &= \delta_{r+2} = \dots = \delta_m = 0, \\
\boldsymbol{\Delta}_* &= \mathrm{diag}(\delta_1, \dots, \delta_r, \delta_{r+1}, \dots, \delta_m) = the\ first\ m\ rows\ of\ \boldsymbol{\Delta}, \\
\mathbf{U}_{n\times n} &= (\mathbf{U}_1 : \mathbf{U}_0), \quad \mathbf{U}_1 \in \mathbb{R}^{n\times r}, \quad \mathbf{U}'\mathbf{U} = \mathbf{U}\mathbf{U}' = \mathbf{I}_n, \\
\mathbf{V}_{m\times m} &= (\mathbf{V}_1 : \mathbf{V}_0), \quad \mathbf{V}_1 \in \mathbb{R}^{m\times r}, \quad \mathbf{V}'\mathbf{V} = \mathbf{V}\mathbf{V}' = \mathbf{I}_m, \\
\mathbf{U}_* &= (\mathbf{u}_1 : \dots : \mathbf{u}_m) = the\ first\ m\ columns\ of\ \mathbf{U}, \quad \mathbf{U}_* \in \mathbb{R}^{n\times m}, \\
\mathbf{V}'\mathbf{A}'\mathbf{A}\mathbf{V} &= \boldsymbol{\Delta}'\boldsymbol{\Delta} = \boldsymbol{\Delta}_*^2 = \begin{pmatrix} \boldsymbol{\Delta}_1^2 & \mathbf{0} \\ \mathbf{0} & \mathbf{0} \end{pmatrix} \in \mathbb{R}^{m\times m}, \\
\mathbf{U}'\mathbf{A}\mathbf{A}'\mathbf{U} &= \boldsymbol{\Delta}\boldsymbol{\Delta}' = \begin{pmatrix} \boldsymbol{\Delta}_*^2 & \mathbf{0} \\ \mathbf{0} & \mathbf{0} \end{pmatrix} \in \mathbb{R}^{n\times n}, \\
\mathbf{u}_i &= \text{i-}th\ left\ singular\ vector\ of\ \mathbf{A} = \text{i-}th\ eigenvector\ of\ \mathbf{A}\mathbf{A}', \\
\mathbf{v}_i &= \text{i-}th\ right\ singular\ vector\ of\ \mathbf{A} = \text{i-}th\ eigenvector\ of\ \mathbf{A}'\mathbf{A}.
\end{aligned}
$$

References

Golub & Van Loan (1996, Section 2.5.3) , Horn & Johnson (1991), Horn & Olkin (1996), Searle (1982, p. 316), Stewart (1993), Stewart (1998, p. 62).

14. The Cauchy–Schwarz Inequality

Theorem 14.1. *Let* $\mathbf{x}$ *and* $\mathbf{y}$ *be* $n \times 1$ *non-null real vectors. Then*

$$(\mathbf{x}'\mathbf{y})^2 \leq \mathbf{x}'\mathbf{x} \cdot \mathbf{y}'\mathbf{y}, \quad \forall\, \mathbf{x}, \mathbf{y} \tag{54}$$

is the vector version of the Cauchy–Schwarz inequality. Equality holds in (54) *if and only if* $\mathbf{x}$ *and* $\mathbf{y}$ *are linearly dependent, i.e.,*

$$(\mathbf{x}'\mathbf{y})^2 = \mathbf{x}'\mathbf{x} \cdot \mathbf{y}'\mathbf{y} \iff \exists\, \lambda \in \mathbb{R}\colon \mathbf{x} = \lambda\mathbf{y}. \tag{55}$$

References

Baksalary & Puntanen (1991), Chipman (1964), Marcus & Minc (1992, p. 61), Watson, Alpargu & Styan (1997).

Acknowledgements

The authors are grateful to Götz Trenkler for helpful comments. The research of the third author was supported in part by the Natural Sciences and Engineering Research Council of Canada.

References

1. Albert, Arthur (1969). Conditions for positive and nonnegative definiteness in terms of pseudoinverses. *SIAM Journal on Applied Mathematics*, 17, 434–440.
2. Anderson, T. W. (1948). On the theory of testing serial correlation. *Skandinavisk Aktuarietidskrift*, 31, 88–116.
3. Baksalary, Jerzy K. (1987). Algebraic characterizations and statistical implications of the commutativity of orthogonal projectors. *Proceedings of the Second International Tampere Conference in Statistics*, (T. Pukkila and S. Puntanen, Eds.), University of Tampere, Tampere, Finland, pp. 113–142.
4. Baksalary, Jerzy K. (2004). An elementary development of the equation characterizing best linear unbiased estimators. *Linear Algebra and Its Applications*, 388, 3–6.
5. Baksalary, Jerzy K. and Puntanen, Simo (1991). Generalized matrix versions of the Cauchy–Schwarz and Kantorovich inequalities. *Aequationes Mathematicae*, 41, 103–110.
6. Bartmann, Flavio C. and Bloomfield, Peter (1981). Inefficiency and correlation. *Biometrika*, 68, 67–71.
7. Ben-Israel, Adi and Greville, Thomas N. E. (2003). *Generalized Inverses: Theory and Applications.* Second Edition. Springer. [1st Ed. 1974]
8. Bhimasankaram, P. and Saha Ray, R. (1997). On a partitioned linear model and some associated reduced models. *Linear Algebra and Its Applications*, 264, 329–339.
9. Bhimasankaram, P. and Sengupta, D. (1996). The linear zero functions approach to linear models. *Sankhyā, Ser. B*, 58, 338–351.
10. Bhimasankaram, P., Shah, K. R. and Saha Ray, R. (1998). On a singular partitioned linear model and some associated reduced models. *Journal of Combinatorics, Information & System Sciences*, 23, 415–421.
11. Bloomfield, Peter and Watson, Geoffrey S. (1975). The inefficiency of least squares. *Biometrika*, 62, 121–128.
12. Carlson, David (1986). What are Schur complements, anyway? *Linear Algebra and Its Applications*, 74, 257–275 [see also 59, 188–193 (1984)].
13. Chipman, John S. (1964). On least squares with insufficient observations. *Journal of the American Statistical Association*, 59, 1078–1111.
14. Chipman, John S. (1998). The contributions of Ragnar Frisch to economics and econometrics. *Economics and Economic Theory in the 20th Century: The Ragnar Frisch Centennial Symposium* (Strøm, Steinar, ed.). Cambridge University Press, 58–108.
15. Cottle, Richard W. (1974). Manifestations of the Schur complement. *Linear Algebra and Its Applications*, 8, 189–211. [See also *Rendiconti del Seminario Matematico e Fisico di Milano*, 45, (1975), 31–40.]
16. Dey, Aloke; Hande, Sayaji and Tiku, M. L. (1994). Statistical proofs of some matrix results. *Linear and Multilinear Algebra*, 38, 109–116.
17. Golub, Gene H. and Van Loan, Charles F. (1996). *Matrix Computations.* Third Edition. Johns Hopkins University Press, Baltimore.
18. Groß, Jürgen (2004). The general Gauss–Markov model with possibly singu-

lar covariance matrix. *Statistical Papers*, 45, 311–336.

19. Groß, Jürgen and Trenkler, Götz (1998). On the product of oblique projectors. *Linear and Multilinear Algebra*, 30, 1–13.
20. Groß, Jürgen and Puntanen, Simo (2000). Estimation under a general partitioned linear model. *Linear Algebra and Its Applications*, 321, 131–144.
21. Halmos, P. R. (1951). *Introduction to Hilbert Space and the Theory of Spectral Multiplicity*. Chelsea, New York. [2nd Ed. 1957]
22. Henderson, Harold V. and Searle, Shayle R. (1981). On deriving the inverse of a sum of matrices. *SIAM Review*, 23, 53–60.
23. Horn, Roger A. and Johnson, Charles R. (1991). *Topics in Matrix Analysis*. Cambridge University Press, New York.
24. Horn, Roger A. and Olkin, Ingram (1996). When does $A^*A = B^*B$ and why does one want to know? *The American Mathematical Monthly*, 103, 470–482.
25. Isotalo, Jarkko; Puntanen, Simo and Styan, George P. H. (2005). Matrix tricks for linear statistical models: our personal Top Fourteen. Third Revised Edition (of report A 345), Dept. of Mathematics, Statistics & Philosophy, Univ. of Tampere, ca. 200 pp., in progress.
26. Magnus, Jan R. and Neudecker, Heinz (1999). *Matrix Differential Calculus with Applications in Statistics and Econometrics*. Revised Edition. Wiley, New York.
27. Marcus, Marcus and Minc, Henryk (1992). *A Survey of Matrix Theory and Matrix Inequalities*. Corrected Reprint Edition. Dover, New York.
28. Marsaglia, George and Styan, George P. H. (1972). When does $\text{rank}(A+B) = \text{rank}(A) + \text{rank}(B)$? *Canadian Mathematical Bulletin*, 15, 451–452.
29. Marsaglia, George and Styan, George P. H. (1974). Equalities and inequalities for ranks of matrices. *Linear and Multilinear Algebra*, 2, 269–292.
30. Nurhonen, Markku and Puntanen, Simo (1992). A property of partitioned generalized regression. *Communications in Statistics: Theory and Methods*, 21, 1579–1583.
31. Ouellette, Diane Valérie (1981). Schur complements and statistics. *Linear Algebra and Its Applications*, 36, 187–295.
32. Puntanen, Simo (1996). Some matrix results related to a partitioned singular linear model. *Communications in Statistics: Theory and Methods*, 25, 269–279.
33. Puntanen, Simo (1997). Some further results related to reduced singular linear models. *Communications in Statistics: Theory and Methods*, 26, 375–385.
34. Puntanen, Simo and Styan, George P. H. (1989). The equality of the ordinary least squares estimator and the best linear unbiased estimator [with comments by Oscar Kempthorne & by Shayle R. Searle and with "Reply" by the authors]. *The American Statistician*, 43, 153–164.
35. Puntanen, Simo & Styan, George P. H. (2005a). Historical introduction: Issai Schur and the early development of the Schur complement. Chapter 0 in *The Schur Complement and Its Applications* (Fuzhen Zhang, ed.), Springer, pp. 1–16.
36. Puntanen, Simo & Styan, George P. H. (2005b). Schur complements in statistics and probability. Chapter 6 in *The Schur Complement and Its Appli-*

cations (Fuzhen Zhang, ed.), Springer, pp. 163–226.

37. Puntanen, Simo, Styan, George P. H. and Werner, Hans Joachim (2000). Two matrix-based proofs that the linear estimator Gy is the best linear unbiased estimator. *Journal of Statistical Planning and Inference*, 88, 173–179.
38. Rao, A. Ramachandra and Bhimasankaram, P. (2000). *Linear Algebra.* Second Edition. Hindustan Book Agency, New Delhi. [1st Ed. 1992]
39. Rao, C. Radhakrishna (1967). Least squares theory using an estimated dispersion matrix and its application to measurement of signals. *Proc. Fifth Berkeley Symposium on Mathematical Statistics and Probability: Berkeley, California, 1965/1966*, vol. 1, L. M. Le Cam and J. Neyman, eds., Univ. of California Press, Berkeley, 355–372.
40. Rao, C. Radhakrishna (1968). A note on a previous lemma in the theory of least squares and some further results. *Sankhyā, Ser. A*, 30, 245–252
41. Rao, C. Radhakrishna (1971). Unified theory of linear estimation. *Sankhyā, Ser. A*, 33, 371–394. [Corrigendum (1972), 34, p. 194 and p. 477.]
42. Rao, C. Radhakrishna (1973a). *Linear Statistical Inference and Its Applications.* Second Edition. Wiley, New York. [1st Ed. 1965]
43. Rao, C. Radhakrishna (1973b). Representations of best linear unbiased estimators in the Gauss–Markoff model with a singular dispersion matrix. *Journal of Multivariate Analysis*, 3, 276–292.
44. Rao, C. Radhakrishna and Mitra, Sujit Kumar (1971). *Generalized Inverse of Matrices and Its Applications.* Wiley, New York.
45. Rao, C. Radhakrishna; Mitra, Sujit Kumar and Bhimasankaram, P. (1972). Determination of a matrix by its subclasses of generalized inverses. *Sankhyā, Ser. A*, 34, 5–8.
46. Rao, C. Radhakrishna and Rao, M. B. (1998). *Matrix Algebra and Its Applications to Statistics and Econometrics.* World Scientific, Singapore.
47. Rao, C. Radhakrishna and Yanai, Haruo (1979). General definition and decomposition of projectors and some applications to statistical problems. *Journal of Statistical Planning and Inference*, 3, 1–17.
48. Searle, Shayle R. (1982). *Matrix Algebra Useful for Statistics.* Wiley, New York.
49. Seber, George A. F. (1980). *The Linear Hypothesis: A General Theory.* Second Edition. Griffin, London. [1st Ed. 1966]
50. Seber, George A. F. and Lee, Alan J. (2003). *Linear Regression Analysis.* Second Edition. Wiley, New York. [1st Ed. 1977]
51. Seshadri, V. and Styan, George P. H. (1980). Canonical correlations, rank additivity and characterizations of multivariate normality. *Analytic Function Methods in Probability Theory: Proc. Colloquium on the Methods of Complex Analysis in the Theory of Probability and Statistics held at the Kossuth L. University, Debrecen, Hungary, August 29–September 2, 1977*, Colloquia Mathematica Societatis János Bolyai, vol. 21, János Bolyai, Budapest and North-Holland, Amsterdam, 331–344.
52. Stewart, G. W. (1993). On the early history of the singular value decomposition. *SIAM Review*, 35, 551–566.
53. Stewart, G. W. (1998). *Matrix Algorithms. Volume I: Basic Decompositions.*

Society for Industrial and Applied Mathematics, Philadelphia.
54. Styan, G. P. H. (1985). Schur complements and linear statistical models. *Proceedings of the First International Tampere Seminar on Linear Statistical Models and their Applications*, 37–75. (T. Pukkila and S. Puntanen, eds.) Department of Mathematical Sciences, University of Tampere, Finland.
55. Takane, Yoshio and Yanai, Haruo (1999). On oblique projectors. *Linear Algebra and Its Applications*, 289, 297–310.
56. Trenkler, Götz (1994). Characterizations of oblique and orthogonal projectors. *Proceedings of the International Conference on Linear Statistical Inference LINSTAT '93.* (T. Caliński and R. Kala, eds.) Kluwer, Amsterdam, 255–270.
57. Watson, Geoffrey S. (1955). Serial correlation in regression analysis, I. *Biometrika*, 42, 327–341.
58. Watson, Geoffrey S., Alpargu, Gülhan and Styan, George P. H. (1997). Some comments on six inequalities associated with the inefficiency of ordinary least squares with one regressor. *Linear Algebra and Its Applications*, 264, 13–53.
59. Werner, Hans Joachim and Yapar, C. (1995). More on partitioned possibly restricted linear regression. *Multivariate Statistics and Matrices in Statistics* (E.-M. Tiit, T. Kollo and H. Niemi, eds.), VSP, Utrecht, Netherlands, and TEV, Vilnius, Lithuania, 57–66.
60. Zhang, Fuzhen (2005). (Ed.) *The Schur Complement and Its Applications.* Springer.
61. Zyskind, George (1967). On canonical forms, non-negative covariance matrices and best and simple least squares linear estimators in linear models. *Annals of Mathematical Statistics*, 38, 1092–1109.
62. Zyskind, George (1969). Parametric augmentations and error structures under which certain simple least squares and analysis of variance procedures are also best. *Journal of the American Statistical Association*, 64, 1353–1368.

On Influence Diagnostics in Multivariate Regression Models under Elliptical Distributions

Shuangzhe Liu
School of Information Sciences and Engineering, University of Canberra, Canberra ACT 2601, Australia
E-mail: Shuangzhe.Liu@canberra.edu.au

In this paper multivariate elliptical regression models are studied by using the local influence method in statistical diagnostics. The results are obtained using an approach advocated by Billor and Loynes (1993). They are found to be more informative when compared with the corresponding results obtained by Liu (2002) using Cook's (1986, 1997) approach for the local influence method.

Keywords: diagnostics; likelihood displacement; elliptical regression models; matrix differential.

AMS Classification: 62J05.

1. Introduction

Elliptical distributions and regression models have been studied and applied for over three decades; see e.g. Fang and Zhang (1990) and Gupta and Varga (1993). In the meantime regression diagnostics and sensitivity analysis have also been studied and applied extensively; see e.g. Chatterjee and Hadi (1988) and Atkinson and Riani (2000). Recently the local influence (LI) method originated by Cook (1986, 1997) has been used in diagnostics for several regression models; a comparison of the LI method with the case deletion and influence function methods is made by Jung, Kim and Kim (1997). Univariate elliptical regression models have been studied by Galea, Paula and Bolfarine (1997) and by Liu (2000). Multivariate regression models under normality have been studied by Kim (1995) and by Fung and Tang (1997), and elliptical regression models by Liu (2002) and by Daz, Galea and Leiva (2003). Cook's approach for LI is used in all these papers.

In the present paper we study the multivariate elliptical regression models using Billor and Loynes' (1993) approach, an alternative to Cook's (1986) approach, for LI, and compare the results obtained in both approaches. In section 2 we briefly introduce elliptical regression models. In section 3 we give Cook's and Billor and Loynes' approaches for LI, and define the key matrices and then two measures, namely the slope and the normal curvature, used in the two approaches to implement LI. In sections 4 through 6 we study three perturbation schemes to find the key matrices for the two measures, by using the standard matrix differential techniques developed by Magnus and Neudecker (1999). In section 7 we discuss and comment on the assessment of LI.

2. Elliptical Regression Models

As in Liu (2002) let us consider

$$U = (u_1, \ldots, u_n)' \sim EM_{np}(0, \Sigma, I_n), \tag{1}$$

$$E(\operatorname{vec} U') = 0, \quad \operatorname{var}(\operatorname{vec} U') = -2\phi(0) I_n \otimes \Sigma, \tag{2}$$

$$\psi(U) = |\Sigma|^{-n/2} g(\operatorname{tr}(U\Sigma^{-1}U')), \tag{3}$$

where U is an $n \times p$ data matrix following an elliptical distribution, $U' = (u_1, \ldots, u_n)$ is the transpose of U, the n vectors $u_1, \ldots, u_n$ can be viewed as a sample from a p-dimensional elliptical population, Σ is a $p \times p$ positive definite scale matrix, I_n is an $n \times n$ identity matrix, vec indicates the vectorisation operator which transforms a matrix into a vector by stacking the columns of the matrix one under the other, ϕ is the derivative of the characteristic generator, $\psi(U)$ is the density based on U and g is its known generator. We study the following multivariate elliptical linear model

$$Y = XB + U, \tag{4}$$

where Y is an $n \times p$ matrix of observations, X is an $n \times m$ model matrix, B is an $m \times p$ matrix of parameters, and U is an $n \times p$ matrix as defined in (1)–(3).

Assuming $\tau(z) = z^{np/2} g(z)$, $z \geq 0$, reaches maximum at $z = z_g > 0$ (see Gupta and Varga, 1993, pp 285–286), we have the maximum likelihood estimators of B and Σ

$$\hat{B} = (X'X)^{-1}X'Y, \tag{5}$$

$$\hat{\Sigma} = \frac{p}{z_g} \hat{U}'\hat{U}, \tag{6}$$

$$\hat{U} = (I_n - X(X'X)^{-1}X')Y. \tag{7}$$

We define the following derivatives of $g(z)$ with respect to z, which are used in sections 4 through 6 to follow:

$$G = G(z) = \frac{\partial \ln g(z)}{\partial z} = \frac{g'(z)}{g(z)},$$
$$F = F(z) = \frac{\partial G(z)}{\partial z}.$$

For g, G and F under the multivariate normal and t distributions, see e.g. Liu (2002).

3. Local Influence

3.1. *Cook's approach*

Useful in sensitivity analysis, LI is introduced for assessing the influence of small perturbations in a general statistical model; see Cook (1986). Let $w = (w_1, \ldots, w_n)'$ be an $n \times 1$ perturbation vector of observations, w_0 be an $n \times 1$ no-perturbation vector (with $w_0 = (0, \ldots, 0)'$, or $w_0 = (1, \ldots, 1)'$, or a third choice, depending on the context), θ be an $r \times 1$ vector of parameters of interest, $L(\theta)$ be the log-likelihood of the postulated (i.e. unperturbed) model, and $L(\theta|w)$ be the log-likelihood of the perturbed model with $L(\theta) = L(\theta|w_0)$. To implement the idea of LI is to investigate the extent to which the inference is affected by the corresponding perturbation. We briefly mention some key concepts and the method as follows. In Cook's approach the likelihood displacement (LD) is chosen to be

$$LD(w) = 2[L(\hat{\theta}) - L(\hat{\theta}_w)],$$

where $\hat{\theta}$ and $\hat{\theta}_w$ are the maximum likelihood estimates under the two models respectively. The influence graph is an $(n+1) \times 1$ vector defined as

$$a(w) = (w', LD(w))'.$$

The LI method is based on studying the local behaviour of $a(w)$. Upon $LD(w)$, the normal curvature in direction l is

$$C_l(\theta) = 2|l'\Delta' H^{-1} \Delta l|, \tag{8}$$

where

$$\Delta = \left. \frac{\partial^2 L(\theta|w)}{\partial \theta \partial w'} \right|_{\theta=\hat{\theta}, w=w_0}, \qquad H = \left. \frac{\partial^2 L(\theta)}{\partial \theta \partial \theta'} \right|_{\theta=\hat{\theta}}.$$

Δ and H are the two derivative matrices evaluated at $\theta = \hat{\theta}$ and $w = w_0$.

$C_l(\theta)$ is the local influence of perturbing the postulated model on the estimation of θ. Large values of $C_l(\theta)$ indicate sensitivity to the induced perturbations in direction l. The maximum $C_{\max}$ of C_l and the corresponding direction vector $l_{\max}$ are given by using the largest eigenvalue of $\Delta' H^{-1} \Delta$ and its associated eigenvector. Large values of those elements of $l_{\max}$ indicate the corresponding observations may be influential.

Actually, for a scalar function $\hat{T}_w$ evaluated under the perturbed model for LI in general (see Cook, 1986), we have

$$C_l(\theta) = \frac{|l'Rl|}{(1 + f'f)^{1/2} l'(I + ff')l}, \tag{9}$$

where $f = \partial \hat{T}_w / \partial w|_{w=w_0}$, $R = \partial^2 \hat{T}_w / \partial w \partial w'|_{w=w_0}$. The maximum $C_{\max}$ of C_l and the corresponding direction vector $l_{\max}$ are given by using the largest eigenvalue of $(1 + f'f)^{-1/2}(I + ff')^{-1/2} R (I + ff')^{-1/2}$ and its associated eigenvector.

3.2. *Billor and Loynes' approach*

In this approach as advocated by Billor and Loynes (1993), we consider

$$\begin{aligned} LD^*(w) &= -2[L(\hat{\theta}) - L(\hat{\theta}_w|w)], \\ a^*(w) &= (w', LD^*(w))'. \end{aligned}$$

A measure for the assessment of local influence is the slope, in the direction of l, of $a^*(w)$:

$$S_l = l'f = l' \left. \frac{\partial LD^*(w)}{\partial w} \right|_{\theta=\hat{\theta}, w=w_0} = 2l' \left. \frac{\partial L(\theta|w)}{\partial w} \right|_{\theta=\hat{\theta}, w=w_0}. \tag{10}$$

With the idea as used in Liu (2004) and in Zhang and King (2005) for their models, we consider in this paper $\hat{T}_w = LD^*(w)$ under perturbation to study S_l and C_l for model (4). This is because we expect both $f \neq 0$ and F to provide useful information for this approach; note that $f \neq 0$ here is expected to be better than $f = 0$ in Cook's approach.

For $\hat{T}_w = LD^*(w)$ we have (see e.g. Liu (2004) and Zhang and King (2005)):

$$f = 2h, \tag{11}$$

$$R = 2(P - \Delta' Q^{-1} \Delta), \tag{12}$$

where

$$h = \left. \frac{\partial L(\theta|w)}{\partial w} \right|_{\theta=\hat{\theta}, w=w_0},$$

$$P = \left. \frac{\partial^2 L(\theta|w)}{\partial w \partial w'} \right|_{\theta=\hat{\theta}, w=w_0},$$

$$Q = \left. \frac{\partial^2 L(\theta|w)}{\partial \theta \partial \theta'} \right|_{\theta=\hat{\theta}, w=w_0}.$$

For model (4), Δ and H are given by Liu (2002). To obtain f and R, and then S_l and C_l, we shall derive h, P and Q in sections 4 through 6. We do so by using the matrix differential techniques developed by Magnus and Neudecker (1999).

4. Perturbation in Case-Weights

Model (4) with (1)–(3) is our postulated model. We now consider model (4) with $U \sim EM_{np}(0, \Sigma, W^{-1})$ to be a perturbed model in case-weights, where W is an $n \times n$ nonsingular diagonal matrix of perturbation. Let $W_0 = I$ be the matrix of no-perturbation such that $L(\theta|W_0) = L(\theta)$. Let $\theta = (b', s')'$ with $b = \operatorname{vec} B$ and $s = \operatorname{vech} \Sigma$, where vec denotes the vectorisation operator which transforms a matrix into a vector by stacking the columns of the matrix one under the other, vech denotes the vectorisation operator which eliminates all supradiagonal elements of the matrix, b is an $mp \times 1$ vector, s is a $(p+1)p/2 \times 1$ vector and θ is an $r \times 1$ vector ($r = mp + (p+1)p/2$). The perturbed log-likelihood of model (4) is

$$L_w = L(\theta|w) = \frac{p}{2} \ln|W| - \frac{n}{2} \ln|\Sigma| + \ln g(z_w),$$

where $z_w = \operatorname{tr}(U'WU\Sigma^{-1})$ and $U = Y - XB$.

We now derive $d_w L_w$ which is the differential of L_w with respect to W, and evaluate it at $\theta = \hat{\theta}$ and $w = w_0$ to get $d_w L_w|_{\theta=\hat{\theta}, w=w_0}$, to find h. We derive $d_w^2 L_w$ to find P. We derive $d_b^2 L_w$, $d_{bs}^2 L_w$ and $d_s^2 L_w$ to find Q.

Taking the differential of L_w with respect to W, we obtain

$$\begin{aligned} d_w L_w &= \frac{p}{2} \operatorname{tr}(W^{-1}(dW)) + G(dz_w) \\ &= \frac{p}{2} \operatorname{tr}(W^{-1}(dW)) + G \operatorname{tr}(U\Sigma^{-1}U')(dW) \\ &= \frac{p}{2} \operatorname{vec}'(W^{-1}) J(dw) + G \operatorname{vec}'(U\Sigma^{-1}U') J(dw), \end{aligned}$$

where J is the $n^2 \times n$ selection matrix such that $\operatorname{vec} W = Jw$ and w is an $n \times 1$ vector containing the diagonal elements of W. For J and its properties, including $J'J = I$ and $J'(M \otimes N)J = (M \odot N)$, where $\otimes$ and $\odot$ denote Kronecker and Hadamard products respectively, see e.g. Liu (1995, section 3.1).

Using $W_0 = I$, $\hat{\Sigma}$ given in (6) and $\hat{U} = (I - X(X'X)^{-1}X')Y$ in (7) we get

$$\begin{aligned} d_w L_w|_{\theta=\hat{\theta},w=w_0} &= \frac{p}{2} \operatorname{vec}'(I)J(dw) + \frac{z_g}{p} G \operatorname{vec}'(\hat{U}(\hat{U}'\hat{U})^{-1}\hat{U}')J(dw), \\ h &= \frac{p}{2}\, J' \operatorname{vec}(I) + \frac{z_g}{p}\, GJ' \operatorname{vec}(\hat{U}(\hat{U}'\hat{U})^{-1}\hat{U}'). \end{aligned}$$

Further

$$\begin{aligned} d_w^2 L_w &= -\frac{p}{2} \operatorname{tr}(W^{-1}(dW)W^{-1}(dW)) + F(dz_w)^2 \\ &= -\frac{p}{2} \operatorname{tr}(W^{-1}(dW)W^{-1}(dW)) + F \operatorname{tr}[(dW)U\Sigma^{-1}U'] \operatorname{tr}[U\Sigma^{-1}U'(dW)] \\ &= -\frac{p}{2}(dw)'J'(W^{-1} \otimes W^{-1})J(dw) \\ &\quad + F(dw)'J' \operatorname{vec}(U\Sigma^{-1}U') \operatorname{vec}'(U\Sigma^{-1}U')J(dw). \end{aligned}$$

Then

$$\begin{aligned} d_w^2 L_w|_{\theta=\hat{\theta},w=w_0} &= -\frac{p}{2}\, (dw)'J'(I \otimes I)J(dw) \\ &\quad + \frac{z_g^2}{p^2} F\, (dw)'J' \operatorname{vec}(\hat{U}(\hat{U}'\hat{U})^{-1}\hat{U}') \operatorname{vec}'(\hat{U}(\hat{U}'\hat{U})^{-1}\hat{U}')J(dw) \\ &= -\frac{p}{2}\, (dw)'I(dw) \\ &\quad + \frac{z_g^2}{p^2} F\, (dw)'J' \operatorname{vec}(\hat{U}(\hat{U}'\hat{U})^{-1}\hat{U}') \operatorname{vec}'(\hat{U}(\hat{U}'\hat{U})^{-1}\hat{U}')J(dw), \\ P &= -\frac{p}{2}\, I + \frac{z_g^2}{p^2} FJ' \operatorname{vec}(\hat{U}(\hat{U}'\hat{U})^{-1}\hat{U}') \operatorname{vec}'(\hat{U}(\hat{U}'\hat{U})^{-1}\hat{U}')J. \end{aligned}$$

To find Q, we first take the differentials of L_w with respect to B to obtain

$$\begin{aligned} d_b L_w &= -2G \operatorname{tr}[\Sigma^{-1}U'WX(dB)], \\ d_b^2 L_w &= 4F \operatorname{tr}[(dB)'X'WU\Sigma^{-1}] \operatorname{tr}[\Sigma^{-1}U'WX(dB)] \\ &\quad + 2G \operatorname{tr}[(dB)'X'WX(dB)\Sigma^{-1}]. \end{aligned}$$

As $W_0 = I$, $\hat{U} = (I - X(X'X)^{-1}X')Y$ and $\hat{U}'X = 0$, we have

$$\begin{aligned} d_b^2 L_w|_{\theta=\hat{\theta},w=w_0} &= 2\hat{G} \operatorname{tr}[(dB)'X'X(dB)\hat{\Sigma}^{-1}] \\ &= 2\hat{G}(d \operatorname{vec} B)'(\hat{\Sigma}^{-1} \otimes X'X)(d \operatorname{vec} B), \end{aligned} \tag{13}$$

where $\hat{G} = G(\hat{z})$ with $\hat{z} = z_g$, and $\hat{\Sigma}$ is the same as in (6).

Taking the differential of $d_b L_w$ with respect to Σ, we obtain

$$\begin{aligned} d^2_{bs} L_w &= 2F \operatorname{tr}[(dB)' X' W U \Sigma^{-1}] \ \operatorname{tr}[\Sigma^{-1} U' W U \Sigma^{-1} (d\Sigma)] \\ &\quad + 2G \operatorname{tr}[(dB)' X' W U \Sigma^{-1} (d\Sigma) \Sigma^{-1}]. \end{aligned}$$

As $W_0 = I$, $\hat{U} = (I - X(X'X)^{-1}X')Y$ and $\hat{U}'X = 0$, we have

$$d^2_{bs} L_w|_{\theta=\hat{\theta}, w=w_0} = 0. \tag{14}$$

Taking the differentials of L_w with respect to Σ, we get

$$\begin{aligned} d_s L &= -\frac{n}{2} \operatorname{tr}[\Sigma^{-1}(d\Sigma)] + G(dz_w) \\ &= -\frac{n}{2} \operatorname{tr}[\Sigma^{-1}(d\Sigma)] - G \operatorname{tr}[\Sigma^{-1} U' W U \Sigma^{-1} (d\Sigma)], \\ d^2_s L &= \frac{n}{2} \operatorname{tr}[(d\Sigma)\Sigma^{-1}(d\Sigma)\Sigma^{-1}] \\ &\quad + F \operatorname{tr}[(d\Sigma)\Sigma^{-1} U' W U \Sigma^{-1}] \ \operatorname{tr}[\Sigma^{-1} U' W U \Sigma^{-1} (d\Sigma)] \\ &\quad + 2G \operatorname{tr}[(d\Sigma)\Sigma^{-1} U' W U \Sigma^{-1} (d\Sigma) \Sigma^{-1}], \end{aligned}$$

so that

$$\begin{aligned} & d^2_s L|_{\theta=\hat{\theta}, w=w_0} \\ &= \frac{n}{2} (d \operatorname{vech} \Sigma)' D' (\hat{\Sigma}^{-1} \otimes \hat{\Sigma}^{-1}) D (d \operatorname{vech} \Sigma) \\ &\quad + \hat{F} (d \operatorname{vech} \Sigma)' D' D \operatorname{vech}(\hat{\Sigma}^{-1} \hat{U}' \hat{U} \hat{\Sigma}^{-1}) \operatorname{vech}'(\hat{\Sigma}^{-1} \hat{U}' \hat{U} \hat{\Sigma}^{-1}) D' D (d \operatorname{vech} \Sigma) \\ &\quad + 2\hat{G} (d \operatorname{vech} \Sigma)' D' (\hat{\Sigma}^{-1} \otimes \hat{\Sigma}^{-1} \hat{U}' \hat{U} \hat{\Sigma}^{-1}) D (d \operatorname{vech} \Sigma) \\ &= (\frac{n}{2} + \frac{2 z_g \hat{G}}{p}) (d \operatorname{vech} \Sigma)' D' (\hat{\Sigma}^{-1} \otimes \hat{\Sigma}^{-1}) D (d \operatorname{vech} \Sigma) \\ &\quad + \frac{z_g^2 \hat{F}}{p^2} (d \operatorname{vech} \Sigma)' D' D \operatorname{vech}(\hat{\Sigma}^{-1}) \operatorname{vech}'(\hat{\Sigma}^{-1}) D' D (d \operatorname{vech} \Sigma), \end{aligned} \tag{15}$$

where $\hat{G} = G(\hat{z})$, $\hat{F} = F(\hat{z})$ and $\hat{U}'\hat{U} = \hat{z}\hat{\Sigma}/p$ with $\hat{z} = z_g$, and D is the $p^2 \times (p+1)p/2$ duplication matrix such that $\operatorname{vec} \Sigma = D \operatorname{vech} \Sigma$. For D and its properties, see Magnus and Neudecker (1999).

Hence, it follows from (13), (14) and (15) that

$$Q = \begin{pmatrix} 2\hat{G}(\hat{\Sigma}^{-1} \otimes X'X) & 0 \\ 0 & Q_s \end{pmatrix},$$

where

$$Q_s = (\frac{n}{2} + \frac{2z_g\hat{G}}{p})D'(\hat{\Sigma}^{-1} \otimes \hat{\Sigma}^{-1})D + \frac{z_g^2\hat{F}}{p^2}D'D\,\mathrm{vech}(\hat{\Sigma}^{-1})\,\mathrm{vech}'(\hat{\Sigma}^{-1})D'D.$$

For the normal distribution case, where $\hat{G} = -\frac{1}{2}$, $\hat{F} = 0$ and $z_g = np$, we obtain

$$Q_{\mathrm{nor}} = -\begin{pmatrix} \hat{\Sigma}^{-1} \otimes X'X & 0 \\ 0 & \frac{n}{2}D'(\hat{\Sigma}^{-1} \otimes \hat{\Sigma}^{-1})D \end{pmatrix}.$$

We see that Q_s and Q_{nor} are the same as H_s and H_{nor} in Liu (2002) respectively. In the case when $p = 1$ (and therefore Σ and $D = 1$ both become scalars), Q_s and Q_{nor} become the results corresponding to the ones given in Galea *et al.* (1997) and Cook (1986) respectively.

We can now make local influence assessment by using Δ defined in (8) and f and R in (11–12), as Δ is given in (31) in Liu (2002) and f and R are available with h, P and Q just established above.

5. Perturbation in Explanatory Variables

If the perturbation in explanatory variables is of interest, the perturbed log-likelihood is constructed with X replaced by $X_w = X + WS$, where $W = (w_{ij})$ is an $n \times m$ matrix of perturbations, $W_0 = 0$, $S = \mathrm{diag}(s_1, \ldots, s_m)$ and s_j $(j = 1, \ldots, m)$ is the scale factor. We use the relevant part of the perturbed log-likelihood

$$L_w = -\frac{n}{2}\ln|\Sigma| + \ln g(z_w),$$

where $\theta = (b', s')'$, $b = \mathrm{vec}\,B$, $s = \mathrm{vech}\,\Sigma$, $z_w = \mathrm{tr}(U_w'U_w\Sigma^{-1})$ and $U_w = Y - (X + WS)B$.

By taking the differentials of L_w with respect to W, B and/or Σ, we get

$$\begin{aligned}
d_w L_w &= -2G\,\mathrm{tr}[\Sigma^{-1}U_w'(dW)SB],\\
d_w^2 L_w &= 4F\,\mathrm{tr}[B'S'(dW)'U_w\Sigma^{-1}]\,\mathrm{tr}[\Sigma^{-1}U_w'(dW)SB]\\
&\quad + 2G\,\mathrm{tr}[\Sigma^{-1}B'S'(dW)'(dW)SB],\\
d_b L_w &= -2G\,\mathrm{tr}[U_w'X_w(dB)\Sigma^{-1}],\\
d_b^2 L_w &= 4F\,\mathrm{tr}[\Sigma^{-1}(dB)'X_w'U_w]\,\mathrm{tr}[U_w'X_w(dB)\Sigma^{-1}]\\
&\quad + 2G\,\mathrm{tr}[(dB)'X_w'X_w(dB)\Sigma^{-1}],\\
d_{bs}^2 L_w &= 2F\,\mathrm{tr}[X_w'U_w\Sigma^{-1}(dB)']\,\mathrm{tr}[U_w'U_w\Sigma^{-1}(d\Sigma)\Sigma^{-1}]\\
&\quad + 2G\,\mathrm{tr}[(dB)'X_w'U_w\Sigma^{-1}(d\Sigma)\Sigma^{-1}],\\
d_s L_w &= -\frac{n}{2}\,\mathrm{tr}[\Sigma^{-1}(d\Sigma) - G]\,\mathrm{tr}[U_w'U_w\Sigma^{-1}(d\Sigma)\Sigma^{-1}],\\
d_s^2 L_w &= \frac{n}{2}\,\mathrm{tr}[\Sigma^{-1}(d\Sigma)\Sigma^{-1}(d\Sigma)\Sigma^{-1}]\\
&\quad + F\,\mathrm{tr}[U_w'U_w\Sigma^{-1}(d\Sigma)\Sigma^{-1}]\,\mathrm{tr}[U_w'U_w\Sigma^{-1}(d\Sigma)\Sigma^{-1}]\\
&\quad + 2G[\mathrm{tr}\,U_w'U_w\Sigma^{-1}(d\Sigma)\Sigma^{-1}(d\Sigma)\Sigma^{-1}].
\end{aligned}$$

Hence

$$\begin{aligned}
d_w L_w|_{\theta=\hat\theta, w=w_0} &= -2\hat G\,\mathrm{vec}'(\hat U\hat\Sigma^{-1}\hat B'S')J\,dw,\\
d_w^2 L_w|_{\theta=\hat\theta, w=w_0} &= 4\hat F(dw)'J'\,\mathrm{vec}(\hat U\hat\Sigma^{-1}\hat B'S')\,\mathrm{vec}'(\hat U\hat\Sigma^{-1}\hat B'S')J\,dw\\
&\quad + 2\hat G(dw)'(S\hat B\hat\Sigma^{-1}\hat B'S' \odot I)dw,\\
d_b^2 L_w|_{\theta=\hat\theta, w=w_0} &= (d\,\mathrm{vec}\,B)'(2\hat G\hat\Sigma^{-1}\otimes X'X)d\,\mathrm{vec}\,B,\\
d_{bs}^2 L_w|_{\theta=\hat\theta, w=w_0} &= 0,\\
d_s^2 L_w|_{\theta=\hat\theta, w=w_0} &= (d\,\mathrm{vech}\,\Sigma)'(\frac{z_g^2}{p^2}\hat F D'D\,\mathrm{vech}\,\hat\Sigma^{-1}\,\mathrm{vech}'\,\hat\Sigma^{-1}D'D)(d\,\mathrm{vech}\,\Sigma)\\
&\quad + (d\,\mathrm{vech}\,\Sigma)'(\frac{n}{2} + \frac{2z_g}{p}\hat G)D'(\hat\Sigma^{-1}\otimes\hat\Sigma^{-1})D(d\,\mathrm{vech}\,\Sigma).
\end{aligned}$$

We then obtain

$$\begin{aligned}
h &= -2\hat G J'\,\mathrm{vec}(U\hat\Sigma^{-1}\hat B'S'),\\
P &= 4\hat F J'\,\mathrm{vec}(\hat U\hat\Sigma^{-1}\hat B'S')\,\mathrm{vec}'(\hat U\hat\Sigma^{-1}\hat B'S')J\\
&\quad + 2\hat G(S\hat B\hat\Sigma^{-1}\hat B'S' \odot I),\\
Q &= \begin{pmatrix} 2\hat G(\hat\Sigma^{-1}\otimes X'X) & 0\\ 0 & Q_s \end{pmatrix},
\end{aligned}$$

where

$$Q_s = \frac{z_g^2}{p^2}\hat{F}D'D\,\mathrm{vech}(\hat{\Sigma}^{-1})\,\mathrm{vech}'(\hat{\Sigma}^{-1})D'D + (\frac{n}{2} + \frac{2z_g}{p}\hat{G})D'(\hat{\Sigma}^{-1}\otimes\hat{\Sigma}^{-1})D,$$

z_g is the same as in (6) and S is the same as defined above.

We can now find f and R by using h, P and Q just established above and using Δ given in (62) in Liu (2002).

6. Perturbation in Response Variables

For the perturbation in response variables, the perturbed log-likelihood is constructed with Y replaced by $Y_w = Y + WS$, where $W = (w_{ij})$ is an $n \times p$ matrix of perturbations, $W_0 = 0$, $S = \mathrm{diag}(s_1, \ldots, s_p)$ and s_j $(j = 1, \ldots, p)$ is the scale factor. We have the relevant part of the perturbed log-likelihood

$$L_w = -\frac{n}{2}\ln|\Sigma| + \ln g(z_w),$$

where $\theta = (b', s')'$, $b = \mathrm{vec}\,B$, $s = \mathrm{vech}\,\Sigma$, $z_w = \mathrm{tr}(U_w'U_w\Sigma^{-1})$ and $U_w = Y + WS - XB$.

By taking the differentials of L_w with respect to W, B and/or Σ, we obtain

$$\begin{aligned}
d_w L_w &= 2G\,\mathrm{tr}[U'(dW)S\Sigma^{-1}],\\
d_w^2 L_w &= 4F\,\mathrm{tr}[\Sigma^{-1}S(dW)'U]\,\mathrm{tr}[U'(dW)S\Sigma^{-1}]\\
&\quad + 2G\,\mathrm{tr}[S(dW)'(dW)S\Sigma^{-1}],\\
d_b L_w &= -2G\,\mathrm{tr}[U'X(dB)\Sigma^{-1}],\\
d_b^2 L_w &= 4F\,\mathrm{tr}[(dB)'X'U\Sigma^{-1}] + 2G\,\mathrm{tr}[(dB)'X'X(dB)\Sigma^{-1}],\\
d_{bs}^2 L_w &= 2F\,\mathrm{tr}[\Sigma^{-1}(dB)'X'U]\,\mathrm{tr}[U'U\Sigma^{-1}(d\Sigma)\Sigma^{-1}]\\
&\quad + 2G\,\mathrm{tr}[U'X(dB)\Sigma^{-1}(d\Sigma)\Sigma^{-1}],\\
d_s L_w &= -\frac{n}{2}\,\mathrm{tr}[\Sigma^{-1}(d\Sigma)] - G\,\mathrm{tr}[U'U\Sigma^{-1}(d\Sigma)\Sigma^{-1}],\\
d_s^2 L_w &= \frac{n}{2}\,\mathrm{tr}[\Sigma^{-1}(d\Sigma)\Sigma^{-1}(d\Sigma)]\\
&\quad + F\,\mathrm{tr}[U'U\Sigma^{-1}(d\Sigma)\Sigma^{-1}]\,\mathrm{tr}[U'U\Sigma^{-1}(d\Sigma)\Sigma^{-1}]\\
&\quad + 2G\,\mathrm{tr}[U'U\Sigma^{-1}(d\Sigma)\Sigma^{-1}(d\Sigma)\Sigma^{-1}].
\end{aligned}$$

Using $W_0 = I$, $\hat{U} = (I - X(X'X)^{-1}X')Y$ and $\hat{U}'X = 0$, we have

$$\begin{aligned}
d_w L_w|_{\theta=\hat{\theta},w=w_0} &= 2G \operatorname{vec}'(U\Sigma^{-1}S)Jdw, \\
d_w^2 L_w|_{\theta=\hat{\theta},w=w_0} &= 4F(dw)'J' \operatorname{vec}(U\Sigma^{-1}S) \operatorname{vec}'(U\Sigma^{-1}S)J(dw) \\
&\quad + 2G(dw)'(S\Sigma^{-1}S \odot I)(dw), \\
d_b^2 L_w|_{\theta=\hat{\theta},w=w_0} &= 2G(d \operatorname{vec} B)'(\Sigma^{-1}S \otimes X'X)(d \operatorname{vec} B), \\
d_{bs}^2 L_w|_{\theta=\hat{\theta},w=w_0} &= 0, \\
d_s^2 L_w|_{\theta=\hat{\theta},w=w_0} & \\
= F(d \operatorname{vech} \Sigma)'D' & \operatorname{vec}(\Sigma^{-1}U'U\Sigma^{-1}) \operatorname{vec}'(\Sigma^{-1}U'U\Sigma^{-1})D(d \operatorname{vech} \Sigma) \\
&\quad + (\frac{n}{2} + \frac{2z_g}{p}G)(d \operatorname{vech} \Sigma)'D'(\Sigma^{-1} \otimes \Sigma^{-1})D(d \operatorname{vech} \Sigma).
\end{aligned}$$

Hence

$$\begin{aligned}
h &= 2GJ' \operatorname{vec}(U\Sigma^{-1}S), \\
P &= 4FJ' \operatorname{vec}(U\Sigma^{-1}S) \operatorname{vec}'(U\Sigma^{-1}S)J + 2G(S\Sigma^{-1}S \odot I), \\
Q &= \begin{pmatrix} 2\hat{G}(\hat{\Sigma}^{-1} \otimes X'X) & 0 \\ 0 & Q_s \end{pmatrix},
\end{aligned}$$

where

$$\begin{aligned}
Q_s = FD' & \operatorname{vec}(\Sigma^{-1}U'U\Sigma^{-1}) \operatorname{vec}'(\Sigma^{-1}U'U\Sigma^{-1})D \\
&+ (\frac{n}{2} + \frac{2z_g}{p}G)D'(\Sigma^{-1} \otimes \Sigma^{-1})D.
\end{aligned}$$

We can now find f and R by using h, P and Q just established above and Δ given in (73) in Liu (2002).

7. Concluding Remarks

We have focussed on elliptical distributions with a finite variance. We have discussed Billor and Loynes' approach for LI to find and examine the slope S_l and the normal curvature C_l of the likelihood displacement LD*, for three perturbation schemes. Local influence assessment can then be made based on the largest slope $S_{\max}$ and the largest normal curvature $C_{\max}$ and their corresponding directions $l_{\max}$. The maximum likelihood estimates involved can be obtained either in a closed form for some simple situations or more often via EM or other numerical algorithms with solutions (see e.g. Liu and Heyde (2003) for examples and discussions on the estimation problems). As seen the results obtained seem to be more informative than those obtained using Cook's approach by Liu (2002).

Regarding the three different perturbation schemes in Billor and Loynes' approach, slightly different indications on which observations are most influential are expected. This may well be similar to the case for Cook's approach; it is reflected by the results in e.g. Galea *et al.* (1997) and Liu (2000).

Acknowledgements

The author is grateful to Heinz Neudecker and Götz Trenkler for their constructive comments on an early version of the paper. He also wishes to acknowledge the support given by the University of Canberra.

References

1. Atkinson, A. and Riani, M. (2000) *Robust Diagnostic Regression Analysis*, Springer, New York.
2. Billor, N. and Loynes, R.M. (1993) Local influence: a new approach, *Comm. Statist. Theory Methods*, 22, 1595–1611.
3. Chatterjee, S. and Hadi, A.S. (1988) *Sensitivity Analysis in Linear Regression*, Wiley, New York.
4. Cook, R.D. (1986) Assessment of local influence (with discussion), *J. R. Statist. Soc. B*, 48, 133–169.
5. Cook, R.D. (1997) Local influence, in *Encyclopedia of Statist. Sciences*, S. Kotz, C.B. Read and D.L. Banks eds., Wiley, New York, Update Vol. 1, 380–385.
6. Daz, J.A., Galea, M. and Leiva, V. (2003) Influence diagnostics for elliptical multivariate linear regression models, *Comm. Statist. Theory Methods*, 32, 625–641.
7. Fang, K.T. and Zhang, Y. (1990) *Generalized Multivariate Analysis*, Science Press, Beijing and Springer, Berlin.
8. Fung, W.K. and Tang, M.K. (1997) Assessment of local influence in multivariate regression analysis, *Comm. Statist. Theory Methods*, 26, 821–837.
9. Galea, M., Paula, G.A. and Bolfarine, H. (1997) Local influence in elliptical linear regression models, *The Statistician*, 46, 71–79.
10. Gupta, A.K. and Varga, T. (1993) *Elliptically Contoured Models in Statistics*, Kluwer Academic, Dordrecht.
11. Jung, K.M., Kim, M.G. and Kim, B.C. (1997) Second order local influence in linear discriminant analysis, *J. Japan. Soc. Comp. Statist.*, 10, 1–11.
12. Kim, M.G. (1995) Local influence in multivariate regression, *Comm. Statist. Theory Methods*, 24, 1271–1278.
13. Liu, S. (1995) *Contributions to Matrix Calculus and Applications in Econometrics*, Thesis Publishers, Amsterdam.

14. Liu, S. (2000) On local influence in elliptical linear regression models, *Statistical Papers*, 41, 211–224.
15. Liu, S. (2002) Local influence in multivariate elliptical linear regression models, *Ninth special issue on linear algebra and statistics. Linear Algebra Appl.*, 354, 159–174.
16. Liu, S. (2004) On diagnostics in conditionally heteroskedastic time series models under elliptical distributions, *J. Appl. Probab.*, 41A, 393–405.
17. Liu, S. and Heyde, C.C. (2003) Some efficiency comparisons for estimators from quasi-likelihood and generalized estimating equations, in *Mathematical Statistics and Applications: Festschrift for Constance van Eeden, IMS Lecture Notes Monogr. Ser. Inst. Math. Statist., Beachwood, OH*, Vol. 42, M. Moor, S. Froda and C. Léger eds., pp 357–371.
18. Magnus, J.R. and Neudecker, H. (1999) *Matrix Differential Calculus with Applications in Statistics and Econometrics*, Revised Ed., Wiley, Chichester.
19. Zhang, X. and King, M.L. (2005) Influence diagnostics in generalized autoregressive conditional heteroscedasticity processes, *J. Bus. Econom. Statist.*, 23, 118–129.

A Necessary Condition for Admissibility of Nonnegative Quadratic Estimators of Variance Components when the Moments are not Necessarily as under Normality

Chang-Yu Lu

Financial Research Center, Shanghai Finance University,
Shanghai 201209, People's Republic of China
E-mail: luchy@shfc.edu.cn

Yu Gao

Zhejiang Ocean University,
Zhoushan 316004, People's Republic of China
E-mail: gaoyu_zjhyxy@21cn.com

Binyan Zhang

Department of Computing, Henan Vocational Technology College,
Zhengzhou 450046, People's Republic of China
E-mail: binyan2003@hotmail.com

Simultaneous estimation of the vector of variance components under general variance components models with respect to the scale quadratic loss function is considered. A new method of constructing a better estimator and how to deal with the admissibility of quadratic estimators of variance components when the covariance matrix may be singular is presented. Using this method, a necessary condition for admissibility of nonnegative quadratic estimators is given.

Keywords: variance component; nonnegative quadratic estimator; kurtosis; admissibility.

AMS Classification: 62C.

1. Introduction

Consider the variance component model:

$$Y = X\beta + U_1\varepsilon_1 + U_2\varepsilon_2 + \cdots + U_k\varepsilon_k \tag{1}$$

where $Y(n \times 1)$ is a vector of observations, the design matrix $X(n \times p)$ and the matrices $U_1(n \times p_1), \ldots, U_k(n \times p_k)$ are known. Suppose that $\varepsilon_1 =$

$(\varepsilon_{11}, \ldots, \varepsilon_{1p_1})', \ldots, \varepsilon_k = (\varepsilon_{k1}, \ldots, \varepsilon_{kp_k})'$ are random vectors with $E(\varepsilon_{ij}) = 0$, $E(\varepsilon_{ij}^2) = \sigma_i^2$, $E(\varepsilon_{ij}^3) = 0$, $E(\varepsilon_{ij}^4) = 3(1+\gamma)\sigma_i^4$, for $i = 1, 2, \ldots, k$ and $j = 1, 2, \ldots, p_i$, and $\varepsilon_{11}, \ldots, \varepsilon_{1p_1}, \ldots, \varepsilon_{k1}, \ldots, \varepsilon_{kp_k}$ are independent. Further $\beta \in \mathbb{R}^p$ and $\sigma^2 = (\sigma_1^2, \ldots, \sigma_k^2) \in \Omega = (0, +\infty) \times (0, +\infty) \times \cdots \times (0, +\infty)$ are unknown parameters. The parameter γ is the so-called kurtosis parameter; its greatest lower bound is $-2/(n+2)$; see, for example, Bentler and Berkane (1986) and Srivastava and Chandra (1985). For the multivariate normal distributions, γ is zero. Throughout this paper, we assume γ is known.

We are interested in estimation of the vector of the variance components $\sigma^2 = (\sigma_1^2, \ldots, \sigma_k^2)$ when $k \geq 2$ and when β is treated as a nuisance parameter. This problem has been intensively investigated. An excellent reference is a monograph by Rao and Kleffe (1988). We focus our attention on admissibility of quadratic estimators of variance components. Most relevant are the papers by Olsen, Seely and Birkes (1976) and by LaMotte (1982) in the case of estimators restricted to satisfying some additional conditions, such as invariance or equivalence or unbiasedness. Their works, which provided seminal results in the characterisation of admissible linear estimators in the general linear model and the admissibility of invariance quadratic estimators of variance components among the class of all invariance quadratic estimators, which is equivalent to the admissibility of linear estimators in some linear model, is of great theoretical importance, but is too complicated to give an explicit characterisation of admissible estimators of variance components. This problem is further developed in detail by, among others, Gnot and Kleffe (1983), Kleffe and Seifert (1986), and Klonecki and Zontek (1987, 1989, 1992). However, the invariance is not so relevant for the estimation procedures for variance components, that is, the invariant quadratic estimator is often not nonnegative. Verdooren (1988) pointed out that nonnegativity must be the first requirement for estimating variance components. Gnot, Kleffe and Zmyslony (1985) investigated the nonnegativity of admissible invariant quadratic estimators. Hartung (1981), Mathew and Sinha (1992) and Chaubey (1984, 1991) investigated nonnegative estimation improvement.

Without the restriction to invariance, the admissibility of quadratic estimators of variance components among the class of all nonnegative quadratic estimators is difficult, and admissible quadratic estimators have been characterised only in special linear models. Wu, Cheng and Li (1981) characterised admissible quadratic estimators in linear models while assuming a covariance matrix of the form $\sigma^2 I$. Lu (1988) accomplished the same characterisation for the covariance matrix of the form $\sigma^2 V$; V may be singular.

For the variance components model (1), while the kurtosis parameter is 0, Lu (1991) gave a sufficient condition for admissibility and a relation between locally best estimators and admissible estimators. Ye (1988) and Lu (1996) gave some conditions for admissibility. In this paper, we discuss the admissibility of nonnegative quadratic estimators of variance components. In section 2, we give a necessary condition for admissibility of nonnegative quadratic estimators.

For a matrix A, let $A \geq 0$, $\lambda_{\max}(A)$, $\mathrm{tr}(A)$ and A^+ denote that A is a nonnegative definite matrix, the largest eigenvalues of A when $A \geq 0$, the trace of A and the Moore-Penrose inverse, respectively. $A \geq B$ means $A - B \geq 0$. Because of the nonnegativity of σ_i^2, we consider the natural class of estimators as follows:

$$\mathcal{D} = \{(Y'A_1Y, \ldots, Y'A_kY) : A_1 \geq 0, \ldots, A_k \geq 0\}.$$

We are concerned with scale quadratic loss function $L(A_1, \ldots, A_k, \sigma^2)$ associated with a vector estimator $(Y'A_1Y, \ldots, Y'A_kY)$ of σ^2:

$$L(A_1, \ldots, A_k, \sigma^2) = \sum_{i=1}^{k} \frac{1}{\sigma_i^4}(Y'A_iY - \sigma_i^2)^2.$$

The risk function is

$$R(A_1, \ldots, A_k, \beta, \sigma_1^2, \ldots, \sigma_k^2) = E(L(A_1, \ldots, A_k, \sigma^2)).$$

The estimator $(Y'A_1Y, \ldots, Y'A_kY)$ is said to be $\mathcal{D}$-admissible for σ^2 if there exists no estimator in $\mathcal{D}$ which is better than $(Y'A_1Y, \ldots, Y'A_kY)$.

2. A Necessary Condition For Admissibility

Lemma 2.1. *Let matrix* $A \geq 0$. *Then* $A \leq (\mathrm{tr}\, A)I$.

Lemma 2.2. *Let matrix*

$$V = \begin{pmatrix} V_{11} & V_{12} \\ V_{21} & V_{22} \end{pmatrix} \geq 0.$$

Then $V_{22} - V_{21}V_{11}^+V_{12} \geq 0$.

Lemma 2.3. *Let*

$$V = \begin{pmatrix} V_{11} & V_{12} \\ V_{21} & V_{22} \end{pmatrix} \geq 0,$$

V_{11} *be a real number and the order of* V_{22} *be* $n-1$.
Then

$$F = x^2 V_{11}\, \mathrm{tr}(WV_{22}) + 2x(V_{12}W\theta)y + (\theta'W\theta)y^2 \geq 0 \tag{2}$$

and

$$G = y^2 \operatorname{tr}(WV_{22}) + 2(V_{12}W\theta)y + (\theta' W\theta) \geq 0 \tag{3}$$

for all $x \in \mathbb{R}$, $y \in \mathbb{R}$, $\theta \in \mathbb{R}^{n-1}$ *and* $W \geq 0$.

The first two Lemmas are well-known, and we prove the third lemma.

Proof. Without loss of generality, we suppose $V_{11} > 0$ and $\operatorname{tr}(WV_{22}) > 0$. F is a quadratic function of x for y, θ and W given. By Lemmas (2.1) and (2.2), the discriminant

$$\begin{aligned}\Delta &= y^2\theta' WV_{21}V_{12}W\theta - y^2V_{11}\operatorname{tr}(WV_{22})(\theta' W\theta)\\ &\leq y^2V_{11}\theta' W(V_{21}V_{11}^{-1}V_{12} - V_{22})W\theta\\ &\leq 0,\end{aligned}$$

therefore (2) holds. The proof of (3) is similar. □

Definition 2.1. The sequence of matrices $\{B_k\}_{k=1}^{+\infty}$ is said to converge to a matrix B, written as $B = \lim_{k\to+\infty} B_k$, if

$$b_{ij} = \lim_{k\to+\infty} b_{ij}^k$$

for all i, j, where b_{ij} and b_{ij}^k denote the (i, j)-th elements of B and B_k, respectively.

Lemma 2.4. *Let the sequence of matrices* $\{B_k\}_{k=1}^{+\infty}$ *satisfy* $0 \leq B_k \leq A$, $\forall\, k \geq 1$, *and* $A \geq 0$. *Then there exists a nonnegative definite matrix* B, *satisfying*

$$B = \lim_{m\to+\infty} B_{k_m}$$

for some subsequence, and

$$0 \leq B \leq A.$$

Proof. Since $0 \leq B_k \leq A$, we have

$$|b_{ij}^k| \leq \frac{b_{ii}^k + b_{jj}^k}{2} \leq \frac{a_{ii} + a_{jj}}{2} \leq \operatorname{tr} A$$

where a_{tt} denotes the (t, t)-th element of A.

Therefore, $\{b_{ij}^k\}_{k=1}^{+\infty}$, $i, j = 1, \ldots, n$, where n is the order of A, are bounded sequences of numbers and the length of the sequence is finite, so $\lim_{m\to+\infty} b_{ij}^{k_m}$ exists for all i, j, for some subsequence $\{k_m\}_{m=1}^{+\infty}$. Write

$$b_{ij} = \lim_{m\to+\infty} b_{ij}^{k_m}.$$

Let $B = (b_{ij})$. Then B satisfies Lemma (2.4). □

Theorem 2.1. *Suppose* $r = (1+\gamma) \leq (n+4)/(n+2)$, *and the nonnegative estimator* $(Y'A_1Y, \dots, Y'A_kY)$ *is* $\mathcal{D}$*-admissible for* σ^2 *under the model* (1). *Then*

$$2\lambda_{\max}(A_iV_i) + \mathrm{tr}(A_iV_i) \leq r^{-1}, \quad i = 1, 2, \dots, k, \tag{4}$$

where $V_i = U_iU_i'$, $i = 1, 2, \dots, k$.

Proof. We need only to show (4) holds for $i = 1$ (and then the same will apply for $i = 2, ..., n$). Suppose $2\lambda_{\max}(A_1V_1) + \mathrm{tr}(A_1V_1) > r^{-1}$. We will distinguish two cases:

(a) V_1 is positive definite and
(b) V_1 is nonnegative definite,

to construct a estimator which is better than $(Y'A_1Y, \dots, Y'A_kY)$, thus being led to a contradiction.

Without loss of generality, assume $r = 1$ in the following.

Case (a). $|V_1| \neq 0$. Since $V_1^{1/2}A_1V_1^{1/2} \geq 0$, we can write

$$V_1^{1/2}A_1V_1^{1/2} = P' \,\mathrm{diag}(\lambda_1, \dots, \lambda_n)P,$$

where $\lambda_1 \geq \lambda_2 \geq \cdots \geq \lambda_n$; λ_i, $i = 1, 2, \dots, n$, are eigenvalues of A_1V_1; $\lambda_1 = \lambda_{\max}(A_1V_1)$; $V_1^{1/2}$ is positive definite such that $V_1^{1/2}V_1^{1/2} = V_1$; and P is an orthogonal matrix.

Let $0 \leq \tau \leq \lambda_1$ satisfy $3\tau + \sum\limits_{i=2}^{n} \lambda_i = \max(1, \sum\limits_{i=2}^{n} \lambda_i)$, and

$$B_1 = V_1^{-1/2}P' \,\mathrm{diag}(\tau, \lambda_2, \lambda_3, \dots, \lambda_n)PV_1^{-1/2}.$$

In the following we prove that $(Y'B_1Y,\ Y'A_2Y,\ \dots,\ Y'A_kY)$ is better than $(Y'A_1Y,\ Y'A_2Y,\ \dots,\ Y'A_kY)$.

Let $W = \frac{1}{\sigma_1^2}\sum_{i=2}^{k} \sigma_i^2 V_i$. By a direct calculation, we have

$$\begin{aligned} &R(A_1, A_2, \dots, A_k, \beta, \sigma_1^2, \dots, \sigma_k^2) - R(B_1, A_2, \dots, A_k, \beta, \sigma_1^2, \dots, \sigma_k^2) \\ &= F_1(\beta, \sigma_1^2) + \frac{1}{\sigma_1^4}F_2(\beta, \sigma_1^2) + 2F_3, \end{aligned} \tag{5}$$

where

$$\begin{aligned}F_1(\beta,\sigma_1^2) &= 2\,\mathrm{tr}(A_1V_1)^2 + [\mathrm{tr}(A_1V_1)-1]^2\\ &\quad + 2\sigma_1^{-2}[(\mathrm{tr}(A_1V_1)-1)\beta'X'A_1X\beta + 2\beta'X'A_1V_1A_1X\beta]\\ &\quad - 2\,\mathrm{tr}(BV_1)^2 - [\mathrm{tr}(BV_1)-1]^2\\ &\quad - 2\sigma_1^2\left[(\mathrm{tr}(BV_1)-1)\beta'X'BX\beta + 2\beta'X'BV_1BX\beta\right],\end{aligned}$$

$$\begin{aligned}F_2(\beta,\sigma_1^2) &= \sigma_1^4\left[2\,\mathrm{tr}(A_1W)^2 + (\mathrm{tr}(A_1W))^2\right] + 2\sigma_1^2\,\mathrm{tr}(A_1W)\beta'X'A_1X\beta\\ &\quad + 4\sigma_1^2\beta'X'A_1WA_1X\beta + (\beta'X'A_1X\beta)^2\\ &\quad - \sigma_1^4\left[2\,\mathrm{tr}(BW)^2 + (\mathrm{tr}(BW))^2\right] - 2\sigma_1^2\,\mathrm{tr}(BW)\beta'X'BX\beta\\ &\quad - 4\sigma_1^2\beta'X'A_1WBX\beta - (\beta'X'BX\beta)^2\end{aligned}$$

and

$$\begin{aligned}F_3 &= [\mathrm{tr}(A_1W)][\mathrm{tr}(A_1V_1)-1] - [\mathrm{tr}(BW)][\mathrm{tr}(BV_1)-1]\\ &\quad + 2\,\mathrm{tr}(A_1V_1A_1W) - 2\,\mathrm{tr}(BV_1BW).\end{aligned}$$

Write

$$\theta = (\theta_1, \theta_2')' = PV_1^{-1/2}X\beta,$$

where θ_1 is a real number and θ_2 is an $(n-1)\times 1$ vector. Denote

$$\mathrm{diag}(\lambda_2,\lambda_3,\ldots,\lambda_n) \widehat{=} \Lambda, \qquad V_1^{-1/2}WV_1^{-1/2} = \begin{pmatrix} V_{11} & V_{12}\\ V_{21} & V_{22}\end{pmatrix},$$

where V_{11} is a nonnegative number. For any $(\beta,\sigma^2)\in\mathbb{R}^n\times\Omega$, we have

$$\begin{aligned}F_1(\beta,\sigma_1^2) &= (\lambda_1-\tau)\left(3\lambda_1+3\tau+2\sum_{i=2}^n\lambda_i-1\right)\\ &\quad + \frac{2}{\sigma_1^2}(\lambda_1-\tau)\left(\left(3\lambda_1+3\tau+\sum_{i=2}^n\lambda_i-1\right)\theta_1^2+\theta_2'\Lambda\theta_2\right)\\ &\geq (\lambda_1-\tau)\left(3\lambda_1+3\tau+2\sum_{i=2}^n\lambda_i-1\right)\\ &\geq \min\left\{\lambda_1\left(2\lambda_1+\mathrm{tr}(A_1V_1)-1\right), \frac{1}{3}\left(2\lambda_1+\mathrm{tr}(A_1V_1)-1\right)^2\right\}\\ &> 0 \end{aligned} \tag{6}$$

and

$$\begin{aligned} F_2(\beta,\sigma_1^2) &= (\lambda_1-\tau)\sigma_1^4\left[(3\lambda_1+\tau)V_{11}^2+2V_{11}\operatorname{tr}(\Lambda V_{22})+4V_{12}\Lambda V_{21}\right] \\ &\quad+(\lambda_1-\tau)\sigma_1^2\left[(6\lambda_1+6\tau)V_{11}+2\operatorname{tr}(\Lambda V_{22})\right]\theta_1^2 \\ &\quad+(\lambda_1-\tau)\sigma_1^2\left[8(V_{12}\Lambda V_{21})\theta_1+2V_{11}\theta_2'\Lambda\theta_2\right] \\ &\quad+(\lambda_1^2-\tau^2)\theta_1^4+2(\lambda_1-\tau)\theta_1^2(\theta_2'\Lambda\theta_2) \\ &\geq 2(\lambda_1-\tau)\left[\sigma_1^4V_{11}\operatorname{tr}(\Lambda V_{22})+2\sigma_1^2(V_{12}\Lambda\theta_2)\theta_1+(\theta_2'\Lambda\theta_2)\theta_1^2\right] \\ &\quad+2(\lambda_1-\tau)\sigma_1^2\left[\operatorname{tr}(\Lambda V_{22})\theta_1^2+2(V_{12}\Lambda\theta_2)\theta_1+V_{11}(\theta_2'\Lambda\theta_2)\right] \\ &= 2(\lambda_1-\tau)g_1(\sigma_1^2,\theta)+2(\lambda_1-\tau)\sigma_1^2g_2(\theta) \end{aligned} \tag{7}$$

where

$$g_1(\sigma_1^2,\theta)=\sigma_1^4V_{11}\operatorname{tr}(\Lambda V_{22})+2\sigma_1^2(V_{12}\Lambda\theta_2)\theta_1+(\theta_2'\Lambda\theta_2)\theta_1^2$$

and

$$g_2(\theta)=\operatorname{tr}(\Lambda V_{22})\theta_1^2+2(V_{12}\Lambda\theta_2)\theta_1+V_{11}(\theta_2'\Lambda\theta_2).$$

By Lemma (2.3), $g_1(\sigma_1^2,\theta)\geq 0$ and $g_2(\theta)\geq 0$ for all σ^2 and all θ, and we also have

$$\begin{aligned} F_3 &= (\lambda_1V_{11}+\operatorname{tr}(\Lambda V_{22}))\,(\lambda_1+\operatorname{tr}\Lambda-1) \\ &\quad-(\tau V_{11}+\operatorname{tr}(\Lambda V_{22}))\,(\tau+\operatorname{tr}\Lambda-1)+2(\lambda_1^2-\tau^2)V_{11} \\ &= (\lambda_1-\tau)\left[(3\lambda_1+3\tau+\operatorname{tr}\Lambda-1)V_{11}+\operatorname{tr}(\Lambda V_{22})\right] \\ &\geq (\lambda_1-\tau)V_{11}(3\lambda_1+3\tau+\operatorname{tr}\Lambda-1) \\ &\geq 0. \end{aligned} \tag{8}$$

Hence, for any $(\beta,\sigma^2)\in\mathbb{R}^n\times\Omega$,

$$\begin{aligned} &R(A_1,A_2,\ldots,A_k,\beta,\sigma_1^2,\ldots,\sigma_k^2)-R(B_1,A_2,\ldots,A_k,\beta,\sigma_1^2,\ldots,\sigma_k^2) \\ &\geq \min\left\{\lambda_1(2\lambda_1+\operatorname{tr}(A_1V_1)-1),\frac{1}{3}(2\lambda_1+\operatorname{tr}(A_1V_1)-1)^2\right\} \\ &> 0. \end{aligned} \tag{9}$$

This means that $(Y'B_1Y,\ Y'A_2Y,\ \ldots,\ Y'A_kY)$ is better than $(Y'A_1Y,\ Y'A_2Y,\ \ldots,\ Y'A_kY)$, which contradicts that $(Y'A_1Y,\ Y'A_2Y,\ \ldots,\ Y'A_kY)$ is $\mathcal{D}$-admissible for σ_1^2.

Case (b). $|V_1|=0$. Set $V(k)=V_1+k^{-1}I$, where k is a natural number. Then $V(k)>V_1$, and so $2\lambda(A_1V(k))+\operatorname{tr}(A_1V_k)>1$. By the proof of case (a), there exists B_k such that $0\leq B_k\leq A_1$, and

$$\begin{aligned}
F(k) &= 2\,\mathrm{tr}\,[A_1(V(k)+W)]^2 + [\mathrm{tr}\,(A_1(V(k)+W)) - 1]^2 \\
&\quad + \sigma_1^{-4}(\beta' X' A_1 X\beta)^2 + 4\sigma_1^{-2}\beta' X' A_1(V(k)+W)A_1 X\beta \\
&\quad + 2\sigma_1^{-2}\,[\mathrm{tr}\,(A_1(V(k)+W)) - 1]\,\beta' X' A_1 X\beta - 2\,\mathrm{tr}\,[B_k(V(k)+W)]^2 \\
&\quad - [\mathrm{tr}\,(B_k(V(k)+W)) - 1]^2 - \sigma_1^{-4}(\beta' X' B_k X\beta)^2 \\
&\quad - 4\sigma_1^{-2}\beta' X' B_k(V(k)+W)B_k X\beta \\
&\quad - 2\sigma_1^{-2}\,[\mathrm{tr}\,(B_k(V(k)+W)) - 1]\,\beta' X' B_k X\beta \\
&\geq \min\left\{\lambda(A_1V(k))\,[2\lambda(A_1V(k)) + \mathrm{tr}(A_1V(k)) - 1]\right., \\
&\qquad\qquad \left.\frac{1}{3}\,[2\lambda(A_1V(k)) + \mathrm{tr}(A_1V(k)) - 1]^2\right\}. \qquad (10)
\end{aligned}$$

Since $0 \leq B(k) \leq A_1$, by Lemma (2.4), B_{k_m} converges to a nonnegative definite matrix B and $0 \leq B \leq A$ for some subsequence $\{k_m\}_{m=1}^{+\infty}$. Let $B_1 = B$; then, from (10), in the limit, we have that, for any $(\beta, \sigma^2) \in \mathbb{R}^n \times \Omega$,

$$\begin{aligned}
&R(A_1, A_2, \ldots, A_k, \beta, \sigma_1^2, \ldots, \sigma_k^2) - R(B_1, A_2, \ldots, A_k, \beta, \sigma_1^2, \ldots, \sigma_k^2) \\
&= \lim_{m\to\infty} F(k) \geq \min\{\lambda_1(2\lambda_1 + \mathrm{tr}(A_1V_1) - 1), \frac{1}{3}(2\lambda_1 + \mathrm{tr}(A_1V_1) - 1)^2\} > 0.
\end{aligned}$$

Hence, $(Y'B_1Y,\ Y'A_2Y,\ \ldots,\ Y'A_kY)$ is better than $(Y'A_1Y,\ Y'A_2Y,\ \ldots,\ Y'A_kY)$, which contradicts that $(Y'A_1Y,\ Y'A_2Y,\ \ldots,\ Y'A_kY)$ is $\mathcal{D}$-admissible for σ^2. The proof is completed. □

Theorem (2.1) is a general result, in which a necessary condition for admissibility of nonnegative quadratic estimators of variance components, when the moments are not necessarily as under normality, is established. For the important case that ε_i $(i = 1, 2, \ldots, k)$ follow a multivariate normal distribution, when γ is zero, by applying Theorem (2.1), we get

Corollary 2.1. *Suppose $\gamma = 0$, and the nonnegative estimator $(Y'A_1Y,\ \ldots,\ Y'A_kY)$ is $\mathcal{D}$-admissible for σ^2. Under model (1), we have*

$$2\lambda_{\max}(A_iV_i) + \mathrm{tr}(A_iV_i) \leq 1,\ i = 1, 2, \ldots, k$$

where $V_i = U_iU_i'$, $i = 1, 2, \ldots, k$.

Acknowledgement

The research of Lu is supported by National Natural Sciences Foundation of P. R. China #10471043.

References

1. Bentler, P.M. and Berkane, M. (1986), Greatest lower bound to the elliptical theory kurtosis parameter. *Biometrika* **73**(1), 240–241.
2. Chaubey, Y.P. (1984), On the comparison of some nonnegative estimator of variance components for two models. *Comm. Statist.–Simulation Comput.* **13**(5), 619–633.
3. Chaubey, Y.P. (1991), A note on nonnegative minimum bias MINQUE in variance components model. *Statistics & Probability Letters* **11**, 395–397.
4. Gnot, S. and Kleffe, J. (1983), Quadratic estimates in mixed linear models with two variance components. *J. Statist. Planning & Inference* **8**, 249–258.
5. Gnot, S., Kleffe, J. and Zmyslony, R. (1985), Nonnegativity of admissible invariant quadratic estimates in mixed linear models with two variance components. *J. Statist. Planning & Inference* **12**, 249–258.
6. Hartung, J. (1981), Nonnegative minimum bias invariant estimation in variance components models. *Ann. Statist.* **9**, 278–292.
7. Kleffe, J. and Seifert, B. (1986), Computation of variance components by the MINQUE method. *J. Multivariate Anal.* **18**, 107–116.
8. Klonecki, W. and Zontek, S. (1987), On admissible invariant estimators of variance components which dominate unbiased invariant estimators. *Statistics* **18**, 483–498.
9. Klonecki, W. and Zontek, S. (1989), Variance components admissible estimators from some unbalanced data: Formulae for the nested design. *Probab. Math. Statist.* **10**, 313–331.
10. Klonecki, W. and Zontek, S. (1992), Admissible estimators of variance components obtained via submodel. *Ann. Statist.* **20**(3), 1454–1467.
11. LaMotte, L.R. (1982), Admissibility in linear estimation. *Ann. Statist.* **10**, 245–256.
12. Lu, C.-Y. (1988), On the question of admissibility of quadratic estimators of error variance in a linear model (in Chinese). *Acta. Sci. Natur. Univ. Jilin.* **3**, 7–13
13. Lu, C.-Y. (1991), Admissibility of nonnegative quadratic estimators for combined variance components (in Chinese). *Chinese Ann. Math.,* **12A**, 699–707.
14. Lu, C.-Y. (1996), Admissibility of simultaneous estimators of variance components (in Chinese). *J. Systems Science and Mathematical Science.* **16**(4), 361–366.
15. Mathew, T. and Sinha, B.K. (1992), Nonnegative estimation of variance components in unbalanced mixed models with two variance components. *J. Multivariate Anal.* **42**, 77–101.
16. Olsen, A., Seely, J. and Birkes, D. (1976), Invariance quadratic estimation for two variance components. *Ann. Statist.* **4**, 878–890.
17. Rao, C.R. and Kleffe, J. (1988), *Estimation of Variance Components and Application.* North-Holland, Amsterdam.
18. Srivastava, V.K. and Chandra, R.(1985), Properties of the mixed regression estimator when disturbances are not necessarily normal. *J. Statistical Planning and Inference* **11**, 15–21.

19. Verdooren, L.R. (1988), Least squares estimators and nonnegative estimators of variance components. *Commun. Statist.* **17A**, 1027–1051.
20. Wu, Q.-G., Cheng, P. and Li, G.-Y. (1981), Admissibility of quadratic form estimators of error variance in a linear model (in Chinese). *Science in Sinica* **7**, 815–825.
21. Ye, C.-N. (1988), Admissibility of simultaneous estimators of two variance components (in Chinese). *J. Chinese App. Prob. Statist.* **4**, 35–43.

Small-Sample Performance of Robust Methods in Logistic Regression

Suraiya Nargis

Population Health Unit, Australian Institute of Health and Welfare, Canberra ACT 2601, Australia
E-mail: suraiya.nargis@aihw.gov.au

Alice Richardson

School of Information Sciences and Engineering, University of Canberra, Canberra ACT 2601, Australia
E-mail: alice.richardson@canberra.edu.au

In this paper we aim to deepen our understanding of the behaviour of robust methods in logistic regression. Firstly we describe a number of approaches to robust estimation in logistic regression. Next we test aspects of their behaviour through a small simulation study of binary data only. Finally, opportunities for further research are presented.

Keywords: generalised linear model; simulation study.

1. Introduction

Linear models have a long history and they provide a rich class of methods for describing the relationship between a dependent variable and one or more explanatory variables. Nelder and Wedderburn (1972) extended the class to include discrete response variables and called their new models Generalised Linear Models (GLMs). McCullagh and Nelder (1989) noted that GLMs are a powerful and popular technique for modelling a large variety of data. In particular GLMs allow researchers to model the relationship between predictors and a response variable which follows a distribution from the exponential family.

Logistic regression is a special case of a GLM. Logistic regression relates a binary response variable to covariates. The possible outcome of a binary response variable might be classified as success or failure and can be represented by 1 and 0. The Bernoulli distribution for binary random variables specifies probabilities $P(y = 1) = \pi$ and $P(y = 0) = 1 - \pi$ for the two outcomes. The probability mass function for a general Bernoulli random variable is as follows:

$$f(y_i; \pi_i) = \pi_i^{y_i}(1 - \pi_i)^{1-y_i}. \tag{1}$$

The natural parameter $\log(\pi/(1 - \pi))$, the log odds of response 1, is called the logit of π.

When linear regression is used for binary data, three problems may arise. First, the variance of the error term is not constant; second, the error term is not normally distributed; third, there is no restriction requiring the predicted values to fall between 0 and 1.

Thus normal distributions for the ordinary least-squares estimators do not apply to binary data and the logistic regression model produces a more fruitful estimator. The logistic regression model is

$$\pi(x) = \frac{\exp(\alpha + \beta x)}{1 + \exp(\alpha + \beta x)}. \tag{2}$$

This model relates $\mu_i = \pi_i$, the mean of y_i, to a linear model via

$$g(\mu_i) = g(\pi_i) = x_i^T \beta = \eta_i$$

where g is the function $g(\pi) = \log(\pi/(1 - \pi))$.

During the past decades researchers have become increasingly aware that some of the most common statistical procedures are excessively sensitive to seemingly minor deviations from the assumptions.

Robust procedures have been proposed as an alternative. The methods used in this paper will be those inspired by the infinitesimal approach of Huber (1981) and Hampel *et al.* (1986). This approach studies the effect of tiny changes in the data on parameter estimates. For example, consider the multiple linear regression model

$$y = X^T \beta + \epsilon$$

where y is an $n \times 1$ vector of observations; X is an $n \times p$ design matrix for the fixed effects; β is a $p \times 1$ vector of fixed effect parameters; and ϵ is an $n \times 1$ vector of random errors with $\epsilon \sim N(0, \sigma^2)$. This model has log-likelihood

$$l(\beta; y) = -\frac{n}{2}\log(2\pi\sigma^2) - \frac{1}{2}\sum_{i=1}^{n}\left(\frac{y_i - x_i^T\beta}{\sigma}\right)^2. \tag{3}$$

So calculating the MLE of β involves differentiating (3) and solving

$$\sum_{i=1}^{n} \left(\frac{x_i}{\sigma}\right) \left(\frac{y_i - x_i^T \beta}{\sigma}\right) = 0. \tag{4}$$

Huber's M-estimation, described in Huber (1981), replaces the linear term in (4) with a function that downweights the effects of large y on the estimating equation. Huber's choice of function for this was

$$\psi(z) = \begin{cases} c\,\mathrm{sgn}(z) \ , & |z| < c, \\ z \ , & |z| \geq c. \end{cases} \tag{5}$$

To achieve a reasonable degree of robustness without compromising efficiency, a typical value of c is 1.345 for the estimation of location parameters.

In Section 2 of this paper we review robust logistic regression. In Section 3.1 we carry out a simulation study on simple logistic regression for both classical and robust cases with no outliers. In Section 3.2 we study the effect of an increase in the proportion of misclassified points in the data set and in Section 3.3 we study the effect of one misclassified point at a time. In Section 4 we conclude with suggestions for further research.

2. Algorithms for Robust Logistic Regression

The term outlier refers to an observation which appears to be inconsistent with the rest of the data, relative to an assumed model. Outliers can occur for a variety of reasons including data entry errors, scoring errors, error in the measurement recording or sample data that are genuinely atypical.

It is not clear how this definition could be extended to binary data because y takes values 0 or 1, neither of which can be considered inconsistent with the other. This makes outliers in logistic regression harder to define and identify than outliers in linear regression. Ruckstuhl and Welsh (2001) consider a contamination model for outliers in binomial data as follows:

$$f(y) = (1 - \epsilon) \binom{m}{y} \pi^y (1 - \pi)^{m-y} + \epsilon g(y)$$

where g is an arbitrary probability function on $\{0, \ldots, m\}$ and $0 \leq \epsilon \leq 1$. They note that no contamination is possible with binary data, because in that case g must be a binary distribution so the linear combination of the two is also binary. Thus the maximum likelihood estimator of π will be robust for binary data, but better estimators exist for binomial data.

We will therefore pursue misclassification rather than contamination in order to define outliers in binary data. This consideration identifies an outlier in binary data as an isolated success/failure that is surrounded by many failures/successes.

In the next three subsections we consider three approaches to robust logistic regression: by Pregibon (1982), by Richardson (1999), and by Cantoni and Ronchetti (2001).

2.1. *Pregibon*

Pregibon (1982) considers a binary outcome y with probability function as shown in (1). The log likelihood is then

$$\sum_{i=1}^{n} \{y_i \log \pi_i + (1-y_i)\log(1-\pi_i)\}$$

and maximising this is the same as minimising the deviance

$$\begin{aligned} D &= -2\left[\sum_{i=1}^{n} \{y_i \log \pi_i + (1-y_i)\log(1-\pi_i)\} \right. \\ &\qquad \left. - \sum_{i=1}^{n} \{y_i \log \hat{\pi}_i + (1-y_i)\log(1-\hat{\pi}_i)\}\right] \\ &= -2\left[\sum_{i=1}^{n} \left\{y_i \log \frac{\pi_i}{\hat{\pi}_i} + (1-y_i)\log\frac{1-\pi_i}{1-\hat{\pi}_i}\right\}\right]. \end{aligned}$$

Pregibon's suggestion is to replace the deviance with a function λ of it that curtails large contributions, namely

$$-2\left[\sum_{i=1}^{n} \lambda\left(y_i \log \frac{\pi_i}{\hat{\pi}_i} + (1-y_i)\log\frac{1-\pi_i}{1-\hat{\pi}_i}\right)\right].$$

The function λ grows more slowly than a straight line; e.g.

$$\lambda(d) = \begin{cases} d\,, & d \le H, \\ 2(dH)^{1/2} - H\,, & \text{otherwise.} \end{cases}$$

2.2. *Richardson*

Richardson (1995) derived robust estimating equations for the parameters of a mixed linear model as follows. Consider the mixed linear model

$$y = X\beta + Zb + \epsilon \tag{6}$$

where y is an $n \times 1$ vector of observations; X is an $n \times p$ design matrix for the fixed effects; β is a $p \times 1$ vector of fixed effect parameters; Z is an

$n \times q$ design matrix for the random effects; b is a $q \times 1$ vector of random effects with $b \sim N(0, \sigma_b^2)$; and ϵ is an $n \times 1$ vector of random errors with $\epsilon \sim N(0, \sigma_\epsilon^2)$ independent of b. Thus $\text{var}(y) = \sigma_b^2 ZZ^T + \sigma_\epsilon^2 I = V$.

Richardson (1995) showed that maximum likelihood estimating equations for the variance parameters of this model are

$$\hat{\sigma}_b^2 = \frac{\sigma_b^2 (y - X\beta)^T V^{-1} ZZ^T V^{-1} (y - X\beta)}{\text{tr}[V^{-1} ZZ^T]}$$
$$\hat{\sigma}_\epsilon^2 = \frac{\sigma_\epsilon^2 (y - X\beta)^T V^{-1} V^{-1} (y - X\beta)}{\text{tr}[V^{-1}]}$$

and that given $\hat{\sigma}_b^2$ and $\hat{\sigma}_\epsilon^2$, the maximum likelihood estimate of β is

$$\hat{\beta} = (X^T \hat{V}^{-1} X)^{-1} X^T \hat{V}^{-1} y.$$

Estimating equations to produce robust M-estimates of Type II of the variance parameters are

$$\hat{\sigma}_b^2 = \frac{\sigma_b^2 \psi(V^{-1/2}(y - X\beta))^T V^{-1/2} ZZ^T V^{-1/2} \psi(V^{-1/2}(y - X\beta))}{k\, \text{tr}[V^{-1} ZZ^T]}$$
$$\hat{\sigma}_\epsilon^2 = \frac{\sigma_\epsilon^2 \psi(V^{-1/2}(y - X\beta))^T V^{-1/2} V^{-1/2} \psi(V^{-1/2}(y - X\beta))}{k\, \text{tr}[V^{-1}]}$$

and that given $\hat{\sigma}_b^2$ and $\hat{\sigma}_\epsilon^2$, the Type II M-estimate of β is $\hat{\beta} =$

$$\beta + (X^T \hat{V}^{-1/2} \psi'(\hat{V}^{-1/2}(y - X\beta)) \hat{V}^{-1/2} X)^{-1} X^T \hat{V}^{-1/2} \psi(\hat{V}^{-1/2}(y - X\beta)).$$

Here $\psi(z)$ is Huber's function in (5), applied elementwise to the vector $V^{-1/2}(y - X\beta)$ and used to create the diagonal entries of $\psi'(\hat{V}^{-1/2}(y - X\beta))$; and k is an appropriate consistency correction. Richardson (1999) used the same estimating equation approach to derive estimators for the generalised linear mixed model. These can be seen as extensions of the estimating equations used in classical maximum likelihood estimation of logistic regression parameters, namely $\hat{\beta} = (X^T W X)^{-1} X^T W z$ where $W = \text{diag}(\hat{\mu}_i(1 - \hat{\mu}_i))$ and $z = \hat{\eta}_i + (y_i - \hat{\mu}_i)\frac{\partial \eta}{\partial \mu}$. This suggests that an appropriate robust estimating equation for the logistic regression model is $\hat{\beta} =$

$$(X^T \hat{W}^{-1/2} \psi'(\hat{W}^{-1/2}(z - X\beta)) \hat{W}^{-1/2} X)^{-1} X^T \hat{W}^{-1/2} \psi(\hat{W}^{-1/2}(z - X\beta)). \tag{7}$$

Splus code for implementing this method is given in the Appendix.

2.3. *Cantoni and Ronchetti*

Cantoni and Ronchetti (2001) suggested Huber quasi-likelihood estimation as a robust alternative to maximum likelihood estimation. They also began with estimating equations

$$\left(\psi(V^{1/2}(y-\mu))W(X)V^{-1/2}\frac{\partial\mu}{\partial\eta}\right) - k = 0 \tag{8}$$

where μ and η are parts of the logistic regression model as in (2); $\psi(z)$ is Huber's function as in (5); W is a function designed to reduce the influence of large X; and k is a consistency correction.

Very few statistical computing packages have implemented such methods. Cantoni and Ronchetti (2001) wrote Splus functions to implement their quasi-likelihood approach. Richardson (1999) also had Splus functions to implement her estimating equation approach. The method will be denoted RII and will be used later in this paper.

Splus has a robust option within the GLM function, but some functions within Splus give much more direct access to the code than is possible with `glm()`. For example, the `varcomp` command has an option `method = winsor` which calls the Splus function `varcomp.fit.winsor`. This function implements the method of Burns (1994) as follows. Initial values of the parameters are found and used to apply Huber ψ functions to the data. A classical fitting method is then used to update the parameter estimates. The algorithm is iterated to convergence. It seems likely that a similar algorithm is used within `glm()`. In fact, Bondell (2005) claims that Splus implements the conditionally unbiased bounded influence estimator of Künsch *et al.* (1989) and the Mallows-type estimator of Carroll and Pederson (1993) with weights defined by a robust Mahalanobis distance. There is also an option to specify that the scale parameter is fixed at 1, i.e. using `glm(y ~ x, family = robust(binomial, scale = 1))` instead of `glm(y ~ x, family = robust(binomial))` which estimates the scale parameter as well.

3. Simulation Study: Rationale

The infinitesimal approach to robustness is based on the following concepts of Huber (1981): firstly, qualitative robustness, meaning that a small perturbation should have small effects; secondly, influence functions, which are a measure of the effects of infinitesimal perturbation; thirdly, breakdown point, which is a measure of how big the perturbation can be before the method under consideration breaks down.

Hampel (1985) used the above concepts and linked them to the stability of an estimator. According to Hampel, the most important robustness requirements (besides qualitative robustness) are a high breakdown point and low gross-error sensitivity. A sensitivity analysis quantifies how changes in the values of the data alter the values of the parameter estimates.

We will follow the approach of Nargis (2005) and study the sensitivity of methods (robust and non-robust) by increasing the outlying-ness of a single point in the data set and studying changes in the resulting parameter estimates. Breakdown is a measure of the instability of an estimator to multiple outliers in the data. Roughly it gives the smallest fraction of data contamination needed to cause an arbitrarily large change in the estimate.

We will study the breakdown of methods (both robust and non-robust) by increasing the proportion of outliers in a data set and then studying changes in the parameter estimates similarly.

3.1. *Simple logistic regression, classical and robust: no outliers*

The algorithm for this part of the simulation study is as follows.

(1) Set up X as 50 values from $N(0,1)$.
(2) Calculate $xbeta = 1 + X$.
(3) Calculate $prob = \exp(xbeta)/(1 + \exp(xbeta))$.
(4) Simulate $Y = Bernoulli(prob)$.
(5) Estimate parameters by classical (Splus and RII) and robust logistic regression (Splus with fixed scale parameter, Splus with free scale parameter, and RII).
(6) Repeat Steps 1 to 5 50 times to produce 50 sets of parameter estimates.

Thus the model is

$$P(Y = 1) = \frac{\exp(1 + X)}{1 + \exp(1 + X)}. \tag{9}$$

The simulation was done using Splus. Table 1 gives the mean and standard deviation (s.d.) for the 50 sets of estimates of the parameters.

We use Q–Q plots in Figure 1 to study the overall shape of the distribution of the estimates.

The classical method produces estimates that are closest to the true values and show the smallest amount of variation. This is because the data contained no outliers to begin with. For the remaining robust methods,

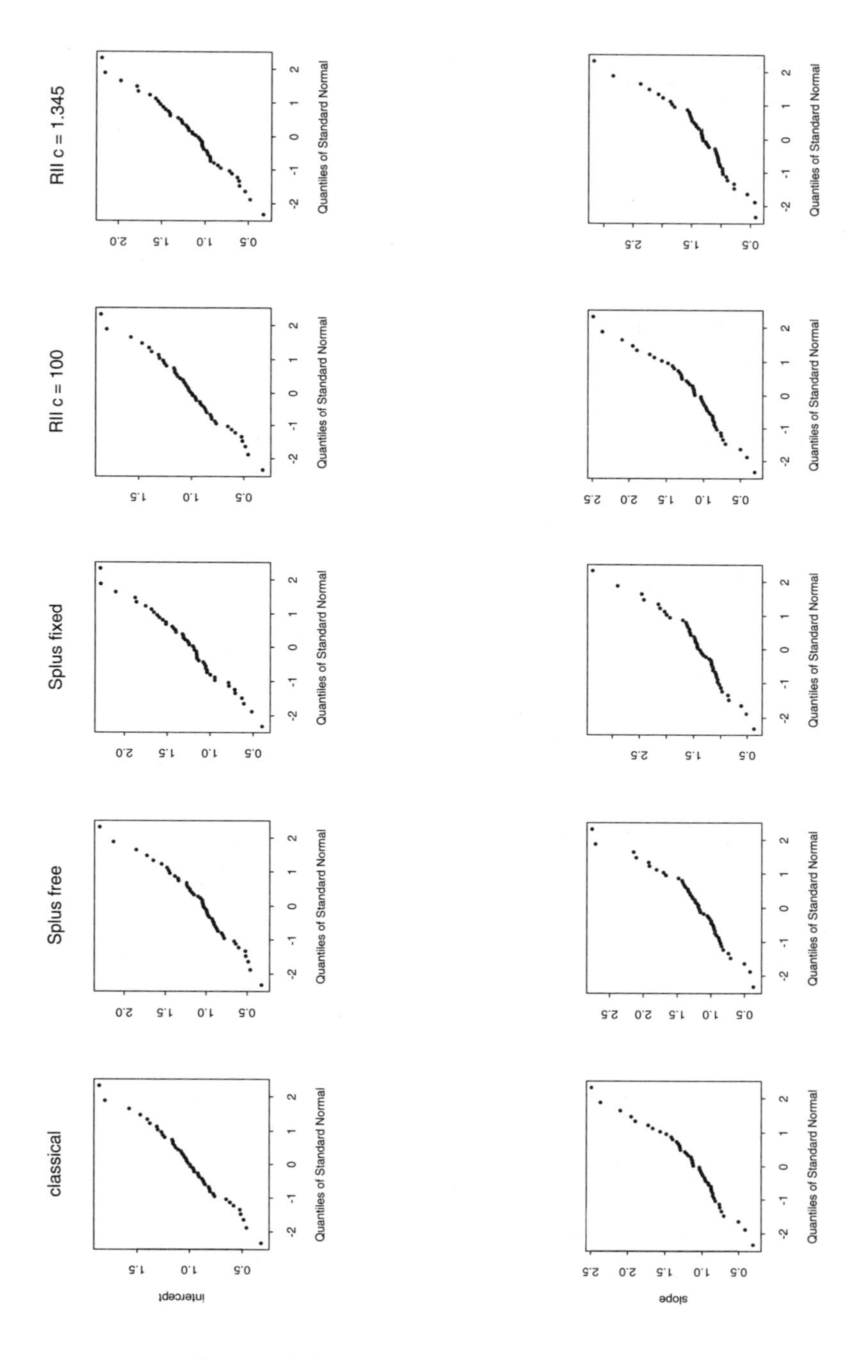

Fig. 1. Q–Q plots of parameter estimates.

Table 1. Mean and s.d. for 50 sets of estimated logistic regression coefficients. True value = 1 in each case.

	classical		Splus fixed		Splus free		RII c = 100		RII c = 1.345	
	β_0	β_1	β_0	β_1	β_0	β_1	β_0	β_1	β_0	β_1
mean	1.02	1.12	1.08	1.19	1.27	1.40	1.02	1.12	1.17	1.30
s.d.	0.37	0.42	0.42	0.48	0.48	0.54	0.38	0.42	0.46	0.53

the means of $\hat{\beta}_0$ and $\hat{\beta}_1$ are all close to the true value of 1 in each case. This shows that the robust methods still work well, even when there are no outliers in the data. The shape of the Q–Q plots shows that the estimates are approximately normally distributed with a slight positive skew.

3.2. *Breakdown analysis: increasing proportion of misclassified points*

The idea of changing a proportion of the responses from 0 to 1 and vice versa has also been employed by Mills *et al.* (2002). The algorithm for this part of the simulation study is as follows.

(1) Set $i = 1$ and set up X as 50 values of $N(0, 1)$.
(2) Calculate $xbeta = 1 + X$.
(3) Calculate $prob = \exp(xbeta)/(1 + \exp(xbeta))$.
(4) Simulate $Y = Bernoulli(prob)$.
(5) Estimate the parameters by classical and robust logistic regression, as before.
(6) Change i^{th} value of y from 0 to 1 or from 1 to 0.
(7) Let $i = i + 1$ and repeat from Step 5 until $i = 50$.

When we change one value we keep the previously changed values as they are. The proportion of misclassified observations will increase gradually and we can study the effect of the change on the parameter estimates.

The simulation was done using Splus. After obtaining the outputs of the simulation with an increasing proportion of outliers, we have plotted the parameter estimates as a function of the proportion of misclassified points.

The values of both intercept and slope estimates gradually decrease as the proportion of outliers increases, until the estimates have essentially turned round, from around 1.0 to around -1.0. The robust parameter estimates are actually slightly more extreme than the non-robust estimates, particularly in the regions where there are either very few or very many misclassified observations.

Therefore both the methods have managed the increasing proportion of outliers in a similar way. There does not seem to be any strong indication of a breakdown point, i.e. a point beyond which the non-robust estimates become infinite or otherwise useless but the robust ones continue to provide useful results.

3.3. *Sensitivity analysis: one misclassified point at a time*

The aim of this part of the simulation study is to investigate the variability in parameter estimates when the proportion of misclassified observations is held constant. The algorithm for this part of the simulation study is as follows.

(1) Set $i = 1$ and set up X as 50 values of $N(0, 1)$.
(2) Calculate $xbeta = 1 + X$.
(3) Calculate $prob = \exp(xbeta)/(1 + \exp(xbeta))$.
(4) Simulate $Y = Bernoulli(prob)$.
(5) Change i^{th} value of y from 0 to 1 or from 1 to 0 (i.e. misclassify it).
(6) Estimate the parameters by classical and robust logistic regression, as before.
(7) Return i^{th} value of y to its original value.
(8) Let $i = i + 1$ and repeat from Step 5 until $i = 50$.

When we change one value we change the previous one back to its original value. The changes will happen one at a time, to see the effect of the change of one value on the parameter estimates. This is akin to considering the points on Figure 2 relating to number of misclassified observations = 1, and changing the observation number which is misclassified. These small numbers of misclassified points were the ones which showed the greatest difference between non-robust and robust fits in Figure 2.

The simulation was done using Splus. After obtaining the outputs of the simulation with an increasing proportion of outliers, we have plotted the parameter estimates as a function of the misclassified point number.

Once again the robust parameter estimates are slightly more extreme than the non-robust ones, but the differences are not large. Hence in Figure 3 we see the gap between the lines for the robust and non-robust fit, matching the gap between the parameter estimates in the extremes of Figure 2. The parameter estimates in Figure 3 vary around the true values of 1.0, suggesting that either method provides a reasonable parameter estimate in this situation.

4. Conclusion

In this paper we have investigated some aspects of the behaviour of a range of methods for robust logistic regression, through the use of a small simulation with binary data. The hypothetical data of Hosmer and Lemeshow (2000) offer another approach to simulation in binary data. They construct a single continuous covariate and seven different sets of twenty binary responses that fit a logistic regression model more or less well. Their example could be extended by adding extra covariates or increasing the sample size.

A simulation study involving binomial data rather than binary data would also be useful as it would overcome the problem of defining contaminated data highlighted by Ruckstuhl and Welsh (2001). An expanded simulation study could also include the Splus routines of Cantoni and Ronchetti (2001), and restricted maximum likelihood methods that match Richardson (1999). It would also be interesting to study the effect of perturbations in the values of X instead of, or as well as, perturbations in y as shown in section 3.2.

Appendix

The Splus code used to implement Richardson's method is given below.

```
PSI <- function(vec, tune = 100) {
out <- vector(length = length(vec))
for(i in 1:length(vec))
out[i] = max(-tune, min(tune, vec[i])
out
}

PSIDASH <- function(vec, tune = 100) {
out <- matrix(0, ncol = length(vec), nrow=length(vec))
for(i in 1:length(vec))
out[i, i] = ifelse(abs(vec[i]) < tune, 1, 0)
out
}

R2 <- function(y, ssize, x, inita, tunea = 100) {
# enter y as a vector, x as matrix
# ML only
# initialise parameter estimates
# let alpha = q by 1 matrix
```

```
alpha <- matrix(inita)
# begin loop to update estimates of everything
iter <- 0
repeat {
iter <- iter + 1
# transform y
eta <- as.vector(x %*% alpha)
mu <- (ssize * exp(eta))/(exp(eta) + 1)
gdashmu <- ssize/((mu + 0.1) * (ssize - mu + 0.1))
r <- eta + (y - mu) * gdashmu
newr <- r
# get jinv
u <- diag(gdashmu)
jinv <- solve(u)
sva <- svd(jinv)
jhalfinv <- sva$u %*% diag(sqrt(sva$d)) %*% t(sva$v)
jhalfresid <- jhalfinv %*% (newr - x %*% alpha)
# update estimates of alpha ROBUSTLY
psia <- PSI(jhalfresid, tunea)
psidash <- PSIDASH(jhalfresid, tunea)
h <- solve(t(x) %*% jhalfinv %*% psidash %*% jhalfinv %*% x)
newalpha <- alpha + h %*% t(x) %*% jhalfinv %*%
   as.matrix(psia)
# check whether parameter estimates are different from last
# time
diff <- c(abs(newalpha - alpha))
alpha <- newalpha
etahat <- as.vector(x %*% alpha)
muhat <- (ssize * exp(etahat))/(exp(etahat) + 1)
print(c(iter, as.vector(alpha)))
if(iter > 99)
break
if(max(diff) < 0.0001)
break
}
# output
list(iter = iter, alpha = as.vector(alpha))
}
```

References

1. Bondell, H.D. (2005) Minimum distance estimation for the logistic regression model. *Biometrika* **92**, 724 – 731.
2. Burns, P. (1994) Winsorised REML estimates of variance components. Technical Report, Statistical Sciences Inc, Seattle.
3. Cantoni, E. and Ronchetti, E. (2001) Robust inference for generalized linear models. *Journal of the American Statistical Association* **96**, 1022 – 1030.
4. Carroll, R.J. and Pederson, S. (1993) On robust estimation in the logistic regression model. *Journal of the Royal Statistical Society, Series B* **55**, 693 – 706.
5. Hampel, F.R. (1985) The breakdown point of the mean combined with some rejection rules. *Technometrics* **27**, 95 - 107.
6. Hampel, F.R., Ronchetti, E.M., Rousseeuw, P.J. and Stahel, W.A. (1986) *Robust Statistics.* New York: Wiley.
7. Hosmer, D.W. and Lemeshow, S. (2000) *Applied Logistic Regression.* 2nd edition. New York: Wiley.
8. Huber, P. J. (1981) *Robust Statistics.* New York: Wiley.
9. Künsch, H.R., Stefanski, L.A. and Carroll, R.J. (1989) Conditionally unbiased bounded influence estimation in general regression models, with applications to generalized linear models. *Journal of the American Statistical Association* **84**, 460 – 466.
10. McCullagh, P.J. and Nelder, J.A. (1989) *Generalised Linear Models.* London: Chapman and Hall.
11. Mills, J.E., Field, C.A. and Dupuis, D.J. (2002) Marginally specified generalized linear mixed models: a robust approach. *Biometrics* **58**, 727 –734.
12. Nargis, S. (2005) *Robust methods in logistic regression.* Unpublished Masters thesis, University of Canberra, Canberra, Australia.
13. Nelder, J.A. and Wedderburn, R.W.M. (1972) Generalised linear models. *Journal of the Royal Statistical Society, Series A* **135**, 370 – 384.
14. Pregibon, D. (1982) Resistant fits for some commonly used logistic models with medical applications. *Biometrics* **38**, 485 – 498.
15. Richardson, A.M. (1995) *Some problems in estimation in mixed linear models.* Unpublished Ph.D. thesis, Australian National University, Canberra, Australia.
16. Richardson, A.M. (1999) *Computer implementation of generalised linear mixed models.* University of Canberra Technical Report ISE R99/108.
17. Ruckstuhl, A.F. and Welsh, A.H. (2001) Robust fitting of the binomial model. *Annals of Statistics* **29**, 1117 – 1136.

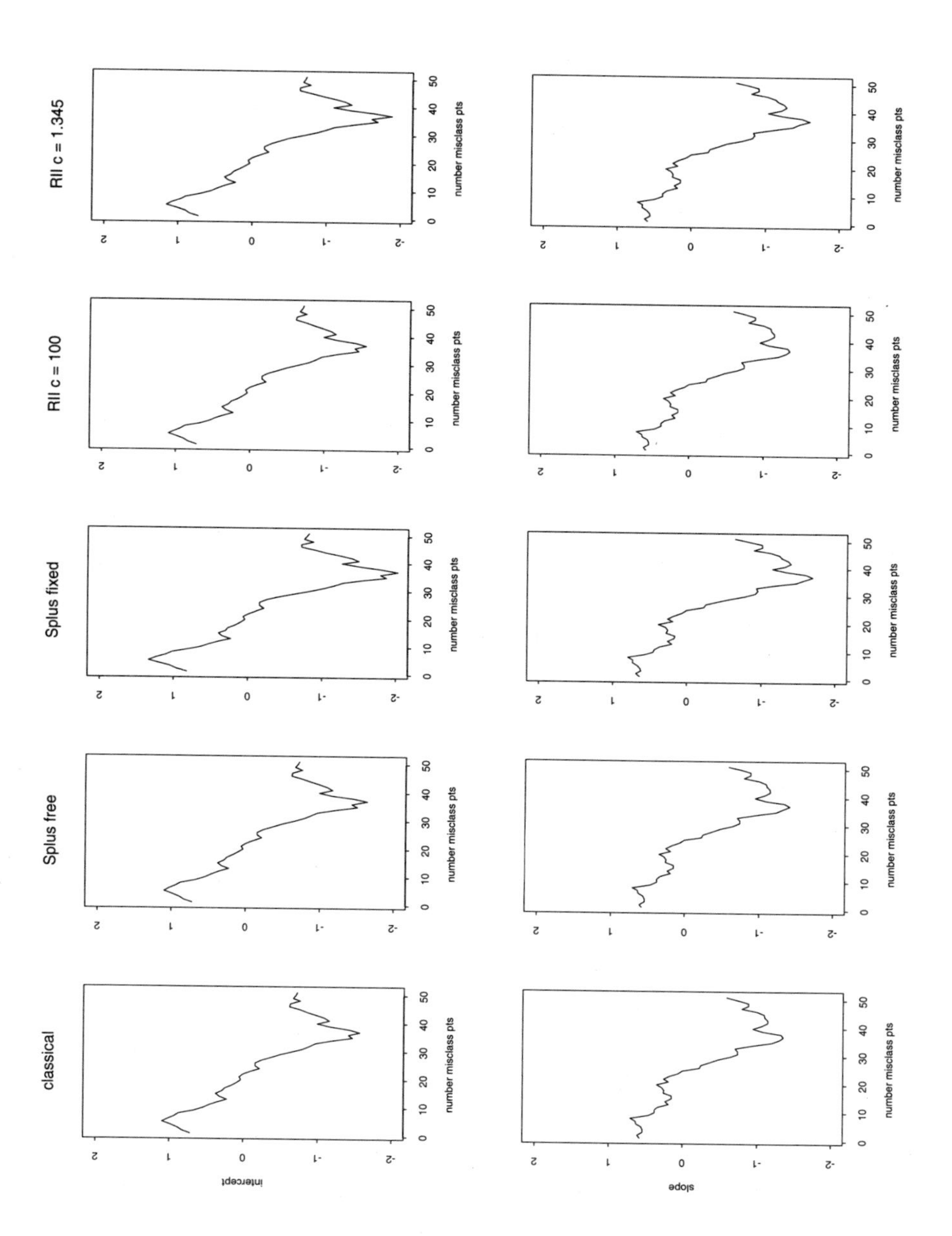

Fig. 2. Parameter estimates under increasing misclassification.

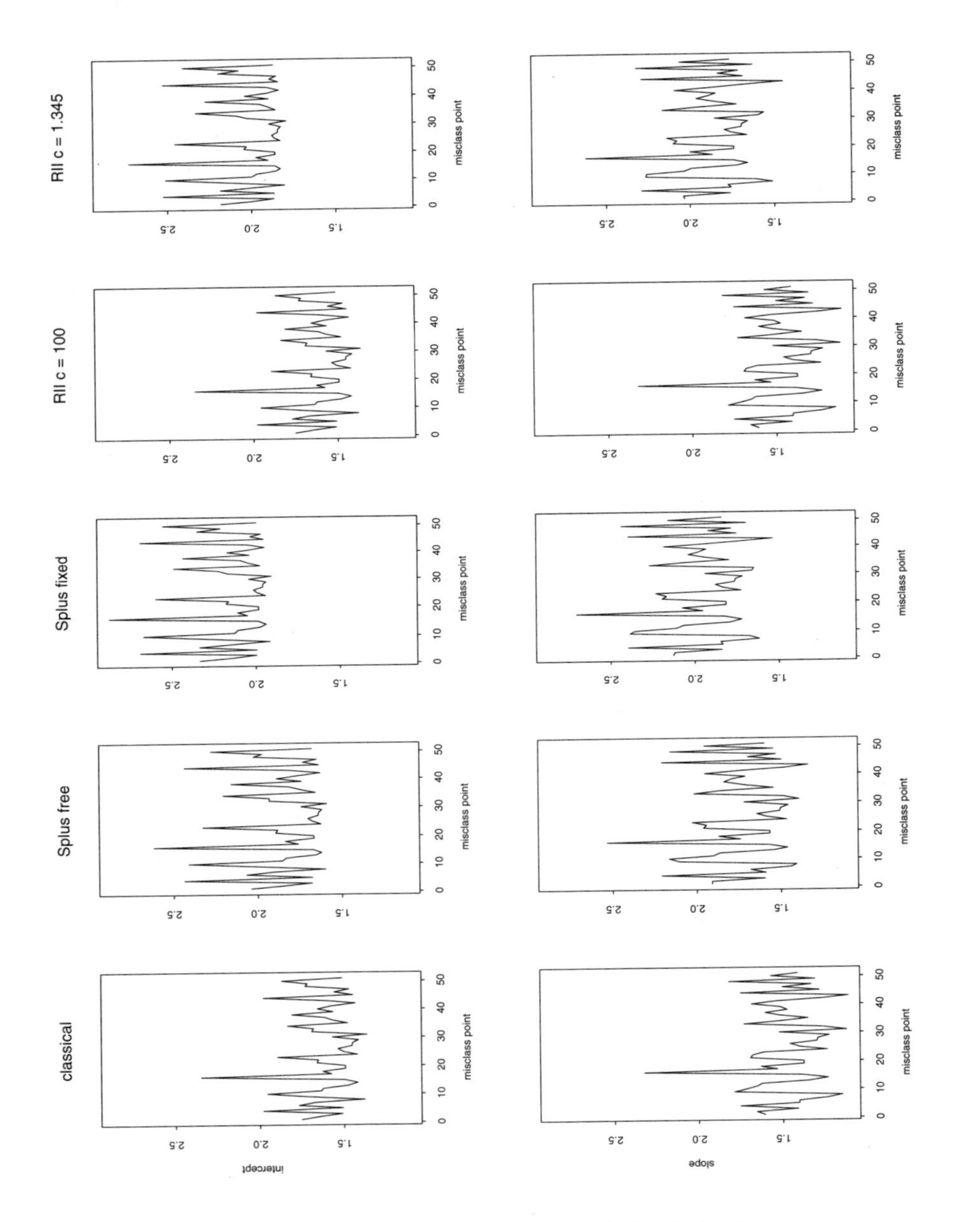

Fig. 3. Parameter estimates for one misclassified point.

On the Asymptotic Distribution of the 'Natural' Estimator of Cronbach's Alpha with Standardised Variates under Nonnormality, Ellipticity and Normality

Heinz Neudecker

School of Economics and Business, University of Amsterdam, 1018 WB Amsterdam, The Netherlands
E-mail: h.neudecker@uva.nl

Following van Zyl, Neudecker and Nel (2000) we consider asymptotic properties of the 'natural' estimator of Cronbach's alpha when the variates are standardised. This means that the population correlation matrix P is the population variance matrix Σ, because now all diagonal elements of Σ are equal to unity. The 'natural' estimator $\hat{\alpha}_s = (p-1)^{-1}p[1 - p(1'R1)^{-1}]$, where R is the sample correlation matrix and p is the number of items (responses). We find the asymptotic distribution of $\hat{\alpha}_s$ under nonnormality, ellipticity and normality. Use is made of a (0,1) 'duplication' matrix $\tilde{D}$. This enables us to switch easily between $\operatorname{vec} A$ and $\mathrm{w}(A)$, where A is a symmetric zero-axial matrix ($A_d = 0$) and $\mathrm{w}(A)$ contains the 'free' elements of A.

Keywords: responses; maximum-likelihood estimator; nonnormality.

1. Introduction

Some time ago van Zyl, Neudecker and Nel (2000) considered the estimation of Cronbach's $\alpha = (p-1)^{-1}p[1 - (1'\Sigma 1)^{-1} \operatorname{tr} \Sigma]$, where Σ is the population variance of a random vector x (of p responses) and 1 is a column p-vector of ones. Further $Ex = 0$. As usual E is the expectation operator.

The 'natural' estimator $\hat{\alpha} = (p-1)^{-1}p[1-(1'S1)^{-1} \operatorname{tr} S]$ was used, where $S = (n-1)^{-1}X'(I-n^{-1}11')X$ is the usual unbiased estimator of Σ. Further $n^{-1}(n-1)S$ is the maximum-likelihood estimator under normality and n is the sample size. The three authors gave the asymptotic distribution of $\hat{\alpha}$ under nonnormality. See their result (21).

In this paper we shall consider the estimation of α when the variates are *standardised.* This means that $\Sigma_d = I$, where $(.)_d$ denotes the diagonal of the matrix $(.)$. Using the notation α_s we have $\alpha_s = (p-1)^{-1}p[1-p(1'P1)^{-1}]$, where P denotes the population correlation matrix.

Clearly in the standardised case $P = \Sigma$ because $\Sigma_d = I$. As estimator we shall adopt $\hat{\alpha}_s = (p-1)^{-1}p[1 - p(1'R1)^{-1}]$, where R is the sample correlation matrix.

2. The Asymptotic Distribution of $\hat{\alpha}_s$ Under Nonnormality

We shall employ the Delta method; see e.g. van Zyl *et al.* (2000). If $\hat{\theta}$ is any estimator of the parameter vector θ such that

$$n^{1/2}(\hat{\theta} - \theta) \xrightarrow{d} N(0, \Phi), \text{ as } n \to \infty \tag{1}$$

and if F is a scalar-valued function of a vector variable z with first-order partial derivatives continuous at $z = \theta$, then

$$n^{1/2}\left(F(\hat{\theta}) - F(\theta)\right) \xrightarrow{d} N(0, h'\Phi h), \text{ as } n \to \infty \tag{2}$$

where $h = \frac{\partial F(z)}{\partial z}|_{z=\theta}$ and $N(\mu, \Phi)$ denotes the normal distribution with mean μ and variance Φ.

It is well-known, see e.g. Magnus (1988) and Neudecker and Wesselman (1990), that

$$n^{1/2}\,\text{vec}(R - P) \xrightarrow{d} N(0, \Psi), \text{ as } n \to \infty \tag{3}$$

where

$$\Psi = [I - L(I \otimes P)K_d](\Sigma_d^{-1/2} \otimes \Sigma_d^{-1/2})V(\Sigma_d^{-1/2} \otimes \Sigma_d^{-1/2})[I - K_d(I \otimes P)L],$$
$$L = \frac{1}{2}(I + K),$$
$$V = E[(x-\mu)(x-\mu)' \otimes (x-\mu)(x-\mu)' - (\text{vec}\,\Sigma)(\text{vec}\,\Sigma)'],$$

K is the commutation matrix: $K\,\text{vec}\,X = \text{vec}\,X'$ and K_d is its diagonal. From this follows easily for $\text{w}(R-P)$, the vector of 'free' elements of $R-P$:

$$n^{1/2}\,\text{w}(R - P) \xrightarrow{d} N(0, \tilde{D}'\Psi\tilde{D}), \tag{4}$$

where $\text{w}(A) = \frac{1}{2}\tilde{D}'\,\text{vec}\,A$ and $\text{vec}\,A = \tilde{D}\,\text{w}(A)$, for A such that $A_d = 0$ and $A' = A$ (In fact $\tilde{D}'$ eliminates the diagonal and supradiagonal elements of A).

Further

$$\tilde{D}'\Psi\tilde{D} = \tilde{D}'[I - (I \otimes P)K_d](\Sigma_d^{-1/2} \otimes \Sigma_d^{-1/2})V(\Sigma_d^{-1/2} \otimes \Sigma_d^{-1/2})[I - K_d(I \otimes P)]\tilde{D},$$

because $L\tilde{D} = \tilde{D}$. See for $\tilde{D}$ and its properties, Neudecker and Satorra (1996). The term involving $\operatorname{vec}\Sigma$ vanishes, because

$$\begin{aligned}
&\tilde{D}'[I - (I \otimes P)K_d](\Sigma_d^{-1/2} \otimes \Sigma_d^{-1/2}) \operatorname{vec}\Sigma \\
&= \tilde{D}'[I - (I \otimes P)K_d] \operatorname{vec} P \\
&= \tilde{D}' \operatorname{vec} P - \tilde{D}'(I \otimes P) \operatorname{vec} P_d \\
&= \tilde{D}' \operatorname{vec} P - \tilde{D}'(I \otimes P) \operatorname{vec} I \\
&= \tilde{D}' \operatorname{vec} P - \tilde{D}' \operatorname{vec} P \\
&= 0. \qquad (5)
\end{aligned}$$

It follows that

$$\begin{aligned}
\tilde{D}'\Psi\tilde{D} = \tilde{D}'[I - (I \otimes P)K_d]E[(\Sigma_d^{-1/2}(x-\mu)(x-\mu)'\Sigma_d^{-1/2}) \otimes \\
(\Sigma_d^{-1/2}(x-\mu)(x-\mu)'\Sigma_d^{-1/2})][I - K_d(I \otimes P)]\tilde{D} \qquad (6)
\end{aligned}$$

which is slightly simpler than the corresponding expression in Magnus' equation (10.23).

From the asymptotic result for $\mathrm{w}(R)$ we get

$$n^{1/2}(\hat{\alpha}_s - \alpha_s) \xrightarrow{d} N\left(0, 4p^4(p-1)^{-2}(1'P1)^{-4}1_*'\tilde{D}'\Psi\tilde{D}1_*\right), \qquad (7)$$

where 1_* has $p_* = \frac{1}{2}p(p-1)$ unit elements.

The asymptotic result for $\hat{\alpha}_s$ given above is actually based on the derivative of $(p-1)^{-1}p[1 - p(1'Z1)^{-1}]$ with respect to $\mathrm{w}(Z)$. Mind that the 'free' elements are the p_* infradiagonal elements of Z. The operator $\tilde{D}'$ eliminates the unit diagonal elements of Z and the zero diagonal elements of the differential dZ. For matrix differential calculus, see Magnus and Neudecker (1999). Clearly this p_*-vector lists the p_* 'free' elements of $Z : Z_d = I, Z' = Z$. We have

$$\begin{aligned}
d(1'Z1)^{-1} &= -(1'Z1)^{-2}1'(dZ)1 \\
&= -(1'Z1)^{-2}(1 \otimes 1)' d \operatorname{vec} Z \\
&= -(1'Z1)^{-2}(1 \otimes 1)'\tilde{D} d\,\mathrm{w}(Z). \qquad (8)
\end{aligned}$$

Hence the derivative of $(p-1)^{-1}p[1 - p(1'Z1)^{-1}]$ is

$$\begin{aligned}
(p-1)^{-1}p^2(1'Z1)^{-2}\tilde{D}(1 \otimes 1) &= 2(p-1)^{-1}p^2(1'Z1)^{-2}\,\mathrm{w}(11') \\
&= 2(p-1)^{-1}p^2(1'Z1)^{-2}1_*. \qquad (9)
\end{aligned}$$

From this result we shall derive the asymptotic distribution of $\hat{\alpha}_s$ in the case of ellipticity.

3. The Asymptotic Distribution of $\hat{\alpha}_s$ Under Ellipticity

In the elliptic case $V = 2(1+\kappa)L(\Sigma\otimes\Sigma)+\kappa(\operatorname{vec}\Sigma)(\operatorname{vec}\Sigma)'$, where κ is the common kurtosis parameter. Hence

$$\begin{aligned}&(\Sigma_d^{-1/2}\otimes\Sigma_d^{-1/2})V(\Sigma_d^{-1/2}\otimes\Sigma_d^{-1/2})\\ &= 2(1+\kappa)(\Sigma_d^{-1/2}\otimes\Sigma_d^{-1/2})L(\Sigma\Sigma_d^{-1/2}\otimes\Sigma\Sigma_d^{-1/2})\\ &\quad+\kappa(\Sigma_d^{-1/2}\otimes\Sigma_d^{-1/2})(\operatorname{vec}\Sigma)(\operatorname{vec}\Sigma)'(\Sigma_d^{-1/2}\otimes\Sigma_d^{-1/2})\\ &= 2(1+\kappa)L(P\otimes P)+\kappa(\operatorname{vec}P)(\operatorname{vec}P)'. \end{aligned}\tag{10}$$

Further

$$\begin{aligned}\Psi &= 2(1+\kappa)[I-L(I\otimes P)K_d]L(P\otimes P)[I-K_d(I\otimes P)L]\\ &\quad+\kappa[I-L(I\otimes P)K_d](\operatorname{vec}P)(\operatorname{vec}P)'[I-K_d(I\otimes P)L]\\ &= 2(1+\kappa)[I-L(I\otimes P)K_d]L(P\otimes P)[I-K_d(I\otimes P)L]\\ &\quad+\kappa[\operatorname{vec}P-L(I\otimes P)\operatorname{vec}I][\operatorname{vec}P-L(I\otimes P)\operatorname{vec}I]'\\ &= 2(1+\kappa)L[I-(I\otimes P)K_d](P\otimes P)[I-K_d(I\otimes P)]L, \end{aligned}\tag{11}$$

because $L(I\otimes P)\operatorname{vec}I = L\operatorname{vec}P = \operatorname{vec}P$. Hence under ellipticity, $\tilde{D}'\Psi\tilde{D} = 2(1+\kappa)\tilde{D}'[I-(I\otimes P)K_d](P\otimes P)[I-K_d(I\otimes P)\tilde{D}$, because $L\tilde{D}=\tilde{D}$.

Finally

$$n^{1/2}(\hat{\alpha}_s-\alpha_s)\xrightarrow{d}N(0,W_e),\tag{12}$$

where $W_e =$ $8(1+\kappa)(p-1)^{-2}p^4(1'P1)^{-4}1_*'\tilde{D}'[I-(I\otimes P)K_d](P\otimes P)[I-K_d(I\otimes P)]\tilde{D}1_*$. The result for normality follows from $\kappa=0$; see Section 4.

4. The Asymptotic Distribution of $\hat{\alpha}_s$ Under Normality

From the preceding section we get under normality

$$n^{1/2}(\hat{\alpha}_s-\alpha_s)\xrightarrow{d}N(0,W_n),\tag{13}$$

where $W_n = 8(p-1)^{-2}p^4(1'P1)^{-4}1_*'\tilde{D}'[I-(I\otimes P)K_d](P\otimes P)[I-K_d(I\otimes P)]\tilde{D}1_*$.

References

1. Magnus, J.R. (1988) *Linear Structures*, Charles Griffin & Company, London and Oxford University Press, New York.
2. Magnus, J.R. and Neudecker, H. (1999) *Matrix Differential Calculus with Applications in Statistics and Econometrics*, Revised Ed., Wiley, Chichester.
3. Neudecker, H. and Satorra, A. (1996) The algebraic equality of two asymptotic tests for the hypothesis that a normal distribution has a specified correlation matrix, *Statist. Probab. Lett.* 30(2), 99–103.
4. Neudecker, H. and Wesselman, A.M. (1990) The asymptotic variance matrix of the sample correlation matrix, *Linear Algebra Appl.* 127, 589–599.
5. van Zyl, J. M., Neudecker, H. and Nel, D.G. (2000) On the distribution of the maximum likelihood estimator of Cronbach's alpha, *Psychometrika* 65(3), 271–280.

On the Approximate Variance of a Nonlinear Function of Random Variables

Heinz Neudecker

School of Economics and Business, University of Amsterdam, 1018 WB Amsterdam, The Netherlands
E-mail: h.neudecker@uva.nl

Götz Trenkler

Department of Statistics, University of Dortmund, D-44221 Dortmund, Germany
E-mail: trenkler@statistik.uni-dortmund.de

In this paper we consider the problem of approximating the variance of a nonlinear function of random variables on the basis of a second degree Taylor series expansion. In contrast to the result achieved by Tiwari and Elston (1999), our approach in addition uses the covariances between the random variables to obtain a better approximation.

Keywords: nonlinear function; second degree approximation of variance; heterozygosity.

1. Introduction

In their paper, Tiwari and Elston (1999) investigated the problem of approximating the variance of a nonlinear function of random variables on the basis of the first three terms of a Taylor series expansion. Unfortunately, when calculating the variance of the second degree Taylor expansion, the authors did not take into account the possible correlation between the variables, which, as their examples show, is indispensable. Subsequently, we resume their analysis and obtain some alternative expressions.

2. Variance of the Second Degree Taylor Series Approximation

As in Tiwari and Elston (1999), we consider a scalar function f of the random vector $\mathbf{y}$, where $\mathbf{y} = (y_1, y_2, \ldots, y_m)'$. Let $E(\mathbf{y}) = \boldsymbol{\mu} = (\mu_1, \ldots, \mu_m)'$

and $\boldsymbol{\Sigma} = D(\mathbf{y}) = E[(\mathbf{y}-\boldsymbol{\mu})(\mathbf{y}-\boldsymbol{\mu})']$ denote the expectation vector and the dispersion matrix, respectively, of $\mathbf{y}$. Our problem is to find the expectation of a scalar function $f(\mathbf{y})$, where we assume that the first and second partial derivatives of f with respect to each y_i $(i = 1, \ldots, m)$ exist in an open neighbourhood containing $\boldsymbol{\mu}$.

Let $\mathbf{y} = \boldsymbol{\mu} + \Delta\mathbf{y}$ and $f(\mathbf{y}) = f(\boldsymbol{\mu}) + \Delta f(\mathbf{y})$. Using Taylor's formula we get the following approximation:

$$\begin{aligned} f(\mathbf{y}) &\approx f(\boldsymbol{\mu}) + \left.\frac{\partial f(\mathbf{y})}{\partial \mathbf{y}'}\right|_{\mathbf{y}=\boldsymbol{\mu}} d\mathbf{y} + \frac{1}{2}(d\mathbf{y})' \left.\frac{\partial^2 f(\mathbf{y})}{\partial \mathbf{y}\partial \mathbf{y}'}\right|_{\mathbf{y}=\boldsymbol{\mu}} d\mathbf{y} \\ &= f(\boldsymbol{\mu}) + \mathbf{a}'d\mathbf{y} + \frac{1}{2}(d\mathbf{y})'\mathbf{A}d\mathbf{y}, \end{aligned}$$

where $\mathbf{a} = \frac{\partial f(\mathbf{y})}{\partial \mathbf{y}'}|_{\mathbf{y}=\boldsymbol{\mu}}$, $\mathbf{A} = \frac{\partial^2 f(\mathbf{y})}{\partial \mathbf{y}\partial \mathbf{y}'}|_{\mathbf{y}=\boldsymbol{\mu}}$ and $d\mathbf{y} = \Delta\mathbf{y} = \mathbf{y} - \boldsymbol{\mu}$.

For the expectation of $f(\mathbf{y})$ we get

$$E[f(\mathbf{y})] \approx f(\boldsymbol{\mu}) + \mathbf{a}'E(d\mathbf{y}) + \frac{1}{2}\operatorname{tr}(\mathbf{A}\boldsymbol{\Sigma}),$$

since $E[(d\mathbf{y})(d\mathbf{y})'] = D(\mathbf{y}) = \boldsymbol{\Sigma}$.

As $E(d\mathbf{y}) = \mathbf{0}$, this reduces to $E[f(\mathbf{y})] \approx f(\boldsymbol{\mu}) + \frac{1}{2}\operatorname{tr}(\mathbf{A}\boldsymbol{\Sigma})$.

Note that Tiwari and Elston (1999) gave the following approximation

$$E[f(\mathbf{y})] \approx f(\boldsymbol{\mu}) + \frac{1}{2}\mathbf{1}'\mathbf{A}_d\boldsymbol{\Sigma}_d\mathbf{1},$$

where generally $\mathbf{X}_d$ is the diagonal matrix featuring the diagonal of a square matrix $\mathbf{X}$.

It is easy to see that $\mathbf{1}'\mathbf{A}_d\boldsymbol{\Sigma}_d\mathbf{1} = \operatorname{tr}(\mathbf{A}\boldsymbol{\Sigma})$ iff $\mathbf{A} = \mathbf{A}_d$ or $\boldsymbol{\Sigma} = \boldsymbol{\Sigma}_d$.

The authors' approximation of $E[f(\mathbf{y})]$ differs from ours. They obviously neglected the correlation between the y_i.

When $f(\mathbf{y}) \approx f(\boldsymbol{\mu}) + \mathbf{a}'d\mathbf{y}$ (first degree Taylor expansion), we then have $E[f(\mathbf{y})] \approx f(\boldsymbol{\mu})$. Further, $D[f(\mathbf{y})] \approx \mathbf{a}'\boldsymbol{\Sigma}\mathbf{a}$, because
$D(\mathbf{a}'d\mathbf{y}) = \mathbf{a}'[D(d\mathbf{y})]\mathbf{a} = \mathbf{a}'[D(\mathbf{y}-\boldsymbol{\mu})]\mathbf{a} = \mathbf{a}'[D(\mathbf{y})]\mathbf{a}$.

Tiwari and Elston (1999) gave as variance $\mathbf{a}'\boldsymbol{\Sigma}_d\mathbf{1}\mathbf{a}$. Because this is a *column vector*, it cannot be the variance. Let us return to the second degree approximation

$$f(\mathbf{y}) \approx f(\boldsymbol{\mu}) + \mathbf{a}'d\mathbf{y} + \frac{1}{2}(d\mathbf{y})'\mathbf{A}d\mathbf{y}.$$

Its variance is

$$\begin{aligned} D[f(\mathbf{y})] &\approx D[\mathbf{a}'d\mathbf{y} + \frac{1}{2}(d\mathbf{y})'\mathbf{A}d\mathbf{y}] \\ &= D(\mathbf{a}'d\mathbf{y}) + \frac{1}{4}D[(d\mathbf{y})'\mathbf{A}d\mathbf{y}] + \operatorname{cov}[\mathbf{a}'d\mathbf{y}, (d\mathbf{y})'\mathbf{A}d\mathbf{y}], \end{aligned}$$

where the last term denotes the covariance between $\mathbf{a}'d\mathbf{y}$ and $(d\mathbf{y})'\mathbf{A}d\mathbf{y}$.

Evaluating the three terms we get

(i) $D(\mathbf{a}'d\mathbf{y}) = \mathbf{a}'\mathbf{\Sigma}\mathbf{a}$

(ii)

$$\begin{aligned}D[(d\mathbf{y})'\mathbf{A}d\mathbf{y}] &= E[(d\mathbf{y})'\mathbf{A}d\mathbf{y}]^2 - [E(d\mathbf{y})'\mathbf{A}d\mathbf{y}]^2 \\ &= E[(d\mathbf{y}\otimes d\mathbf{y})'(\mathbf{A}\otimes\mathbf{A})(d\mathbf{y}\otimes d\mathbf{y})] - [\operatorname{tr}\mathbf{A}E(d\mathbf{y})(d\mathbf{y})']^2 \\ &= \operatorname{tr}\{(\mathbf{A}\otimes\mathbf{A})E[(d\mathbf{y})(d\mathbf{y})'\otimes(d\mathbf{y})(d\mathbf{y})']\} - [\operatorname{tr}(\mathbf{A}\mathbf{\Sigma})^2] \\ &= \operatorname{tr}[(\mathbf{A}\otimes\mathbf{A})\mathbf{\Psi}] - [\operatorname{tr}(\mathbf{A}\mathbf{\Sigma})]^2,\end{aligned}$$

where $\mathbf{\Psi} = E[(d\mathbf{y})(d\mathbf{y})'\otimes(d\mathbf{y})(d\mathbf{y})']$ and $\otimes$ denotes Kronecker product.

(iii)

$$\begin{aligned}\operatorname{cov}[\mathbf{a}'d\mathbf{y},(d\mathbf{y})'\mathbf{A}d\mathbf{y}] &= \mathbf{a}'E[(d\mathbf{y})(d\mathbf{y})'\mathbf{A}d\mathbf{y}] \\ &= \mathbf{a}'E\operatorname{vec}[(d\mathbf{y})(d\mathbf{y})'\mathbf{A}d\mathbf{y}] \\ &= \mathbf{a}'E[d\mathbf{y}\otimes(d\mathbf{y})(d\mathbf{y})']'\operatorname{vec}\mathbf{A} \\ &= \mathbf{a}'\mathbf{\Phi}'\operatorname{vec}\mathbf{A} \\ &= (\operatorname{vec}\mathbf{A})'\mathbf{\Phi}\mathbf{a},\end{aligned}$$

where $\mathbf{\Phi} = E[d\mathbf{y}\otimes(d\mathbf{y})(d\mathbf{y})']$ and 'vec' denotes the vectorisation operator which transforms a matrix into a vector by stacking the columns of the matrix one under the other.

Collecting terms we arrive at the result:

$$D[f(\mathbf{y})] \approx \mathbf{a}'\mathbf{\Sigma}\mathbf{a} + \frac{1}{4}\operatorname{tr}[(\mathbf{A}\otimes\mathbf{A})\Psi] - \frac{1}{4}[\operatorname{tr}(\mathbf{A}\mathbf{\Sigma})]^2 + (\operatorname{vec}\mathbf{A})'\mathbf{\Phi}\mathbf{a}.$$

With the notations introduced above we get

Theorem 2.1.

(i) $E_a[f(\mathbf{y})] = f(\boldsymbol{\mu}) + \frac{1}{2}\operatorname{tr}(\mathbf{A}\mathbf{\Sigma})$

(ii) $D_a[f(\mathbf{y})] = \mathbf{a}'\mathbf{\Sigma}\mathbf{a} + \frac{1}{4}\operatorname{tr}[(\mathbf{A}\otimes\mathbf{A})\mathbf{\Psi}] - \frac{1}{4}[\operatorname{tr}(\mathbf{A}\mathbf{\Sigma})]^2 + (\operatorname{vec}\mathbf{A})'\mathbf{\Phi}\mathbf{a}$,

where $E_a[f(\mathbf{y})]$ *and* $D_a[f(\mathbf{y})]$ *are the approximate expectation and variance, respectively, of* $f(\mathbf{y}) = f(\boldsymbol{\mu}) + \mathbf{a}'d\mathbf{y} + \frac{1}{2}(d\mathbf{y})'\mathbf{A}d\mathbf{y}$. *When the higher derivatives are zero, both formulae are exact.*

3. Application to Genetics

As in Tiwari and Elston (1999) we consider a panmitic population in Hardy-Weinberg equilibrium. Suppose that there are m dominant alleles segregating at a locus. Let p_i be the frequency of the i-th allele in the population for $i = 1, \ldots, m$. Assume a sample consists of n alleles chosen at random from the population. The random variables X_i denote the number of i-th alleles in the sample. Then the set $X_1, \ldots, X_m$ has a joint multinomial distribution given by

$$P(x_1, \ldots, x_m) = \frac{n!}{x_1! \ldots x_m!} p_1^{x_1} \ldots p_m^{x_m},$$

where each X_i may take on the values $0, \ldots, n$, $\sum_{i=1}^{m} x_i = n$ and $\sum_{i=1}^{m} p_i = 1$.

Let $\widehat{\mathbf{p}} = (\widehat{p}_1, \ldots, \widehat{p}_m)'$ be the random vector of sample allele frequencies, where $\widehat{p}_i = X_i/n$. Then $E(\widehat{p}_i) = p_i$, so that $E(\widehat{\mathbf{p}}) = \mathbf{p}$, where $\mathbf{p} = (p_1, \ldots, p_m)'$ is the unknown vector of allele frequencies. It follows that

$$\mathbf{\Sigma} = D(\widehat{\mathbf{p}}) = \frac{1}{n}(\mathbf{P} - \mathbf{p}\mathbf{p}'),$$

where $\mathbf{P} = \text{diag}(p_i)$ is the $m \times m$ diagonal matrix having the p_i as its diagonal elements; see Tiwari and Elston (1999) or Bickel and Doksum (2001, A13).

Tiwari and Elston (1999) investigated the problem of estimating heterozygosity, which is defined as the probability that a randomly chosen individual from the population is heterozygous at a locus. Heterozygosity is defined as

$$\text{Het} = 1 - \sum_{i=1}^{m} p_i^2 = 1 - \mathbf{p}'\mathbf{p}.$$

Then a reasonable estimator for the parameter function Het should be

$$H_1 = \widehat{\text{Het}} = 1 - \widehat{\mathbf{p}}'\widehat{\mathbf{p}} = 1 - \sum_{i=1}^{m} \widehat{p}_i^2.$$

The statistic H_1 is not unbiased, since by our Theorem in the setup of Section 2 we have $f(\mathbf{y}) = 1 - \mathbf{y}'\mathbf{y} = f(\boldsymbol{\mu}) + \mathbf{a}'d\mathbf{y} + \frac{1}{2}(d\mathbf{y})'\mathbf{A}d\mathbf{y}$ with $\mathbf{y} = \widehat{\mathbf{p}}$, $\boldsymbol{\mu} = \mathbf{p} = E(\widehat{\mathbf{p}})$, $A = -2\mathbf{I}$ and $\mathbf{a} = -2\mathbf{p}$. Hence $f(\mathbf{y})$ coincides with its second degree Taylor expansion, and consequently the approximations of our Theorem are exact. This implies

$$E(H_1) = \text{Het} - \text{Het}/n = \frac{n-1}{n} \text{Het}.$$

This identity may also be shown directly. Since

$$\begin{aligned} E(\widehat{\mathbf{p}}'\widehat{\mathbf{p}}) &= \operatorname{tr} E(\widehat{\mathbf{p}}\widehat{\mathbf{p}}') \\ &= \operatorname{tr}[D(\widehat{\mathbf{p}}) + (E\widehat{\mathbf{p}})(E\widehat{\mathbf{p}})'] \\ &= \operatorname{tr}\left[\frac{1}{n}(\mathbf{P} - \mathbf{p}\mathbf{p}') + \mathbf{p}\mathbf{p}'\right] \\ &= \frac{1}{n}\operatorname{tr}\mathbf{P} - \frac{1}{n}\mathbf{p}'\mathbf{p} + \mathbf{p}'\mathbf{p} \\ &= \frac{1}{n} + \frac{n-1}{n}\mathbf{p}'\mathbf{p}, \end{aligned}$$

it follows that $E(1 - \widehat{\mathbf{p}}'\widehat{\mathbf{p}}) = (n-1)/n$ Het.

Observe that Tiwari and Elston (1999) gave an incorrect expression for $E(H_1)$. An unbiased estimator for Het is

$$H_2 = \frac{n}{n-1}(1 - \widehat{\mathbf{p}}'\widehat{\mathbf{p}}).$$

Furthermore, our Theorem yields

$$D_a(H_1) = D(H_1) = 4\mathbf{p}'\mathbf{\Sigma}\mathbf{p} + \operatorname{tr}\mathbf{\Psi} - (\operatorname{tr}\mathbf{\Sigma})^2 + 4(\operatorname{vec}\mathbf{I})'\mathbf{\Phi}\mathbf{p},$$

where

$$\mathbf{\Phi} = E[(\widehat{\mathbf{p}} - \mathbf{p}) \otimes (\widehat{\mathbf{p}} - \mathbf{p})(\widehat{\mathbf{p}} - \mathbf{p})']$$

and

$$\mathbf{\Psi} = E[(\widehat{\mathbf{p}} - \mathbf{p})(\widehat{\mathbf{p}} - \mathbf{p})' \otimes (\widehat{\mathbf{p}} - \mathbf{p})(\widehat{\mathbf{p}} - \mathbf{p})'].$$

The variance approximation given in Tiwari and Elston (1999, formula 12) is

$$\operatorname{var}_D(H_1) = \frac{4}{n}[\sum_{i=1}^{m} p_i^3 - (\mathbf{p}'\mathbf{p})^2],$$

which can be written as

$$\begin{aligned} \operatorname{var}_D(H_1) &= \frac{4}{n}[\mathbf{1}'\mathbf{P}^3\mathbf{1} - (\mathbf{p}'\mathbf{p})^2] \\ &= \frac{4}{n}[\mathbf{p}'\mathbf{P}\mathbf{p} - (\mathbf{p}'\mathbf{p})^2] \\ &= 4\mathbf{p}'\mathbf{\Sigma}\mathbf{p}. \end{aligned}$$

Hence we obtain

$$D_a(H_1) - \operatorname{var}_D(H_1) = \operatorname{tr}\mathbf{\Psi} - (\operatorname{tr}\mathbf{\Sigma})^2 + 4(\operatorname{vec}\mathbf{I})'\mathbf{\Phi}\mathbf{p}.$$

We note in passing that our Theorem is applicable also to the estimation of the so-called polymorphism information content measure (PIC), defined as

$$\text{PIC} = 1 - \sum_{i=1}^{m} p_i^2 - \sum_{i \neq j} p_i^2 p_j^2.$$

However, we shall not pursue this possibility any further.

References

1. Bickel, P.J. and Doksum, K.A. (2001) *Mathematical Statistics. Basic Ideas and Selected Topics.* Vol. 1. Prentice Hall, New Jersey.
2. Tiwari, H.K. and Elston, R.C. (1999) The approximate variance of a function of random variables. *Biometrical Journal* 41, 351–357.

On Oblique and Orthogonal Projectors

Götz Trenkler

Department of Statistics, University of Dortmund,
D-44221 Dortmund, Germany
E-mail: trenkler@statistik.uni-dortmund.de

Projection matrices with real elements play an important role in statistics, for instance the distribution of quadratic forms. In recent years the scope of interest has extended to a more general point of view, namely projectors with possibly complex entries. Basing on a powerful representation of matrices related to singular value decomposition, new results concerning Moore-Penrose inverse, group inverse, and algebraic transformations of projectors are presented.

Keywords: oblique and orthogonal projectors; singular value decomposition; Moore-Penrose inverse; group inverse; algebraic transformations.

1. Introduction

In Trenkler (1994), several characterisations of oblique and orthogonal projection matrices with real entries were given. In the following we extend these investigations to the complex case by using an approach by Hartwig and Spindelböck (1984).

We start from a singular value decomposition (SVD) which exists for every complex matrix. For our purpose it suffices to consider a square matrix $\boldsymbol{A}$ from $\mathbb{C}_{n\times n}$. Then $\boldsymbol{A}$ can be written in the form

$$\boldsymbol{A} = \boldsymbol{U} \begin{pmatrix} \boldsymbol{\Sigma} & \boldsymbol{0} \\ \boldsymbol{0} & \boldsymbol{0} \end{pmatrix} \boldsymbol{V}^*,$$

where $\boldsymbol{U}$ and $\boldsymbol{V}$ are unitary matrices, and

$$\boldsymbol{\Sigma} = \mathrm{diag}(\sigma_1 \boldsymbol{I}_{r_1}, \ldots, \sigma_t \boldsymbol{I}_{r_t})$$

is the diagonal matrix of singular values of $\boldsymbol{A}$, $\sigma_1 > \sigma_2 > \cdots > \sigma_t > 0$ and $r_1 + r_2 + \cdots + r_t = r = \mathrm{rank}(\boldsymbol{A})$ (see Zhang, 1999, p.66).

Using this decomposition, Hartwig and Spindelböck (1984, Corollary 6) derived the following result. Let $\boldsymbol{A} \in \mathbb{C}_{n\times n}$; then

$$\boldsymbol{A} = \boldsymbol{U}\begin{pmatrix} \boldsymbol{\Sigma K} & \boldsymbol{\Sigma L} \\ \boldsymbol{0} & \boldsymbol{0} \end{pmatrix}\boldsymbol{U}^*,$$

where $\boldsymbol{U}$ is unitary; $\boldsymbol{KK}^* + \boldsymbol{LL}^* = \boldsymbol{I}_r$; $\boldsymbol{\Sigma} = \mathrm{diag}(\sigma_1\boldsymbol{I}_{r_1}, \ldots, \sigma_t\boldsymbol{I}_{r_t})$; $r_1 + r_2 + \cdots + r_t = r = \mathrm{rank}(\boldsymbol{A})$; and $\sigma_1 > \sigma_2 > \cdots > \sigma_t > 0$.

Basing on this representation of any square matrix with complex entries we may state:

(i) $\boldsymbol{A} = \boldsymbol{U}\begin{pmatrix} \boldsymbol{\Sigma K} & \boldsymbol{\Sigma L} \\ \boldsymbol{0} & \boldsymbol{0} \end{pmatrix}\boldsymbol{U}^*$ is an oblique projector if and only if $\boldsymbol{\Sigma K} = \boldsymbol{I}_r$. Hence any oblique projector can always be written as

$$\boldsymbol{A} = \boldsymbol{U}\begin{pmatrix} \boldsymbol{I}_r & \boldsymbol{H} \\ \boldsymbol{0} & \boldsymbol{0} \end{pmatrix}\boldsymbol{U}^*.$$

(ii) $\boldsymbol{A} = \boldsymbol{U}\begin{pmatrix} \boldsymbol{\Sigma K} & \boldsymbol{\Sigma L} \\ \boldsymbol{0} & \boldsymbol{0} \end{pmatrix}\boldsymbol{U}^*$ is an orthogonal projector (i.e. idempotent and hermitian) if and only if $\boldsymbol{\Sigma K} = \boldsymbol{I}_r$ and $\boldsymbol{L} = \boldsymbol{0}$. Thus an orthogonal projector can always be written as

$$\boldsymbol{A} = \boldsymbol{U}\begin{pmatrix} \boldsymbol{I}_r & \boldsymbol{0} \\ \boldsymbol{0} & \boldsymbol{0} \end{pmatrix}\boldsymbol{U}^*.$$

In the following we consider oblique and orthogonal projectors in the representations from above, where without loss of generality we assume $\boldsymbol{U} = \boldsymbol{I}_n$.

2. Characterisation of oblique projectors

Consider the oblique projector

$$\boldsymbol{P} = \begin{pmatrix} \boldsymbol{I}_r & \boldsymbol{H} \\ \boldsymbol{0} & \boldsymbol{0} \end{pmatrix}.$$

Due to its simple form, we easily obtain its Moore-Penrose inverse and related transforms.

Theorem 2.1. *Let us be given the oblique projector*

$$\boldsymbol{P} = \begin{pmatrix} \boldsymbol{I}_r & \boldsymbol{H} \\ \boldsymbol{0} & \boldsymbol{0} \end{pmatrix}$$

with $\mathrm{rank}(\boldsymbol{P}) = r$. *Then*

(i) $\boldsymbol{P}^+ = \begin{pmatrix} \boldsymbol{E} & \boldsymbol{0} \\ \boldsymbol{H}^*\boldsymbol{E} & \boldsymbol{0} \end{pmatrix}$, *where* $\boldsymbol{E} = (\boldsymbol{I}_r + \boldsymbol{HH}^*)^{-1}$

(ii) $\boldsymbol{PP}^+ = \begin{pmatrix} \boldsymbol{I}_r & \boldsymbol{0} \\ \boldsymbol{0} & \boldsymbol{0} \end{pmatrix} = \boldsymbol{P}(\boldsymbol{P}+\boldsymbol{P}^* - \boldsymbol{I}_n)^{-1}\boldsymbol{P}^*$

(iii) $\boldsymbol{P}^+\boldsymbol{P} = \begin{pmatrix} \boldsymbol{E} & \boldsymbol{EH} \\ \boldsymbol{H}^*\boldsymbol{E} & \boldsymbol{H}^*\boldsymbol{EH} \end{pmatrix}$

(iv) $(\boldsymbol{I}-\boldsymbol{P})^+ = \begin{pmatrix} \boldsymbol{0} & \boldsymbol{0} \\ -\boldsymbol{H}^*\boldsymbol{E} & \boldsymbol{E} \end{pmatrix}$

(v) $(\boldsymbol{P}-\boldsymbol{P}^*)^+ = \begin{pmatrix} \boldsymbol{0} & -\boldsymbol{H}^{*+} \\ \boldsymbol{H}^+ & \boldsymbol{0} \end{pmatrix}$

(vi) $\boldsymbol{P}(\boldsymbol{P}-\boldsymbol{P}^*)^+\boldsymbol{P}^* = \boldsymbol{0}$

(vii) $(\boldsymbol{P}+\boldsymbol{P}^*-\boldsymbol{I}_n)^{-1} = \boldsymbol{PP}^+ - (\boldsymbol{I}_n - \boldsymbol{P}^+\boldsymbol{P}) = \begin{pmatrix} \boldsymbol{E} & \boldsymbol{EH} \\ \boldsymbol{H}^*\boldsymbol{E} & \boldsymbol{H}^*\boldsymbol{EH} - \boldsymbol{I}_{n-r} \end{pmatrix}$

(viii) *All generalised inverses of* $\boldsymbol{P}$ *are given by* $\boldsymbol{P} = \begin{pmatrix} \boldsymbol{I}_r - \boldsymbol{HY} & \boldsymbol{X} \\ \boldsymbol{Y} & \boldsymbol{Z} \end{pmatrix}$, *where* $\boldsymbol{X}$, $\boldsymbol{Y}$ *and* $\boldsymbol{Z}$ *are arbitrary conformable matrices.*

Proof. Conditions (i) – (vii) are straightforward. For condition (vii) note that $\boldsymbol{PP}^+$ is the orthogonal projector on the column space of $\boldsymbol{P}$, whereas $\boldsymbol{I}_n - \boldsymbol{P}^+\boldsymbol{P}$ is the orthogonal projector on the null space of $\boldsymbol{P}$. Representation (viii) follows from Theorem 9.2 in Harville (1997). Although Harville's book deals with real matrices only, the result we need also holds in the complex case. □

From Theorem 2.1 it is obvious that for an oblique projector $\boldsymbol{P}$ the conjugate transpose and the Moore-Penrose inverse commute, since $\boldsymbol{P}^*\boldsymbol{P}^+ = \boldsymbol{P}^+ = \boldsymbol{P}^+\boldsymbol{P}^*$. Thus any oblique projector is star-dagger; see Hartwig and Spindelböck (1984).

Interestingly, as we shall see in the following theorem, each of the conditions $\boldsymbol{A}^*\boldsymbol{A}^+ = \boldsymbol{A}^+$ and $\boldsymbol{A}^+\boldsymbol{A}^* = \boldsymbol{A}^+$ is necessary and sufficient for $\boldsymbol{A}$ to be an oblique projector.

Furthermore, $\mathrm{rank}(\boldsymbol{P}) = \mathrm{rank}(\boldsymbol{P}^2)$, showing that an oblique projector has a group inverse.

Recall that the group inverse $\boldsymbol{A}^\#$ of a square matrix $\boldsymbol{A}$, if it exists, is uniquely defined by the conditions

(i) $\boldsymbol{AA}^\#\boldsymbol{A} = \boldsymbol{A}$
(ii) $\boldsymbol{A}^\#\boldsymbol{AA}^\# = \boldsymbol{A}^\#$
(iii) $\boldsymbol{A}^\#\boldsymbol{A} = \boldsymbol{AA}^\#$

(see Campbell and Meyer, 1979, p.124). We call a square matrix $\boldsymbol{A}$ GP if $\boldsymbol{A}$ has a group inverse. This happens if and only if $\mathrm{rank}(\boldsymbol{A}) = \mathrm{rank}(\boldsymbol{A}^2)$.

Basing on the representation

$$A = \begin{pmatrix} \Sigma K & \Sigma L \\ 0 & 0 \end{pmatrix} \tag{1}$$

discussed in the introduction, we may state the following characterisations of an oblique projector.

Theorem 2.2. *Assume that the group inverse $A^{\#}$ of A exists. Then the following conditions are equivalent:*

(i) A *is an oblique projector*
(ii) $A = AA^{\#}$
(iii) A^* *is an oblique projector*
(iv) $A^+ = A^*A^+$
(v) $A^+ = A^+A^*$
(vi) $A^{\#}$ *is an oblique projector*
(vii) $A^{\#} = AA^{\#}$
(viii) $\Sigma K = I_r$.

Proof. We show only (viii) ⇔ (iv) and (viii) ⇔ (vii). The other equivalences are proved similarly.
(viii) ⇔ (iv): It is readily established that

$$A^+ = \begin{pmatrix} K^*\Sigma^{-1} & 0 \\ L^*\Sigma^{-1} & 0 \end{pmatrix}$$

and

$$A^*A^+ = \begin{pmatrix} K^*\Sigma K^*\Sigma^{-1} & 0 \\ L^*\Sigma K^*\Sigma^{-1} & 0 \end{pmatrix}.$$

Hence $A^+ = A^*A^+$ is equivalent to $K^* = K^*\Sigma K^*$ and $L^* = L^*\Sigma K^*$. Multiplication with K and L, respectively, from the left gives $KK^* = KK^*\Sigma K^*$ and $LL^* = LL^*\Sigma K^*$. Using $KK^* + LL^* = I_r$ yields $\Sigma K^* = I_r$. Since Σ is a real diagonal matrix, we get $K^* = \Sigma^{-1} = K$. It follows that $\Sigma K = I_r$ is necessary and sufficient for $A^+ = A^*A^+$.
(viii) ⇔ (vii): From the representation (2.1) we obtain

$$A^{\#} = \begin{pmatrix} K^{-1}\Sigma^{-1} & K^{-1}\Sigma^{-1}K^{-1}L \\ 0 & 0 \end{pmatrix},$$

where we use Corollary 6 in Hartwig and Spindelböck (1984), saying that A is GP if and only if K^{-1} exists. It follows that

$$AA^{\#} = \begin{pmatrix} I_r & K^{-1}L \\ 0 & 0 \end{pmatrix}.$$

Hence $\boldsymbol{A}^{\#} = \boldsymbol{A}\boldsymbol{A}^{\#}$ if and only if $\boldsymbol{K}^{-1}\boldsymbol{\Sigma}^{-1} = \boldsymbol{I}_r$, or equivalently $\boldsymbol{\Sigma}\boldsymbol{K} = \boldsymbol{I}_r$. □

According to Marathe (1956), a matrix $\boldsymbol{B}$ is called quasi-idempotent if $\boldsymbol{B}(\boldsymbol{I} - \boldsymbol{B})^k = \boldsymbol{0}$ for some positive integer k. Of course, if $k = 1$, the notions idempotency (i.e. being an oblique projector) and quasi-idempotency coincide. Are there other conditions for this?

Theorem 2.3. *For $\boldsymbol{B} \in \mathbb{C}_{n\times n}$ the following statements are equivalent:*

(i) $\boldsymbol{B}$ is an oblique projector
(ii) $\boldsymbol{B}$ is quasi-idempotent and $\boldsymbol{I}_n - \boldsymbol{B}$ is GP.

Proof. We show the nontrivial direction (ii) ⇒ (i).
Put $\boldsymbol{A} = \boldsymbol{I} - \boldsymbol{B}$ and write

$$\boldsymbol{A} = \begin{pmatrix} \boldsymbol{\Sigma}\boldsymbol{K} & \boldsymbol{\Sigma}\boldsymbol{L} \\ \boldsymbol{0} & \boldsymbol{0} \end{pmatrix}.$$

Then

$$\boldsymbol{A}^k = \begin{pmatrix} (\boldsymbol{\Sigma}\boldsymbol{K})^k & (\boldsymbol{\Sigma}\boldsymbol{K})^{k-1}\boldsymbol{\Sigma}\boldsymbol{L} \\ \boldsymbol{0} & \boldsymbol{0} \end{pmatrix}.$$

By Hartwig and Spindelböck (1984, Corollary 6), $\boldsymbol{A}$ is GP if and only if $\boldsymbol{K}^{-1}$ exists. Since $\boldsymbol{B}$ is quasi-idempotent, we have $\boldsymbol{B}(\boldsymbol{I}_n - \boldsymbol{B})^k = \boldsymbol{0}$ so that $(\boldsymbol{I}_n - \boldsymbol{A})\boldsymbol{A}^k = \boldsymbol{0}$. Hence $(\boldsymbol{I}_r - \boldsymbol{\Sigma}\boldsymbol{K})(\boldsymbol{\Sigma}\boldsymbol{K})^k = \boldsymbol{0}$, which implies $\boldsymbol{\Sigma}\boldsymbol{K} = \boldsymbol{I}_r$. Consequently

$$\boldsymbol{A} = \begin{pmatrix} \boldsymbol{I}_r & \boldsymbol{\Sigma}\boldsymbol{L} \\ \boldsymbol{0} & \boldsymbol{0} \end{pmatrix}$$

is an oblique projector. □

From the representation $\begin{pmatrix} \boldsymbol{I}_r & \boldsymbol{H} \\ \boldsymbol{0} & \boldsymbol{0} \end{pmatrix}$ it is clear that for any oblique projector $\boldsymbol{A}$ we have $\mathrm{tr}(\boldsymbol{A}) = \mathrm{rank}(\boldsymbol{A})$, where $\mathrm{tr}(\cdot)$ denotes the trace of a matrix. We now state a related result characterising an oblique projector by rank and trace.

Theorem 2.4. *Let $\boldsymbol{A} \in \mathbb{C}_{n\times n}$ with $\mathrm{rank}(\boldsymbol{A}) = r$. Then the following statements are equivalent:*

(i) $\boldsymbol{A}$ is an oblique projector
(ii) $\mathrm{rank}(\boldsymbol{I}_n - \boldsymbol{A}) = n - \mathrm{rank}(\boldsymbol{A})$

(iii) $\text{rank}(\boldsymbol{A}) = \text{tr}(\boldsymbol{A})$ *and* $\text{rank}(\boldsymbol{I}_n - \boldsymbol{A}) = \text{tr}(\boldsymbol{I}_n - \boldsymbol{A})$
(iv) $\text{rank}(\boldsymbol{A}) \leq \text{tr}(\boldsymbol{A})$ *and* $\text{rank}(\boldsymbol{I}_n - \boldsymbol{A}) \leq \text{tr}(\boldsymbol{I}_n - \boldsymbol{A})$.

Proof. Write

$$\boldsymbol{A} = \begin{pmatrix} \boldsymbol{\Sigma K} & \boldsymbol{\Sigma L} \\ \boldsymbol{0} & \boldsymbol{0} \end{pmatrix}$$

to obtain

$$\boldsymbol{I}_n - \boldsymbol{A} = \begin{pmatrix} \boldsymbol{I}_r - \boldsymbol{\Sigma K} & -\boldsymbol{\Sigma L} \\ \boldsymbol{0} & \boldsymbol{I}_{n-r} \end{pmatrix}.$$

(i) $\Rightarrow$ (ii): If $\boldsymbol{A}$ is an oblique projector, then $\boldsymbol{\Sigma K} = \boldsymbol{I}_r$, and thus (ii) is satisfied.
(ii) $\Rightarrow$ (iii): From (ii) we get $\boldsymbol{I}_r - \boldsymbol{\Sigma K} = \boldsymbol{0}$, and (iii) is valid.
(iii) $\Rightarrow$ (iv): This implication is trivial.
(iv) $\Rightarrow$ (i): The inequality $\text{rank}(\boldsymbol{A}) \leq \text{tr}(\boldsymbol{A})$ yields $\text{rank}(\boldsymbol{I}_r - \boldsymbol{\Sigma K}) + n - r \leq n - \text{tr}(\boldsymbol{\Sigma K})$, so that $\text{rank}(\boldsymbol{I}_r - \boldsymbol{\Sigma K}) \leq r - \text{tr}(\boldsymbol{\Sigma K}) \leq 0$. Hence $\boldsymbol{\Sigma K} = \boldsymbol{I}_r$ and $\boldsymbol{A}$ is an oblique projector. □

We note that the equivalence (i) $\Leftrightarrow$ (ii) is well-known in the literature (see Sibuya, 1970).

Theorem 2.5. *Let* $\boldsymbol{A}$ *be a square matrix with complex elements. Then* $\boldsymbol{A}$ *is an oblique projector if and only if* $\boldsymbol{A}$ *is similar to* $\boldsymbol{A}\boldsymbol{A}^+$*, the orthogonal projector on the column space of* $\boldsymbol{A}$.

Proof. Let $\boldsymbol{A} \sim (n, n)$ be an oblique projector in the form

$$\boldsymbol{A} = \begin{pmatrix} \boldsymbol{I}_r & \boldsymbol{H} \\ \boldsymbol{0} & \boldsymbol{0} \end{pmatrix}.$$

Consider the matrix

$$\boldsymbol{R} = \begin{pmatrix} \boldsymbol{I}_r & -\boldsymbol{H} \\ \boldsymbol{0} & \boldsymbol{I}_{n-r} \end{pmatrix}.$$

Then $\boldsymbol{R}$ is nonsingular,

$$\boldsymbol{AR} = \begin{pmatrix} \boldsymbol{I}_r & \boldsymbol{0} \\ \boldsymbol{0} & \boldsymbol{0} \end{pmatrix}$$

and $\boldsymbol{RAR} = \boldsymbol{AR}$. Then $\boldsymbol{P} = \boldsymbol{AR} = \boldsymbol{AA}^+$ is the orthogonal projector on the column space of $\boldsymbol{A}$ satisfying $\boldsymbol{RP} = \boldsymbol{AR}$. Hence $\boldsymbol{A} = \boldsymbol{RPR}^{-1}$, and $\boldsymbol{A}$ is similar to $\boldsymbol{P}$. The other direction is trivial. □

Consider now the identity $\boldsymbol{A}^k = \boldsymbol{A}^{k+1}$ for some integer k. It is easy to verify that in this case $\boldsymbol{A}^k$ is an oblique projector. A more interesting result is given subsequently.

Theorem 2.6. *The following two statements are equivalent:*

(i) $\boldsymbol{P}$ *is an oblique projector*
(ii) $\boldsymbol{P}$ *has a group inverse and* $\boldsymbol{P}^k = \boldsymbol{P}^{k+1}$ *for some integer* k

Proof. We prove the nontrivial direction (ii) $\Rightarrow$ (i). Let

$$\boldsymbol{P} = \begin{pmatrix} \boldsymbol{\Sigma K} & \boldsymbol{\Sigma L} \\ \boldsymbol{0} & \boldsymbol{0} \end{pmatrix}.$$

Since $\boldsymbol{P}$ has a group inverse, by Corollary 6 in Hartwig and Spindelböck (1984), $\boldsymbol{K}$ is nonsingular. Consequently, since $\boldsymbol{\Sigma}$ is nonsingular, $(\boldsymbol{\Sigma K})^{-1}$ exists. From $\boldsymbol{P}^k = \boldsymbol{P}^{k+1}$ we obtain $(\boldsymbol{\Sigma K})^k = (\boldsymbol{\Sigma K})^{k+1}$ which implies $\boldsymbol{\Sigma K} = \boldsymbol{I}_r$. Hence

$$\boldsymbol{P} = \begin{pmatrix} \boldsymbol{I}_r & \boldsymbol{\Sigma L} \\ \boldsymbol{0} & \boldsymbol{0} \end{pmatrix}$$

is an oblique projector. □

3. Characterisation of orthogonal projectors

Recall that an orthogonal projector $\boldsymbol{P}$ can be written in the form

$$\boldsymbol{P} = \begin{pmatrix} \boldsymbol{I}_r & \boldsymbol{0} \\ \boldsymbol{0} & \boldsymbol{0} \end{pmatrix},$$

where $r = \text{rank}(\boldsymbol{P})$. Thus, if $\boldsymbol{A} \in \mathbb{C}_{n\times n}$ is given in the representation

$$\boldsymbol{A} = \begin{pmatrix} \boldsymbol{\Sigma K} & \boldsymbol{\Sigma L} \\ \boldsymbol{0} & \boldsymbol{0} \end{pmatrix}, \tag{2}$$

$\boldsymbol{A}$ is an orthogonal projector if and only if

$$\boldsymbol{\Sigma K} = \boldsymbol{I}_r \text{ and } \boldsymbol{L} = \boldsymbol{0}. \tag{3}$$

Using this fact and representation (2) we obtain the following theorem.

Theorem 3.1. *The following statements are necessary and sufficient for* $\boldsymbol{A}$ *to be an orthogonal projector:*

(i)	$\boldsymbol{A} = \boldsymbol{A}\boldsymbol{A}^*$	*(xiv)*	$\boldsymbol{A}^+ = \boldsymbol{A}\boldsymbol{A}^\#$
(ii)	$\boldsymbol{A} = \boldsymbol{A}\boldsymbol{A}^+$	*(xv)*	$\boldsymbol{A}^+ = \boldsymbol{A}^+\boldsymbol{A}$
(iii)	$\boldsymbol{A} = \boldsymbol{A}^*\boldsymbol{A}$	*(xvi)*	$\boldsymbol{A}^+ = \boldsymbol{A}^\#\boldsymbol{A}^*$
(iv)	$\boldsymbol{A} = \boldsymbol{A}^+\boldsymbol{A}$	*(xvii)*	$\boldsymbol{A}^+ = \boldsymbol{A}^\#\boldsymbol{A}^+$
(v)	$\boldsymbol{A}^* = \boldsymbol{A}\boldsymbol{A}^*$	*(xviii)*	$\boldsymbol{A}^+ = \boldsymbol{A}^\#\boldsymbol{A}^\#$
(vi)	$\boldsymbol{A}^* = \boldsymbol{A}\boldsymbol{A}^+$	*(xix)*	$\boldsymbol{A}^\# = \boldsymbol{A}\boldsymbol{A}^+$
(vii)	$\boldsymbol{A}^* = \boldsymbol{A}\boldsymbol{A}^\#$	*(xx)*	$\boldsymbol{A}^\# = \boldsymbol{A}^*\boldsymbol{A}^+$
(viii)	$\boldsymbol{A}^* = \boldsymbol{A}^*\boldsymbol{A}$	*(xxi)*	$\boldsymbol{A}^\# = \boldsymbol{A}^+\boldsymbol{A}$
(ix)	$\boldsymbol{A}^* = \boldsymbol{A}^*\boldsymbol{A}^\#$	*(xxii)*	$\boldsymbol{A}^\# = \boldsymbol{A}^+\boldsymbol{A}^*$
(x)	$\boldsymbol{A}^* = \boldsymbol{A}^+\boldsymbol{A}$	*(xxiii)*	$\boldsymbol{A}^\# = \boldsymbol{A}^+\boldsymbol{A}^+$
(xi)	$\boldsymbol{A}^* = \boldsymbol{A}^\#\boldsymbol{A}$	*(xxiv)*	$\boldsymbol{A}^\# = \boldsymbol{A}^+\boldsymbol{A}^\#$
(xii)	$\boldsymbol{A}^* = \boldsymbol{A}^\#\boldsymbol{A}^*$	*(xxv)*	$\boldsymbol{A}^\# = \boldsymbol{A}^\#\boldsymbol{A}^+.$
(xiii)	$\boldsymbol{A}^+ = \boldsymbol{A}\boldsymbol{A}^+$		

Proof. We prove only the equivalence (xxii) $\Leftrightarrow$ (3); the others follow by a similar reasoning. Recall from Section 2 that for given

$$\boldsymbol{A} = \begin{pmatrix} \boldsymbol{\Sigma K} & \boldsymbol{\Sigma L} \\ \boldsymbol{0} & \boldsymbol{0} \end{pmatrix},$$

we have

$$\boldsymbol{A}^+ = \begin{pmatrix} \boldsymbol{K}^*\boldsymbol{\Sigma}^{-1} & \boldsymbol{0} \\ \boldsymbol{L}^*\boldsymbol{\Sigma}^{-1} & \boldsymbol{0} \end{pmatrix}$$

and

$$\boldsymbol{A}^\# = \begin{pmatrix} \boldsymbol{K}^{-1}\boldsymbol{\Sigma}^{-1} & \boldsymbol{K}^{-1}\boldsymbol{\Sigma}^{-1}\boldsymbol{K}^{-1}\boldsymbol{L} \\ \boldsymbol{0} & \boldsymbol{0} \end{pmatrix},$$

the latter identity being valid only if $\text{rank}(\boldsymbol{A}) = \text{rank}(\boldsymbol{A}^2)$. Then

$$\boldsymbol{A}^+\boldsymbol{A}^* = \begin{pmatrix} \boldsymbol{K}^*\boldsymbol{\Sigma}^{-1}\boldsymbol{K}^*\boldsymbol{\Sigma} & \boldsymbol{0} \\ \boldsymbol{L}^*\boldsymbol{\Sigma}^{-1}\boldsymbol{K}^*\boldsymbol{\Sigma} & \boldsymbol{0} \end{pmatrix}.$$

Consequently, $\boldsymbol{A}^\# = \boldsymbol{A}^+\boldsymbol{A}^*$ is equivalent to $\boldsymbol{L} = \boldsymbol{0}$ and $\boldsymbol{K}^{-1}\boldsymbol{\Sigma}^{-1} = \boldsymbol{K}^*\boldsymbol{\Sigma}^{-1}\boldsymbol{K}^*\boldsymbol{\Sigma}$. Note that if $\boldsymbol{L} = \boldsymbol{0}$, we have $\boldsymbol{K}^{-1} = \boldsymbol{K}^*$. The former conditions are equivalent to $\boldsymbol{L} = \boldsymbol{0}$ and $\boldsymbol{K}^*\boldsymbol{\Sigma} = \boldsymbol{I}_r$, that is, $\boldsymbol{L} = \boldsymbol{0}$ and $\boldsymbol{\Sigma K} = \boldsymbol{I}_r$. □

In many cases, idempotency of a matrix is a strong feature of characterising an orthogonal projector, and only some additional conditions are needed to guarantee this property. Typical examples are provided by the next result.

Theorem 3.2. *The matrix $\boldsymbol{A}$ is an orthogonal projector if and only if one of the following conditions is satisfied:*

(i) $\boldsymbol{A}$ *and* $\boldsymbol{AA}^*$ *are oblique projectors*
(ii) $\boldsymbol{A}$ *and* $(\boldsymbol{A}+\boldsymbol{A}^*)/2$ *are oblique projectors*
(iii) $\boldsymbol{A}$ *is an oblique projector and* $\boldsymbol{AA}^* + \boldsymbol{A}^*\boldsymbol{A} = \boldsymbol{A}+\boldsymbol{A}^*$
(iv) $\boldsymbol{A}$ *is an oblique projector and* $\boldsymbol{A} = \boldsymbol{A}^+$
(v) $\boldsymbol{A}$ *is an oblique projector and* $\boldsymbol{AA}^* = \boldsymbol{AA}^+$
(vi) $\boldsymbol{A} = \boldsymbol{A}^+$ *and* $\boldsymbol{A}^2 = \boldsymbol{A}^*$
(vii) $\boldsymbol{A} = \boldsymbol{A}^+$ *and* $\boldsymbol{A}^2 = \boldsymbol{A}^+$
(viii) $\boldsymbol{A}$ *and* $\boldsymbol{A}^+$ *are oblique projectors*
(ix) $\boldsymbol{A}$ *is an oblique projector and* $\boldsymbol{AA}^*$ *is a generalised inverse of* $\boldsymbol{A}$
(x) $\boldsymbol{A}$ *is an oblique projector and* $(\boldsymbol{A}+\boldsymbol{A}^*)/2 = (\boldsymbol{AA}^*)^{1/2}$
(xi) $\boldsymbol{A}$ *is an oblique projector and* $(\boldsymbol{A}+\boldsymbol{A}^+)/2 = \boldsymbol{AA}^+$
(xii) $\boldsymbol{A}$ *and* $\boldsymbol{A}+\boldsymbol{A}^* - \boldsymbol{AA}^*$ *are oblique projectors*
(xiii) $\boldsymbol{A}$ *and* $\boldsymbol{A}+\boldsymbol{A}^+ - \boldsymbol{AA}^+$ *are oblique projectors.*

Proof. We show only nontrivial implications. Recall that for an oblique projector $\boldsymbol{A}$ we have

$$\boldsymbol{A} = \begin{pmatrix} \boldsymbol{I}_r & \boldsymbol{H} \\ \boldsymbol{0} & \boldsymbol{0} \end{pmatrix}$$

and

$$\boldsymbol{A}^+ = \begin{pmatrix} \boldsymbol{E} & \boldsymbol{0} \\ \boldsymbol{H}^*\boldsymbol{E} & \boldsymbol{0} \end{pmatrix},$$

where $\boldsymbol{E} = (\boldsymbol{I}_r + \boldsymbol{HH}^*)^{-1}$. It is clear that the conditions (i) – (xiii) are necessary. Hence we only show sufficiency.

Condition (i): If $\boldsymbol{A}$ and $\boldsymbol{AA}^*$ are oblique projectors, then

$$\boldsymbol{AA}^* = \begin{pmatrix} \boldsymbol{E}^{-1} & \boldsymbol{0} \\ \boldsymbol{0} & \boldsymbol{0} \end{pmatrix} = \begin{pmatrix} \boldsymbol{E}^{-2} & \boldsymbol{0} \\ \boldsymbol{0} & \boldsymbol{0} \end{pmatrix} = (\boldsymbol{AA}^*)^2$$

such that $\boldsymbol{I}_r + \boldsymbol{HH}^* = \boldsymbol{I}_r$, which implies $\boldsymbol{H} = \boldsymbol{0}$. Hence $\boldsymbol{A} = \begin{pmatrix} \boldsymbol{I}_r & \boldsymbol{0} \\ \boldsymbol{0} & \boldsymbol{0} \end{pmatrix}$ is an orthogonal projector.

Condition (iv): When $\boldsymbol{A}$ is an oblique projector and $\boldsymbol{A} = \boldsymbol{A}^+$, we get $\boldsymbol{H}^*\boldsymbol{E} = \boldsymbol{0}$ and thus $\boldsymbol{H} = \boldsymbol{0}$.

Condition (vi): From Theorem 2.2, we know that for a matrix

$$\boldsymbol{A} = \begin{pmatrix} \boldsymbol{\Sigma K} & \boldsymbol{\Sigma L} \\ \boldsymbol{0} & \boldsymbol{0} \end{pmatrix}$$

we have

$$\boldsymbol{A}^+ = \begin{pmatrix} \boldsymbol{K}^*\boldsymbol{\Sigma}^{-1} & \mathbf{0} \\ \boldsymbol{L}^*\boldsymbol{\Sigma}^{-1} & \mathbf{0} \end{pmatrix}.$$

Hence $\boldsymbol{A} = \boldsymbol{A}^+$ means $\boldsymbol{L} = \mathbf{0}$, $\boldsymbol{K}^* = \boldsymbol{K}^{-1}$ and therefore $(\boldsymbol{\Sigma K})^2 = \boldsymbol{I}$. On the other hand, $\boldsymbol{A}^2 = \boldsymbol{A}^*$ implies $(\boldsymbol{\Sigma K})^2 = \boldsymbol{K}^*\boldsymbol{\Sigma}$ so that $\boldsymbol{\Sigma K} = \boldsymbol{I}$, and $\boldsymbol{A}$ is an orthogonal projector.

Condition (x): When $\boldsymbol{A}$ is an oblique projector, it follows that

$$\frac{1}{2}(\boldsymbol{A} + \boldsymbol{A}^*) = \frac{1}{2}\begin{pmatrix} 2\boldsymbol{I} & \boldsymbol{H} \\ \boldsymbol{H}^* & \mathbf{0} \end{pmatrix}$$

and

$$(\boldsymbol{A}\boldsymbol{A}^*)^{1/2} = \begin{pmatrix} \boldsymbol{E}^{-1/2} & \mathbf{0} \\ \boldsymbol{H}^* & \mathbf{0} \end{pmatrix}.$$

The equality of the preceding matrices yields $\boldsymbol{H} = \mathbf{0}$ and $\boldsymbol{E}^{-1/2} = \boldsymbol{I}$. Thus $\boldsymbol{A}$ is an orthogonal projector.

Condition (xii): Since $\boldsymbol{A}$ is an oblique projector, we easily get

$$\boldsymbol{A} + \boldsymbol{A}^* - \boldsymbol{A}\boldsymbol{A}^* = \begin{pmatrix} 2\boldsymbol{I}_r - \boldsymbol{E}^{-1} & \boldsymbol{H} \\ \boldsymbol{H}^* & \mathbf{0} \end{pmatrix}.$$

Idempotency of this matrix then implies $\boldsymbol{H} = \mathbf{0}$, and hence $\boldsymbol{A}$ is an orthogonal projector. Since the other parts of the proof are similar we can omit them. □

Let us now assume that $\boldsymbol{P} = \begin{pmatrix} \boldsymbol{I}_r & \boldsymbol{H} \\ \mathbf{0} & \mathbf{0} \end{pmatrix}$ is a given oblique projector. Then the following result is important, insofar as it provides weaker conditions than $\boldsymbol{P}$ being hermitian, to identify $\boldsymbol{P}$ nevertheless as an orthogonal projector.

Theorem 3.3. *Let $\boldsymbol{P}$ be an oblique projector. Then the following statements are equivalent:*

(i) *$\boldsymbol{P}$ is an orthogonal projector*
(ii) *$\boldsymbol{P}$ is hermitian (i.e. $\boldsymbol{P} = \boldsymbol{P}^*$)*
(iii) *$\boldsymbol{P}$ is normal (i.e. $\boldsymbol{P}\boldsymbol{P}^* = \boldsymbol{P}^*\boldsymbol{P}$)*
(iv) *$\boldsymbol{P}$ is EP (i.e. $\mathcal{R}(\boldsymbol{P}) = \mathcal{R}(\boldsymbol{P}^*)$, where $\mathcal{R}(\cdot)$ denotes the column space of a matrix)*
(v) *$\boldsymbol{P}$ is a partial isometry (i.e. $\boldsymbol{P}^* = \boldsymbol{P}^+$)*
(vi) *$\boldsymbol{P}$ is bi-normal (i.e. $\boldsymbol{P}\boldsymbol{P}^*\boldsymbol{P}^*\boldsymbol{P} = \boldsymbol{P}^*\boldsymbol{P}\boldsymbol{P}\boldsymbol{P}^*$)*

(vii) $\boldsymbol{P}$ *is bi-EP (i.e.* $\boldsymbol{PP^+P^+P} = \boldsymbol{P^+PPP^+}$*)*
(viii) $\boldsymbol{P}$ *is bi-dagger (i.e.* $(\boldsymbol{P})^2 = (\boldsymbol{P}^2)^+$*).*

Proof. We show only (i) ⇔ (vii) and (i) ⇔ (viii).
(i) ⇔ (vii): Necessity is trivial. Since $\boldsymbol{P^+PPP^+} = \boldsymbol{P^+PP^+} = \boldsymbol{P^+}$, we have to prove that $\boldsymbol{PP^+P^+P} = \boldsymbol{P^+}$ implies $\boldsymbol{P}^2 = \boldsymbol{P}$. From Theorem 2.1 we conclude

$$\boldsymbol{PP^+P^+P} = \begin{pmatrix} \boldsymbol{E} & \boldsymbol{EH} \\ \boldsymbol{0} & \boldsymbol{0} \end{pmatrix}$$

and

$$\boldsymbol{P^+} = \begin{pmatrix} \boldsymbol{E} & \boldsymbol{0} \\ \boldsymbol{H^*E} & \boldsymbol{0} \end{pmatrix},$$

so that $\boldsymbol{PP^+P^+P} = \boldsymbol{P^+}$ entails $\boldsymbol{H} = \boldsymbol{0}$.
(i) ⇔ (viii): Necessity is obvious. Let $\boldsymbol{P^+P^+} = \boldsymbol{P^+}$. Then $\boldsymbol{E}^{-2} = \boldsymbol{E}^{-1}$ which implies $\boldsymbol{H} = \boldsymbol{0}$. □

Theorem 3.4. *Let* $\boldsymbol{P}$ *be an oblique projector. Then the following statements are equivalent:*

(i) $\boldsymbol{P}$ *is an orthogonal projector*
(ii) $\boldsymbol{PP^*} - \boldsymbol{P^*P} = \boldsymbol{P} - \boldsymbol{P^*}$
(iii) $\boldsymbol{PP^*} + \boldsymbol{P^*P} = \boldsymbol{P} + \boldsymbol{P^*}$
(iv) $\boldsymbol{PP^+} - \boldsymbol{P^+P} = \boldsymbol{P} - \boldsymbol{P^+}$
(v) $\boldsymbol{PP^+} + \boldsymbol{P^+P} = \boldsymbol{P} + \boldsymbol{P^+}$
(vi) $\boldsymbol{P} + \boldsymbol{P^*}$ *is nonnegative definite*
(vii) $\boldsymbol{P} - \boldsymbol{P^*}$ *is an oblique projector*
(viii) $\boldsymbol{P} - \boldsymbol{P^+}$ *is an oblique projector*
(ix) $\frac{1}{2}(\boldsymbol{P} + \boldsymbol{P^*})$ *is an oblique projector*
(x) $\frac{1}{2}(\boldsymbol{P} + \boldsymbol{P^+})$ *is an oblique projector.*

Proof. We show only (i) ⇔ (iv) and (i) ⇔ (vi).
(i) ⇔ (iv): Necessity is obvious. Assume now $\boldsymbol{PP^+} - \boldsymbol{P^+P} = \boldsymbol{P} - \boldsymbol{P^+}$. From Theorem 2.1 we get

$$\boldsymbol{PP^+} - \boldsymbol{P^+P} = \begin{pmatrix} \boldsymbol{I}_r - \boldsymbol{E} & -\boldsymbol{EH} \\ -\boldsymbol{H^*E} & -\boldsymbol{H^*EH} \end{pmatrix}$$

and

$$\boldsymbol{P} - \boldsymbol{P^+} = \begin{pmatrix} \boldsymbol{I}_r - \boldsymbol{E} & \boldsymbol{H} \\ -\boldsymbol{H^*E} & \boldsymbol{0} \end{pmatrix}.$$

Hence $\boldsymbol{PP}^+ - \boldsymbol{P}^+\boldsymbol{P} = \boldsymbol{P} - \boldsymbol{P}^+$ entails $\boldsymbol{H} = \boldsymbol{0}$, and thus $\boldsymbol{P}$ is orthogonal.
(i) $\Leftrightarrow$ (vi): We have

$$\boldsymbol{P} + \boldsymbol{P}^* = \begin{pmatrix} 2\boldsymbol{I}_r & \boldsymbol{H} \\ \boldsymbol{H}^* & \boldsymbol{0} \end{pmatrix}.$$

According to Theorem 6.13 in Zhang (1999), $\boldsymbol{P}+\boldsymbol{P}^*$ is nonnegative definite if and only if $-\boldsymbol{H}^*\boldsymbol{H}$ is nonnegative definite. This is equivalent to $\boldsymbol{H} = \boldsymbol{0}$. □

There is also a trace characterisation of orthogonal projectors.

Theorem 3.5. *Let* $\boldsymbol{P}$ *be an oblique projector. Then the following statements are equivalent:*

(i) $\boldsymbol{P}$ *is an orthogonal projector*
(ii) $\mathrm{tr}(\boldsymbol{PP}^*) = \mathrm{tr}(\boldsymbol{P})$
(iii) $\mathrm{tr}(\boldsymbol{P}^+) = \mathrm{rank}(\boldsymbol{P})$.

Proof. We show (ii) $\Rightarrow$ (i) and (iii) $\Rightarrow$ (i). The other implications are trivial.
(ii) $\Rightarrow$ (i): Let

$$\boldsymbol{P} = \begin{pmatrix} \boldsymbol{I}_r & \boldsymbol{H} \\ \boldsymbol{0} & \boldsymbol{0} \end{pmatrix}.$$

Then $\mathrm{tr}(\boldsymbol{PP}^*) = \mathrm{tr}(\boldsymbol{I}_r + \boldsymbol{HH}^*) = r + \mathrm{tr}(\boldsymbol{HH}^*)$, and $\mathrm{tr}(\boldsymbol{PP}^*) = \mathrm{tr}(\boldsymbol{P})$ requires $\boldsymbol{H} = \boldsymbol{0}$.
(iii) $\Rightarrow$ (i): $\mathrm{tr}(\boldsymbol{P}^+) = \mathrm{tr}(\boldsymbol{E}) = \sum_{j=1}^{r}(1/(1+\lambda_j))$, where λ_j are the eigenvalues of $\boldsymbol{HH}^*$. The condition $\mathrm{tr}(\boldsymbol{P}^+) = \mathrm{rank}(\boldsymbol{P}) = r$ then implies $\lambda_j = 0$ for all j, so that $\boldsymbol{HH}^* = \boldsymbol{0}$ and finally $\boldsymbol{H} = \boldsymbol{0}$. □

4. Parallel Sum

According to Rao and Mitra (1971, Sec. 10.1.6), for a pair of matrices $\boldsymbol{A}$ and $\boldsymbol{B}$ of the same order, the parallel sum of $\boldsymbol{A}$ and $\boldsymbol{B}$, denoted by the symbol $\boldsymbol{A} : \boldsymbol{B}$, is defined by

$$\boldsymbol{A} : \boldsymbol{B} = \boldsymbol{A}(\boldsymbol{A} + \boldsymbol{B})^- \boldsymbol{B},$$

where $(\boldsymbol{A} + \boldsymbol{B})^-$ is any g-inverse of $\boldsymbol{A} + \boldsymbol{B}$.

A pair of matrices $\boldsymbol{A}$ and $\boldsymbol{B}$ is said to be parallel summable if $\boldsymbol{A} : \boldsymbol{B}$ is invariant under the choices of $(\boldsymbol{A} + \boldsymbol{B})^-$. This is the case if and only if $\mathcal{R}(\boldsymbol{A}) \subset \mathcal{R}(\boldsymbol{A} + \boldsymbol{B})$ and $\mathcal{R}(\boldsymbol{A}^*) \subset \mathcal{R}(\boldsymbol{A}^* + \boldsymbol{B}^*)$.

Subsequently we consider the pair $\boldsymbol{A} = \boldsymbol{P}$ and $\boldsymbol{B} = \boldsymbol{P}^*$, where $\boldsymbol{P}$ is an oblique projector.

Theorem 4.1. *Let* $\boldsymbol{P} = \begin{pmatrix} \boldsymbol{I} & \boldsymbol{H} \\ \boldsymbol{0} & \boldsymbol{0} \end{pmatrix}$ *be an oblique projector. Then the following results hold:*

(i) $(\boldsymbol{P}+\boldsymbol{P}^*)^+ = \begin{pmatrix} \frac{1}{2}(\boldsymbol{I}_r - \boldsymbol{H}\boldsymbol{H}^+) & \boldsymbol{H}^{+*} \\ \boldsymbol{H}^+ & -2(\boldsymbol{H}^*\boldsymbol{H})^+ \end{pmatrix}$

(ii) $\mathcal{R}(\boldsymbol{P}) \subset \mathcal{R}(\boldsymbol{P}+\boldsymbol{P}^*)$ *and* $\mathcal{R}(\boldsymbol{P}^*) \subset \mathcal{R}(\boldsymbol{P}+\boldsymbol{P}^*)$

(iii) $\mathcal{R}(\boldsymbol{P}+\boldsymbol{P}^*) = \mathcal{R}(\boldsymbol{P}) + \mathcal{R}(\boldsymbol{P}^*)$ *and* $\mathcal{N}(\boldsymbol{P}+\boldsymbol{P}^*) = \mathcal{N}(\boldsymbol{P}) \cap \mathcal{N}(\boldsymbol{P}^*)$

(iv) $\boldsymbol{P}(\boldsymbol{P}+\boldsymbol{P}^*)^-\boldsymbol{P}^*$ *is invariant under the choice of* $(\boldsymbol{P}+\boldsymbol{P}^*)^-$

(v) $2(\boldsymbol{P} : \boldsymbol{P}^*)$ *is the orthogonal projector on* $\mathcal{R}(\boldsymbol{P}) \cap \mathcal{R}(\boldsymbol{P}^*)$.

Proof.

(i) $\boldsymbol{P}+\boldsymbol{P}^*$ is readily seen to be $\boldsymbol{P}+\boldsymbol{P}^* = \begin{pmatrix} 2\boldsymbol{I}_r & \boldsymbol{H} \\ \boldsymbol{H}^* & \boldsymbol{0} \end{pmatrix}$.
Using Theorem 3.5.2 from Campbell and Meyer (1979), we obtain the MP-inverse of $\boldsymbol{P}+\boldsymbol{P}^*$ as in statement (i).

(ii) From (i) we get
$(\boldsymbol{P}+\boldsymbol{P}^*)^+(\boldsymbol{P}+\boldsymbol{P}^*) = (\boldsymbol{P}+\boldsymbol{P}^*)(\boldsymbol{P}+\boldsymbol{P}^*)^+ = \begin{pmatrix} \boldsymbol{I}_r & \boldsymbol{0} \\ \boldsymbol{0} & \boldsymbol{H}^+\boldsymbol{H} \end{pmatrix}$.
Since $\begin{pmatrix} \boldsymbol{I}_r & \boldsymbol{0} \\ \boldsymbol{0} & \boldsymbol{H}^+\boldsymbol{H} \end{pmatrix} \begin{pmatrix} \boldsymbol{I}_r & \boldsymbol{H} \\ \boldsymbol{0} & \boldsymbol{0} \end{pmatrix} = \begin{pmatrix} \boldsymbol{I}_r & \boldsymbol{H} \\ \boldsymbol{0} & \boldsymbol{0} \end{pmatrix}$
and $\begin{pmatrix} \boldsymbol{I}_r & \boldsymbol{0} \\ \boldsymbol{0} & \boldsymbol{H}^+\boldsymbol{H} \end{pmatrix} \begin{pmatrix} \boldsymbol{I}_r & \boldsymbol{0} \\ \boldsymbol{H}^* & \boldsymbol{0} \end{pmatrix} = \begin{pmatrix} \boldsymbol{I}_r & \boldsymbol{0} \\ \boldsymbol{H}^* & \boldsymbol{0} \end{pmatrix}$,
we arrive at $\mathcal{R}(\boldsymbol{P}) \subset \mathcal{R}(\boldsymbol{P}+\boldsymbol{P}^*)$ and $\mathcal{R}(\boldsymbol{P}^*) \subset \mathcal{R}(\boldsymbol{P}+\boldsymbol{P}^*)$.

(iii) From (ii) we get $\mathcal{R}(\boldsymbol{P}) + \mathcal{R}(\boldsymbol{P}^*) \subset \mathcal{R}(\boldsymbol{P}+\boldsymbol{P}^*)$. The other inclusion $\mathcal{R}(\boldsymbol{P}+\boldsymbol{P}^*) \subset \mathcal{R}(\boldsymbol{P}) + \mathcal{R}(\boldsymbol{P}^*)$ is trivial. Furthermore, we have, since $\boldsymbol{P}+\boldsymbol{P}^*$ is hermitian, $\mathcal{N}(\boldsymbol{P}+\boldsymbol{P}^*) = [\mathcal{R}(\boldsymbol{P}+\boldsymbol{P}^*)]^\perp = [\mathcal{R}(\boldsymbol{P})+\mathcal{R}(\boldsymbol{P}^*)]^\perp = \mathcal{R}(\boldsymbol{P})^\perp \cap \mathcal{R}(\boldsymbol{P}^*)^\perp = \mathcal{N}(\boldsymbol{P}^*) \cap \mathcal{N}(\boldsymbol{P})$.

(iv) This statement follows from (ii).

(v) Some straightforward calculations show that

$$2\boldsymbol{P}(\boldsymbol{P}+\boldsymbol{P}^*)^+\boldsymbol{P}^* = \begin{pmatrix} \boldsymbol{I}_r - \boldsymbol{H}\boldsymbol{H}^+ & \boldsymbol{0} \\ \boldsymbol{0} & \boldsymbol{0} \end{pmatrix}$$

is an orthogonal projector. Consulting Theorem 10.1.8 in Rao and Mitra (1971), we conclude that its range is $\mathcal{R}(\boldsymbol{P}) \cap \mathcal{R}(\boldsymbol{P}^*)$. □

Note that for complex numbers we have $r : s = (rs)/(r+s)$, provided $r + s \neq 0$, and for $r \neq 0$, $r : r = r/2$. For a complex matrix $\boldsymbol{A}$ consider its spectral norm $||\boldsymbol{A}||$, defined by the square root of the maximal eigenvalue of $\boldsymbol{AA}^*$. Observe that for oblique projectors $\boldsymbol{P}$ we get $||\boldsymbol{P}|| \geq 1$.

Theorem 4.2. *Let $\boldsymbol{P}$ be an oblique projector. Then the following hold:*

(i) $||\boldsymbol{P} : \boldsymbol{P}^*|| \leq ||\boldsymbol{P}|| : ||\boldsymbol{P}^*||$ *with equality if and only if $\boldsymbol{P}$ is an orthogonal projector*

(ii) $\mathrm{tr}(\boldsymbol{P} : \boldsymbol{P}^*) \leq \mathrm{tr}(\boldsymbol{P}) : \mathrm{tr}(\boldsymbol{P}^*)$ *with equality if and only if $\boldsymbol{P}$ is an orthogonal projector.*

Proof. If $\boldsymbol{P} = \boldsymbol{0}$, nothing has to be proved. Let now $\boldsymbol{P} \neq \boldsymbol{0}$.

(i) Trivially, $||\boldsymbol{P}|| : ||\boldsymbol{P}^*|| = \frac{1}{2}||\boldsymbol{P}||$ and $\boldsymbol{P} : \boldsymbol{P}^* = \frac{1}{2}\begin{pmatrix} \boldsymbol{I}_r - \boldsymbol{HH}^+ & \boldsymbol{0} \\ \boldsymbol{0} & \boldsymbol{0} \end{pmatrix}$ so that $||\boldsymbol{P} : \boldsymbol{P}^*|| = \frac{1}{2} \leq \frac{1}{2}||\boldsymbol{P}|| = ||\boldsymbol{P}|| : ||\boldsymbol{P}^*||$. If $||\boldsymbol{P}|| = 2||\boldsymbol{P} : \boldsymbol{P}^*||$, then $1 = \lambda_{\max}(\boldsymbol{I}_r + \boldsymbol{HH}^*)$ which entails $\boldsymbol{H} = \boldsymbol{0}$.

(ii) Obviously, $\mathrm{tr}(\boldsymbol{P}) : \mathrm{tr}(\boldsymbol{P}^*) = \frac{1}{2}r$, where $r = \mathrm{rank}(\boldsymbol{P})$ and $\mathrm{tr}(\boldsymbol{P} : \boldsymbol{P}^*) = \frac{1}{2}(r - \mathrm{rank}(\boldsymbol{H})) \leq \frac{r}{2}$. Equality happens if and only if $\mathrm{rank}\,\boldsymbol{H} = 0$, i.e. $\boldsymbol{H} = \boldsymbol{0}$. □

References

1. Campbell, S.L. and Meyer, C.D. (1979), *Generalized Inverses of Linear Transformations.* Pitman, London.
2. Hartwig, R.E. and Spindelböck, K. (1984), Matrices for which $\boldsymbol{A}^*$ and $\boldsymbol{A}^+$ commute. *Linear Algebra and its Applications* 14, 241 – 256.
3. Harville, D.A. (1997), *Matrix Algebra from a Statistician's Perspective.* Springer, New York.
4. Marathe, C.R. (1956), A note of quasi-idempotent matrices. *American Mathematical Monthly* 63, 632 – 635.
5. Rao, C.R. and Mitra, S.K. (1971), *Generalized Inverse of Matrices and Its Applications.* Wiley, New York.
6. Sibuya, M. (1970), Subclasses of generalized inverses of matrices. *Annals of the Institute of Statistical Mathematics* 22, 543 – 556.
7. Trenkler, G. (1994), Characterizations of oblique and orthogonal projectors. In: *Proceedings of the International Conference of Linear Statistical Inference LINSTAT '93* ed. by T. Calinski and R. Kala, Kluwer Academic Press, Dordrecht, 255 – 270.
8. Zhang, F. (1999), *Matrix Theory. Basic Results and Techniques.* Springer, New York.

Fiducial Inference of Means and Variances from Normal Populations under Order Restrictions

Baoxue Zhang

School of Mathematics and Statistics, North East Normal University, Changchun 130024, People's Republic of China
E-mail: bxzhang@nenu.edu.cn

Xingzhong Xu

Department of Mathematics, Beijing Institute of Technology, Beijing 100081, People's Republic of China
E-mail: xuxz@bit.edu.cn

Xin Li

School of Software, Beijing Institute of Technology, Beijing 100081, People's Republic of China
E-mail: xinli1974@bit.edu.cn

For k independent normal populations with unknown means θ_i and unknown variances σ_i^2, $i = 1, \ldots, k$, this paper studies the fiducial estimation of $\theta = (\theta_1, \ldots, \theta_k)$ and $\sigma^2 = (\sigma_1^2, \ldots, \sigma_k^2)$ subject to the condition that the θ_i and σ_i^2 are restricted by a given partial ordering; for example, simple order restrictions: $\theta_1 \leq \theta_2 \leq \cdots \leq \theta_k$ and $\sigma_1 \geq \sigma_2 \geq \cdots \geq \sigma_k > 0$. It is well known that it is quite difficult to find the interval estimation of parameters for finite samples. But it is quite simple by using fiducial methods. Not only does the fiducial point estimation given by using fiducial methods appear to be more reasonable than MLEs, since they satisfy order restrictions automatically, but also the fiducial interval estimation can be constructed. This article mainly provides the explicit form for the fiducial distribution in the particular case that $k = 2$ and variances are known. For the general case, an algorithm to simulate the fiducial distribution of Θ and Σ with order restrictions is proposed.

Keywords: pivotal model; fiducial model; order restrictions; algorithm.

AMS Classification: 62B05, 62H12, 62F25.

1. Introduction

Since the first paper on discussion of fiducial distribution was published by Fisher (1930a), a considerable amount of work has been done, especially on how to derive a fiducial distribution of a parameter. For some of the contributions in this area, one may refer to Fisher (1930a, 1930b, 1950, 1973) and Fraser (1961a, 1961b, 1968); one may also refer to Pederson (1978), Zabell (1992) and Barnard (1995) for detailed discussion of results on fiducial argument. Dawid and Stone (1982) introduced the general concept of function models, that is, expressing a sufficient statistic $T(X)$ where X takes values in a sample space $\mathcal{X}$ as a function of an unknown parameter Θ and an error variable E with a known distribution. They found the fiducial distribution of a parameter Θ by giving the solution to the function models, but the range for which the fiducial distribution can be obtained is limited since the solution to the function models sometimes does not exist. Following Dawid and Stone (1982), Xu and Li (2003) first gave a new definition of the fiducial model by using a pivotal family of a distribution, which made larger the scope for which the fiducial distribution can be found. To be specific, let $T(X)$ denote a sufficient statistic whose distribution depends on parameters; i.e. suppose $T(X) \sim P_\theta$, $\theta \in \Omega$, where Ω is non-degenerate. $\{P_\theta : \theta \in \Omega\}$ is a pivotal family of distribution on $T(\cdot)$. $T(X) \stackrel{d}{=} h(\theta, E)$ is a pivotal model in which the pivotal random element E is distributed as a known distribution Q. Define a distance $d(\cdot, \cdot)$ on $T(\cdot)$, given x, the observation of X, and let $\widehat{\theta}_{T(x)}(e)$ denote the unique minimum point of $d(T(x), h(\theta, e))$ (usually take the Euclidean distance) on Ω; that is,

$$\widehat{\theta}_{T(x)}(e) = \arg\min_{\theta \in \Omega} d(T(x), h(\theta, e)). \qquad (1.1)$$

Then $\Theta = \widehat{\theta}_{T(x)}(E)$ is called a *fiducial model*, and the distribution of $\widehat{\theta}_{T(x)}(E)$ under Q is called a *fiducial distribution*. Obviously, if a minimum point of $d(T(x), h(\theta, e))$ does not exist, then the fiducial model does not exist, either. The fiducial cumulative distribution function (FCDF) is denoted by $H_{T(x)}(\theta)$.

Problems concerning estimation and testing hypotheses of parameters of distribution, when it is known *a priori* which are subject to order restrictions, are of considerable interest and are investigated in the literature. Most of the work on order restricted inference was reviewed by Barlow *et al.* (1972) and later by Robertson *et al.* (1988); in particular, the maximum likelihood estimators (MLEs) of parameters which satisfy certain order restrictions for normal distribution with order means and common variance are discussed by Robertson *et al.* (1988). Furthermore, Shi (1994, 1998) has studied some properties of the MLEs of unknown means and unknown variances under certain order restrictions, and proposed an algorithm to obtain the MLEs of unknown means and unknown variances under certain order restrictions. Shi, Gao and Zhang (2001) and Zhang, Gao and Shi (2003) gave an algorithm which extended the well-known pool adjacent violators algorithm (PAVA) for finding the maximum points of the objective concave function under certain order restrictions, which is applied to a numerical example from grouped samples. However, the isotonic maximum likelihood estimates are usually not strictly increasing. For example, suppose $\theta_1 \leq \theta_2 \leq \theta_3$ are the means of three normal distributions with common variances and suppose $\bar{x}_1 = 30$, $\bar{x}_2 = 20$, and $\bar{x}_3 = 50$. If the three sample sizes are equal, then the isotonic maximum likelihood estimates are $\theta_1^{IMLE} = 25$, $\theta_2^{IMLE} = 25$ and $\theta_3^{IMLE} = 50$. Wright (1978) proposed a weighted average of the isotonic estimators to break these ties. Furthermore, it is well known that finding interval estimation of parameters is quite difficult for finite samples under order restrictions, but it may be easy by using fiducial methods. The isotonic fiducial estimators will automatically be strictly monotone. Throughout this paper, $N(\theta, \sigma^2)$ denotes the normal distribution with mean θ and variance σ^2; the normal distribution with $\theta = 0$ and $\sigma^2 = 1$ is known as the standard normal distribution, whose density function and distribution function will be denoted by $\phi(x)$ and $\Phi(x)$ respectively.

In this paper, we mainly consider the fiducial estimations of isotonic normal means and variances by the above fiducial model. In detail, let x_{ij}, $i = 1, \ldots, k$, $j = 1, \ldots, n_i$, be observations from the i-th normal population with unknown mean θ_i and variance σ_i^2, $i = 1, \ldots, k$. Assume $\theta = (\theta_1, \ldots \theta_k)$ and $\sigma = (\sigma_1, \ldots, \sigma_k)$ are restricted by some partial orderings; for instance, the simple order

$$\theta_1 \leq \theta_2 \leq \cdots \leq \theta_k \tag{1.2}$$

and

$$\sigma_1 \geq \sigma_2 \geq \cdots \geq \sigma_k > 0. \tag{1.3}$$

The general pivotal models are given by

$$\bar{X}_i \stackrel{d}{=} \theta_i + \frac{\sigma_i}{\sqrt{n_i}} E_i \tag{1.4}$$

$$S_i \stackrel{d}{=} \sigma_i F_i \tag{1.5}$$

where $\bar{X}_i$ denotes sample mean from the i-th normal population, $S_i^2 = \sum_{j=1}^{n_i}(X_{ij} - \bar{X}_i)^2$, E_i and F_i are mutually independent, the distributions of E_i are standard normal, and the distributions of F_i are $\chi_{(n_i-1)}$ with freedom $n_i - 1$, $i = 1, \ldots, k$ respectively.

In Section 3.1, we present the analytic expressions for the fiducial distribution of θ_i for $k = 2$ when variances are known. In section 3.2, an algorithm to obtain samples of the fiducial distribution of Θ and Σ is proposed. In section 4, a numerical example is given. Proofs are given in the Appendix.

2. Existence of the Fiducial Model

For our problems, (1.1) becomes

$$d(\cdot,\cdot) = \frac{1}{2}\sum_{i=1}^{k}[(\bar{x}_i - \theta_i - \frac{\sigma_i}{\sqrt{n_i}}e_i)^2 + (s_i - \sigma_i f_i)^2] \ . \tag{2.1}$$

We will show that d in (2.1) is a strictly convex function. It suffices to check that the Hessian matrix of $d(\cdot,\cdot)$ is positive definite (see Stoer and Witzgall (1970, Chap. 4); and Horot, Pardalos and Thoai (1995, Chap. 1)). Let the matrix be denoted by H. It can be shown that

$$H = \begin{bmatrix} H_{11} & H_{12} \\ H_{21} & H_{22} \end{bmatrix},$$

where H_{11}, H_{12}, H_{21}, and H_{22} are diagonal matrices, respectively,

$$H_{11} = I \ ,$$

$$H_{12} = H_{21} = \mathrm{diag}\{\frac{e_1}{\sqrt{n_1}}, \ldots, \frac{e_k}{\sqrt{n_k}}\} \ ,$$

$$H_{22} = \mathrm{diag}\{f_1^2 + \frac{e_1^2}{n_1}, \ldots, f_k^2 + \frac{e_k^2}{n_k}\} \ .$$

It is clear that H_{11} and H_{22} are positive definite matrices. Denote the determinant of the $(k+i)$-order principal minor matrix of H by H_{k+i} for $i = 1, \ldots, k$. We need only to prove that $H_{k+i} > 0$. It is easily verified that

$$H_{k+i} = f_1^2 f_2^2 \cdots f_i^2 > 0$$

and hence (2.1) is a strictly convex function. If we let $g_i(\theta) = \theta_i - \theta_{i-1}$ and $l_i(\sigma) = \sigma_{i-1} - \sigma_i$, then the solution to problem (2.1) exists and is unique under restrictions (1.2) and (1.3), which also implies the existence of the fiducial model.

3. The Fiducial Distribution of Θ and Σ

3.1. *The fiducial distribution of Θ when $k = 2$*

It is known that if variances are given, the restriction (1.3) of variance and the pivotal (1.5) are vacuous. At first, we develop the explicit form of the fiducial distribution of Θ for $k = 2$ when variances are known. For simplicity, we assume that the pivotal model (1.4) is as follows:

$$\bar{X} \stackrel{d}{=} \theta_1 + \frac{1}{\sqrt{m}} E_1 \tag{3.1.1}$$

$$\bar{Y} \stackrel{d}{=} \theta_2 + \frac{1}{\sqrt{n}} E_2 \tag{3.1.2}$$

where $\bar{X}$ and $\bar{Y}$ denote sample means from populations $N(\theta_1, 1)$ and $N(\theta_2, 1)$, respectively. E_1 and E_2 are independent, identically distributed as $N(0, 1)$.

To find the fiducial distribution of Θ under order restrictions, we need only to find the minimum points of $[(\bar{x} - \theta_1 - \frac{1}{\sqrt{m}} e_1)^2 + (\bar{y} - \theta_2 - \frac{1}{\sqrt{n}} e_2)^2]$ over $\Omega = \{(\theta_1, \theta_2) : \theta_1 \leq \theta_2\}$, i.e.

$$\min_{\theta \in \Omega} \; [(\bar{x} - \theta_1 - \frac{1}{\sqrt{m}} e_1)^2 + (\bar{y} - \theta_2 - \frac{1}{\sqrt{n}} e_2)^2] \tag{3.1.3}$$

which leads to the following Theorem:

Theorem 3.1. *The fiducial models of Θ_1 and Θ_2 are*

$$\Theta_1 = \begin{cases} \bar{x} - \frac{1}{\sqrt{m}} E_1, & \bar{x} - \frac{1}{\sqrt{m}} E_1 < \bar{y} - \frac{1}{\sqrt{n}} E_2 \,, \\ \frac{1}{2}(\bar{x} + \bar{y} - \frac{1}{\sqrt{m}} E_1 - \frac{1}{\sqrt{n}} E_2), & \bar{x} - \frac{1}{\sqrt{m}} E_1 \geq \bar{y} - \frac{1}{\sqrt{n}} E_2 \,; \end{cases} \tag{3.1.4}$$

$$\Theta_2 = \begin{cases} \bar{y} - \frac{1}{\sqrt{n}} E_2, & \bar{x} - \frac{1}{\sqrt{m}} E_1 < \bar{y} - \frac{1}{\sqrt{n}} E_2 \,, \\ \frac{1}{2}(\bar{x} + \bar{y} - \frac{1}{\sqrt{m}} E_1 - \frac{1}{\sqrt{n}} E_2), & \bar{x} - \frac{1}{\sqrt{m}} E_1 \geq \bar{y} - \frac{1}{\sqrt{n}} E_2 \,. \end{cases} \tag{3.1.5}$$

From Theorem 3.1, we have the following Corollary:

Corollary 3.1. *The fiducial joint distribution of (Θ_1, Θ_2) is* $H_{\bar{x},\bar{y}}(\theta_1, \theta_2)$

$$= \begin{cases} (1 - \Phi(\sqrt{m}(\bar{x} - \theta_1)))(\Phi(\sqrt{n}(\bar{y} - \theta_1)) - \Phi(\sqrt{n}(\bar{y} - \theta_2))) \\ +(1 - \Phi(\sqrt{n}(\bar{y} - \theta_1)) \\ - \int_{\sqrt{n}(\bar{y}-\theta_1)}^{+\infty} \Phi(\sqrt{m}(\bar{x} + \bar{y} - 2\theta_1 - \frac{1}{\sqrt{n}} e_1))\phi(e_1)\, de_1, & \theta_2 \geq \theta_1, \\ \\ 1 - \Phi(\sqrt{m}(\bar{y} - \theta_2)) \\ - \int_{\sqrt{n}(\bar{y}-\theta_2)}^{+\infty} \Phi(\sqrt{m}(\bar{x} + \bar{y} - 2\theta_2 - \frac{1}{\sqrt{n}} e_1))\phi(e_1)\, de_1, & \theta_2 < \theta_1. \end{cases}$$

Furthermore, the marginal fiducial distributions of Θ_i, $i = 1, 2$, are as follows, respectively:

$$F_{\Theta_1}(\theta_1) = 1 - \Phi(\sqrt{m}(\bar{x} - \theta_1))\Phi(\sqrt{n}(\bar{y} - \theta_1)) - \int_{\sqrt{n}(\bar{y}-\theta_1)}^{+\infty} \Phi(\sqrt{m}(\bar{x} + \bar{y} - 2\theta_1 - \frac{1}{\sqrt{n}}e_1))\phi(e_1)\, de_1 \,,$$

$$F_{\Theta_2}(\theta_2) = 1 - \Phi(\sqrt{n}(\bar{y} - \theta_2)) - \int_{\sqrt{n}(\bar{y}-\theta_2)}^{+\infty} \Phi(\sqrt{m}(\bar{x} + \bar{y} - 2\theta_2 - \frac{1}{\sqrt{n}}e_1))\phi(e_1)\, de_1 \,.$$

In general, we use the expectation of Θ_i as the fiducial estimations denoted by $\bar{\theta}_i$. By using expressions (3.1.4) and (3.1.5), we get the following Corollary:

Corollary 3.2.
(a)

$$\bar{\theta}_1 = E\Theta_1 = \frac{\bar{x} + \bar{y}}{2} - \frac{(\bar{y} - \bar{x})}{2}\Phi\left(\frac{\bar{y} - \bar{x}}{\sqrt{\frac{1}{m} + \frac{1}{n}}}\right) - \frac{1}{2}\sqrt{\frac{1}{m} + \frac{1}{n}}\,\phi\left(\frac{\bar{x} - \bar{y}}{\sqrt{\frac{1}{m} + \frac{1}{n}}}\right)$$

$$\bar{\theta}_2 = E\Theta_2 = \frac{\bar{x} + \bar{y}}{2} + \frac{(\bar{y} - \bar{x})}{2}\Phi\left(\frac{\bar{y} - \bar{x}}{\sqrt{\frac{1}{m} + \frac{1}{n}}}\right) + \frac{1}{2}\sqrt{\frac{1}{m} + \frac{1}{n}}\,\phi\left(\frac{\bar{x} - \bar{y}}{\sqrt{\frac{1}{m} + \frac{1}{n}}}\right)$$

(b)
$$E\Theta_1 \le \frac{\bar{x} + \bar{y}}{2} \le E\Theta_2$$

(c)
$$E\Theta_1 \le \bar{x}, \qquad E\Theta_2 \ge \bar{y}$$

3.2. *The proposed algorithm*

For $k > 2$, it is difficult to obtain the analytic expressions for the fiducial distributions of Θ and Σ; however, the fiducial (or pivotal) model plays an important role in finding the fiducial distribution. In this section, an algorithm to simulate the fiducial distribution is proposed based on the the fiducial (or pivotal) model. The following Theorems 3.2 and 3.3 are the basis of the algorithm given in this section.

Theorem 3.2. *Consider the problem*

$$\min_{\theta \in \Theta} \sum_{i=1}^{k} f_i(\theta_i) \tag{3.2.1}$$

where $\Theta = \{(\theta_1, \ldots, \theta_k) : \theta_1 \leq \cdots \leq \theta_k\}$ *is a convex set in* $\mathbb{R}^k$, $f_i(\theta_i)$ *is any convex function having the second partial derivative defined on* Θ_i, *which is a nonempty convex set in* $\mathbb{R}^1, i = 1, \ldots, k$. *For given* $1 \leq s \leq t \leq \lambda$, *let* $\hat{\theta}_{s,t}$ *denote the solution of the following equation:*

$$\sum_{i=s}^{t} \frac{\partial f_i(\vartheta)}{\partial \vartheta} = 0,$$

and let $\bar{\theta}_i$ *be the global minimum point of the optimisation problem (3.2.1). Then*

$$\bar{\theta}_i = \min_{t \geq i} \max_{s \leq i} \hat{\theta}_{s,t} = \max_{s \leq i} \min_{t \geq i} \hat{\theta}_{s,t}. \tag{3.2.2}$$

Remark 3.1. Let $f_i(\theta_i) = (x_i - \theta_i)^2 w_i$.
Then $\bar{\theta}_i = \min_{t \geq i} \max_{s \leq i} \left(\sum_{i=s}^{t} w_i x_i \Big/ \sum_{i=s}^{t} w_i \right)$ is called the isotonic regression (Robertson *et al.*, 1988, P_{24}).

Theorem 3.3. *Consider the problem*

$$\min_{\theta \in D,\ \sigma \in G} \sum f_i(\theta_i, \sigma_i) \tag{3.2.3}$$

where $D = \{\theta = (\theta_1, \ldots, \theta_k) : \theta_1 \leq \cdots \leq \theta_k\}$, $G = \{\sigma = (\sigma_1, \ldots, \sigma_k) : \sigma_1 \geq \cdots \geq \sigma_k > 0\}$, *and* $D \otimes G$ *are convex sets in* $\mathbb{R}^k$, $\mathbb{R}^k$, *and* $\mathbb{R}^k \otimes \mathbb{R}^k$ *respectively.* $f_i(\theta_i, \sigma_i)$ *is any convex function having the second partial derivative defined on* D, *which is a nonempty convex set in* $\mathbb{R}^k$ *when* σ_i *is fixed; similarly,* $f_i(\theta_i, \sigma_i)$ *is any convex function having the second partial derivative defined on* G, *which is a nonempty convex set in* $\mathbb{R}^k$ *when* θ_i *is fixed. Then the local minimum point of the optimisation problem (3.2.3) is given by the following iterative algorithm based on Theorem 3.2:*

step 0: *Let* $\theta = \theta^{(0)}$ *and* $\sigma = \sigma^{(0)}$
step n: (1) *Find* $\hat{\theta}_i^{(n)} = \min_{t \geq i} \max_{s \leq i} \hat{\theta}_{s,t}^{(n)}$
where $\hat{\theta}_{s,t}^{(n)}$ *is the solution of the following equation:*

$$\sum_{i=s}^{t} \frac{\partial f_i(\vartheta, \sigma_i^{(n-1)})}{\partial \vartheta} = 0 \qquad 1 \leq s \leq t \leq k$$

(2) *Find* $\hat{\sigma}_i^{(n)} = \min_{t \geq i} \max_{s \leq i} \hat{\sigma}_{s,t}^{(n)}$
where $\hat{\sigma}_{s,t}$ *is the solution of the following equation:*

$$\sum_{i=s}^{t} \frac{\partial f_i(\theta_i^{(n)}, \nu)}{\partial \nu} = 0 \qquad 1 \leq s \leq t \leq k.$$

Furthermore, if $f(\theta,\sigma)$ is a convex function defined on $D \otimes G$, then the sequence $\{\theta_i^{(n)}, \sigma_i^{(n)}\}$ converges to the global minimum point $(\bar{\theta}, \bar{\sigma})$.

Remark 3.2. For our problem, we have

$$\hat{\theta}_{s,t}^{(n)} = \frac{\sum_{i=s}^{t}(\bar{x}_i - \frac{\sigma_i^{(n-1)}}{\sqrt{n}_i}e_i)}{t-s+1}, \tag{3.2.4}$$

$$\hat{\sigma}_{s,t}^{(n)} = \frac{\sum_{i=s}^{t}(\bar{x}_i - \theta_i^{(n)}) + \sum_{i=s}^{t} s_i f_i}{\sum_{i=s}^{t} f_i^2 + \sum_{i=s}^{t}\frac{e_i^2}{n_i}}. \tag{3.2.5}$$

The fiducial interval estimates, and the fiducial estimates of θ_i, σ_i can be computed using the following algorithm.

ALGORITHM

For a given data set $x_{ij}, i = 1, \ldots, k,\ j = 1, \ldots, n_i$,

compute $\bar{x}_i = \frac{1}{n_i}\sum_{j=1}^{n_i} x_{ij}$ and $s_i^2 = \sum_{j=1}^{n_i}(x_{ij} - \bar{x}_i)^2$.

For $j = 1$ to m,

generate $E_{ij} \overset{\text{i.i.d.}}{\sim} N(0,1)$ and $F_{ij} \overset{\text{i.i.d.}}{\sim} \chi_{(n_i-1)}$,

set $\{\theta_{ij}, \sigma_{ij}\} = \{\hat{\theta}_{ij}^{(n)}, \hat{\sigma}_{ij}^{(n)}\}$ which is given by step n (1) and (2), (3.2.4) and (3.2.5).

(end j loop)

$\frac{1}{m}\sum_{j=1}^{m}\theta_{ij}$ and $\frac{1}{m}\sum_{j=1}^{m}\sigma_{ij}$ are Monte Carlo estimates of θ_i and σ_i.

The $100(\frac{\alpha}{2})$ centile and $100(1-\frac{\alpha}{2})$ centile of $\theta_{i1}, \ldots, \theta_{im}$ and $\sigma_{i1}, \ldots, \sigma_{im}$, denoted by $\theta_i(\frac{\alpha}{2})$, $\theta_i(1-\frac{\alpha}{2})$, $\sigma_i(\frac{\alpha}{2})$, $\sigma_i(1-\frac{\alpha}{2})$, are Monte Carlo estimates of the $100(\frac{\alpha}{2})\%$ and the $100(1-\frac{\alpha}{2})\%$ fiducial lower limit and upper limit for θ_i and σ_i. Moreover, $[\theta_i(\frac{\alpha}{2}),\ \theta_i(1-\frac{\alpha}{2})]$ and $[\sigma_i(\frac{\alpha}{2}),\ \sigma_i(1-\frac{\alpha}{2})]$ are fiducial level $100(1-\alpha)\%$ fiducial interval estimation for θ_i and σ_i, respectively.

4. A Numerical Example

For illustration, the proposed algorithm is used to treat the data in Table 1 taken from Shi (1994). There are five districts in Jilin Province of China: Liaoyuan, Qianfu, Changchun, Tonghua, and Jilin. Group 1, Group 2, Group 3, Group 4 and Group 5 represent the scores of 100 students

Table 1. The Examination Scores of 500 Students.

	Group 1		Group 2		Group 3		Group 4		Group 5	
1	405	400	434	325	388	358	403	332	440	449
2	444	435	476	459	336	358	422	438	421	442
3	326	354	498	420	459	366	386	460	386	294
4	306	348	246	450	369	418	422	368	451	517
5	387	314	456	382	424	352	342	370	493	436
6	365	242	397	364	405	419	373	347	448	388
7	351	486	462	391	448	434	332	397	378	457
8	429	261	266	388	432	217	377	473	493	401
9	345	376	389	433	472	334	508	348	391	409
10	345	503	488	448	400	397	421	379	400	411
11	471	424	277	346	312	347	520	415	369	336
12	354	370	479	501	375	461	306	364	396	368
13	166	399	288	494	373	396	497	364	396	368
14	438	310	453	336	446	473	489	363	519	348
15	379	413	329	423	238	305	493	358	462	432
16	435	387	479	402	434	406	408	240	478	405
17	428	242	384	486	372	323	357	407	458	465
18	328	312	299	436	409	429	534	383	435	442
19	377	384	415	473	416	398	469	416	298	259
20	380	430	368	458	400	420	397	446	468	300
21	414	437	297	391	345	412	444	448	438	424
22	412	463	315	356	287	341	492	460	478	426
23	350	414	241	414	500	383	351	483	413	462
24	528	380	407	448	348	443	355	404	383	327
25	305	434	344	461	322	434	429	390	454	408
26	420	419	389	321	418	425	434	486	484	388
27	413	278	374	353	417	428	420	398	442	414
28	450	469	358	248	413	387	397	460	384	458
29	442	447	424	437	405	372	323	409	354	419
30	448	443	208	385	298	440	509	441	419	466
31	418	433	363	302	487	414	447	377	446	394
32	380	305	393	378	471	202	419	369	507	342
33	391	427	369	467	543	408	386	433	444	269
34	402	347	431	286	488	472	338	346	233	515
35	381	395	275	359	490	496	422	406	445	351
36	429	406	414	313	451	372	464	397	427	405
37	323	307	267	425	413	452	454	349	468	300
38	393	377	231	271	445	462	325	478	341	469
39	470	488	423	280	301	432	433	493	532	215
40	404	417	314	407	397	546	333	342	219	508
41	444	436	433	515	362	380	428	274	383	271
42	238	397	438	381	470	320	434	324	402	544
43	406	448	379	234	419	533	306	362	462	396
44	392	433	457	439	307	417	336	355	535	465
45	269	336	455	413	240	464	350	397	436	419
46	431	417	424	276	483	378	329	384	521	437
47	363	388	503	405	420	375	304	299	478	430
48	416	437	467	416	331	355	374	490	359	390
49	250	399	362	370	470	338	331	273	393	527
50	362	457	302	225	404	327	322	335	496	456
$\bar{x}$	388.270		384.610		398.000		395.170		418.010	
σ^2	4013.917		5354.438		4269.380		3582.821		4928.749	

in each district obtained in the National Matriculation Examination held in 1992, respectively. It is verified that the examination scores follow normal distributions using MATLAB. Let X_i denote the examination score of Group i; then $X_i \sim N(\theta_i, \sigma_i^2)$ for $i = 1, \ldots, 5$.

Prior information tells us

$$\theta_1 \leq \theta_2 \leq \cdots \leq \theta_5 \qquad \text{and} \qquad \sigma_1 \geq \sigma_2 \geq \cdots \geq \sigma_5 > 0$$

A computer program for the proposed algorithm is written using S-plus. The computed results are given in Table 2. The iteration is terminated when the conditions

$$\max_i |\theta_i^{(n-1)} - \theta_i^{(n)}| < 10^{-5} \qquad \text{and} \qquad \max_i |\sigma_i^{(n-1)} - \sigma_i^{(n)}| < 10^{-5}$$

are satisfied. The fiducial interval estimates obtained are based on the algorithm with 100,000 runs.

Table 2. The estimates of θ and σ

θ	θ_1	θ_2	θ_3	θ_4	θ_5
MLE	386.440	386.440	396.583	396.583	418.010
FE	385.149	387.323	395.650	397.898	418.010
FIE	[374.17, 394.75]	[377.43, 396.93]	[386.31, 405.08]	[388.75, 408.12]	[405.43, 430.67]
σ	σ_1	σ_2	σ_3	σ_4	σ_5
MLE	68.466	68.466	65.356	65.244	65.244
FE	69.398	69.198	66.442	64.639	64.554
FIE	[63.87, 76.43]	[63.77, 75.92]	[61.12, 72.26]	[59.35, 69.86]	[59.24, 69.78]

MLE and FE denote maximum likelihood estimate and Fiducial estimate.
FIE denotes Fiducial interval estimate with the Fiducial level 95%.

From the numerical results in Table 2, it is clear that we can not only get the fiducial estimates of θ_i and σ_i, but also obtain the fiducial interval estimates for θ_i and σ_i. Furthermore, it seems to be more reasonable than MLEs of θ_i and σ_i given in Shi (1994), since they increase successively.

Acknowledgements

The authors wish to thank a referee for many helpful comments. The project was supported by the Science Foundation for Young Teachers of Northeast Normal University, China Mathematics Tian Yuan Youth Foundation, the National Natural Science Youth Foundation of China and the National Natural Science Foundation of China.

Appendix

The proof of Corollary 3.1: It can be seen from (3.1.4):

$$
\begin{aligned}
&H_{\bar{x},\bar{y}}(\theta_1,\theta_2) = \widetilde{P}\,(\Theta_1 \le \theta_1, \Theta_2 \le \theta_2)\\
&= Q(\bar{x} - \frac{1}{\sqrt{m}}E_1 \le \theta_1, \bar{y} - \frac{1}{\sqrt{n}}E_2 \le \theta_2, \bar{x} - \frac{1}{\sqrt{m}}E_1 \le \bar{y} - \frac{1}{\sqrt{n}}E_2)\\
&+ Q(\frac{1}{2}(\bar{x}+\bar{y} - \frac{1}{\sqrt{m}}E_1 - \frac{1}{\sqrt{n}}E_2) \le \min(\theta_1,\theta_2), \bar{x} - \frac{1}{\sqrt{m}}E_1 > \bar{y} - \frac{1}{\sqrt{n}}E_2)\\
&= Q(E_1 \ge \sqrt{m}(\bar{x}-\theta_1), E_2 \ge \sqrt{n}(\bar{y}-\theta_2), \frac{1}{\sqrt{m}}E_1 - \frac{1}{\sqrt{n}}E_2 \ge \bar{x}-\bar{y})\\
&+ Q(\frac{1}{\sqrt{m}}E_1 + \frac{1}{\sqrt{n}}E_2 \ge \ \bar{x}+\bar{y} - 2\min(\theta_1,\theta_2), \frac{1}{\sqrt{n}}E_2 - \frac{1}{\sqrt{m}}E_1 > \bar{y}-\bar{x}).
\end{aligned} \tag{1}
$$

For $\theta_1 \le \theta_2$, the first term on the right-hand side of (1) equals

$$
\begin{aligned}
&\int_{\sqrt{n}(\bar{y}-\theta_1)}^{+\infty}\int_{\sqrt{m}(\bar{x}-\bar{y}-\theta_1+\frac{1}{\sqrt{n}}e_2)}^{+\infty} \phi(e_1)\phi(e_2)\,de_1de_2\\
&\qquad + \int_{\sqrt{n}(\bar{y}-\theta_2)}^{\sqrt{n}(\bar{y}-\theta_1)}\int_{\sqrt{m}(\bar{x}-\theta_1)}^{+\infty} \phi(e_1)\phi(e_2)\,de_1de_2\\
&= \int_{\sqrt{n}(\bar{y}-\theta_1)}^{+\infty} (1-\Phi(\sqrt{m}(\bar{x}-\bar{y}+\frac{1}{\sqrt{n}}e_2)))\phi(e_2)\,de_2\\
&\qquad + (1-\Phi(\sqrt{m}(\bar{x}-\theta_1)))(\Phi(\sqrt{n}(\bar{y}-\theta_1)) - \Phi(\sqrt{m}(\bar{y}-\theta_2)))\\
&= 1-\Phi(\sqrt{n}(\bar{y}-\theta_1)) - \int_{\sqrt{n}(\bar{y}-\theta_1)}^{+\infty} \Phi(\sqrt{m}(\bar{x}-\bar{y}+\frac{1}{\sqrt{n}}e_2))\phi(e_2)\,de_2\\
&\qquad + (1-\Phi(\sqrt{m}(\bar{x}-\theta_1)))(\Phi(\sqrt{n}(\bar{y}-\theta_1)) - \Phi(\sqrt{m}(\bar{y}-\theta_2)))
\end{aligned} \tag{2}
$$

and the second term on the right-hand side of (1) equals

$$
\begin{aligned}
&\int_{\sqrt{n}(\bar{y}-\theta_1)}^{+\infty}\int_{\sqrt{m}(\bar{x}+\bar{y}-2\theta_1-\frac{1}{\sqrt{n}}e_2)}^{\sqrt{m}(\bar{x}-\bar{y}+\frac{1}{\sqrt{n}}e_2)} \phi(e_1)\phi(e_2)\,de_1de_2\\
&= \int_{\sqrt{n}(\bar{y}-\theta_1)}^{+\infty} [\Phi(\sqrt{m}(\bar{x}-\bar{y}+\frac{1}{\sqrt{n}}e_2))\\
&\qquad\qquad - \Phi(\sqrt{m}(\bar{x}+\bar{y}-2\theta_1-\frac{1}{\sqrt{n}}e_2))]\ \phi(e_2)\,de_2\\
&= \int_{\sqrt{n}(\bar{y}-\theta_1)}^{+\infty} [\Phi(\sqrt{m}(\bar{x}-\bar{y}+\frac{1}{\sqrt{n}}e_1))\\
&\qquad\qquad - \Phi(\sqrt{m}(\bar{x}+\bar{y}-2\theta_1-\frac{1}{\sqrt{n}}e_1))]\ \phi(e_1)\,de_1\ .
\end{aligned} \tag{3}
$$

Substituting (2) and (3) for (1), we have

$$H_{\bar{x},\bar{y}}(\theta_1,\theta_2) = (1-\Phi(\sqrt{m}(\bar{x}-\theta_1)))(\Phi(\sqrt{n}(\bar{y}-\theta_1))-\Phi(\sqrt{n}(\bar{y}-\theta_2)))$$
$$+(1-\Phi(\sqrt{n}(\bar{y}-\theta_1)))-\int_{\sqrt{n}(\bar{y}-\theta_1)}^{+\infty}\Phi(\sqrt{m}(\bar{x}+\bar{y}-2\theta_1-\frac{1}{\sqrt{n}}e_1))\phi(e_1)\,de_1.$$

Similarly, for $\theta_1 > \theta_2$, we have

$$H_{\bar{x},\bar{y}}(\theta_1,\theta_2) = 1-\Phi(\sqrt{m}(\bar{y}-\theta_2))$$
$$-\int_{\sqrt{n}(\bar{y}-\theta_2)}^{+\infty}\Phi(\sqrt{m}(\bar{x}+\bar{y}-2\theta_2-\frac{1}{\sqrt{n}}e_1))\phi(e_1)\,de_1.$$

Furthermore, when $\theta_2 \to +\infty$, we obtain the marginal fiducial distribution of Θ_1:

$$F_{\Theta_1}(\theta_1) = 1-\Phi(\sqrt{m}(\bar{x}-\theta_1))\Phi(\sqrt{n}(\bar{y}-\theta_1))$$
$$-\int_{\sqrt{n}(\bar{y}-\theta_1)}^{+\infty}\Phi(\sqrt{m}(\bar{x}+\bar{y}-2\theta_1-\frac{1}{\sqrt{n}}e_1))\phi(e_1)\,de_1.$$

Similarly, when $\theta_1 \to +\infty$, we also get the marginal fiducial distribution of Θ_2:

$$F_{\Theta_2}(\theta_2) = 1-\Phi(\sqrt{n}(\bar{y}-\theta_2))-\int_{\sqrt{n}(\bar{y}-\theta_2)}^{+\infty}\Phi(\sqrt{m}(\bar{x}+\bar{y}-2\theta_2-\frac{1}{\sqrt{n}}e_1))\phi(e_1)\,de_1.$$

Thus, we complete the proof of Corollary 3.1

The proof of Corollary 3.2 is based on the following two Lemmas:

Lemma A.1: Suppose that $X \sim N(0,1)$; then

(*a*)
$$E\Phi(a+bX) = \Phi\left(\frac{a}{\sqrt{1+b^2}}\right)$$

(*b*)
$$E\phi(a+bX) = \frac{1}{\sqrt{1+b^2}}\,\phi\left(\frac{a}{\sqrt{1+b^2}}\right)$$

Proof: Since

$$E\Phi(a+bX) = \int_{-\infty}^{+\infty}\Phi(a+bx)\phi(x)\,dx = \int_{-\infty}^{+\infty}\int_{-\infty}^{a+bx}\phi(x)\phi(y)\,dy\,dx$$
$$= P(Y \le a+bX)$$

where $X, Y \overset{i.i.d.}{\sim} N(0,1)$, and $Z = Y - bX - a \sim N(-a, 1+b^2)$, we have

$$E\Phi(a+bX) = P(Z \le 0) = P(\frac{Z+a}{\sqrt{1+b^2}} \le \frac{a}{\sqrt{1+b^2}}) = \Phi\left(\frac{a}{\sqrt{1+b^2}}\right),$$

which implies (a) holds.

Noting that

$$x^2 + (a + bx)^2 = (b^2 + 1)\left(x + \frac{ab}{b^2+1}\right)^2 + \frac{a^2}{b^2+1},$$

we have

$$\begin{aligned}
&E\phi(a + bX) \\
&= \int_{-\infty}^{+\infty} \frac{1}{\sqrt{2\pi}}\, e^{-x^2/2} \frac{1}{\sqrt{2\pi}}\, e^{-(a+bx)^2/2} dx \\
&= \int_{-\infty}^{+\infty} \frac{1}{2\pi}\, e^{-[x^2+(a+bx)^2]/2} dx \\
&= \int_{-\infty}^{+\infty} \frac{1}{2\pi}\, e^{-\frac{(x+ab/(b^2+1))^2}{2/(b^2+1)}}\, e^{-a^2/(2(b^2+1))} dx \\
&= \frac{1}{\sqrt{2\pi}}\sqrt{\frac{1}{b^2+1}}\, e^{-a^2/(2(b^2+1))} \int_{-\infty}^{+\infty} \frac{1}{\sqrt{2\pi}\sqrt{\frac{1}{b^2+1}}}\, e^{-\frac{(x+ab/(b^2+1))^2}{2/(b^2+1)}} dx \\
&= \frac{1}{\sqrt{2\pi}}\sqrt{\frac{1}{b^2+1}}\, e^{-a^2/(2(b^2+1))} \\
&= \frac{1}{\sqrt{b^2+1}}\, \phi\left(\frac{a}{\sqrt{b^2+1}}\right),
\end{aligned}$$

which leads to (b).

Lemma A.2: For all $x \in \mathbb{R}^1$, the following inequality holds:

$$\phi(x) + x\Phi(x) \geq 0$$

Proof: Let $L(x) = \phi(x) + x\Phi(x)$: then $L^{'}(x) = \Phi(x)$, $L^{''}(x) = \phi(x) \geq 0$, which asserts that $L(x)$ is a convex function. It attains global minimum at $x = -\infty$, so we have $L(x) \geq L(-\infty) = 0$. This completes the proof.

The proof of Corollary 3.2: From (3.1.4), we have

$$\begin{aligned}
E\Theta_1 = &\iint\limits_{\bar{x}-\frac{1}{\sqrt{m}}e_1<\bar{y}-\frac{1}{\sqrt{n}}e_2} (\bar{x} - \frac{1}{\sqrt{m}}e_1)\phi(e_1)\phi(e_2)\, de_1 de_2 \\
&+ \iint\limits_{\bar{x}-\frac{1}{\sqrt{m}}e_1\geq\bar{y}-\frac{1}{\sqrt{n}}e_2} \frac{1}{2}(\bar{x} + \bar{y} - \frac{1}{\sqrt{m}}e_1 - \frac{1}{\sqrt{n}}e_2)\phi(e_1)\phi(e_2)\, de_1 de_2 \ \ldots
\end{aligned}$$

$$= \bar{x} - \frac{\bar{x}}{2} \iint\limits_{\bar{x}-\frac{1}{\sqrt{m}}e_1 \geq \bar{y}-\frac{1}{\sqrt{n}}e_2} \phi(e_1)\phi(e_2)\, de_1 de_2$$

$$+ \frac{1}{2\sqrt{m}} \iint\limits_{\bar{x}-\frac{1}{\sqrt{m}}e_1 \geq \bar{y}-\frac{1}{\sqrt{n}}e_2} e_1\phi(e_1)\phi(e_2)\, de_1 de_2$$

$$+ \frac{\bar{y}}{2} \iint\limits_{\bar{x}-\frac{1}{\sqrt{m}}e_1 \geq \bar{y}-\frac{1}{\sqrt{n}}e_2} \phi(e_1)\phi(e_2)\, de_1 de_2$$

$$- \frac{1}{2\sqrt{n}} \iint\limits_{\bar{x}-\frac{1}{\sqrt{m}}e_1 \geq \bar{y}-\frac{1}{\sqrt{n}}e_2} e_2\phi(e_1)\phi(e_2)\, de_1 de_2. \quad (4)$$

From Lemma A.1 (a), we have

$$\iint\limits_{\bar{x}-\frac{1}{\sqrt{m}}e_1 \geq \bar{y}-\frac{1}{\sqrt{n}}e_2} \phi(e_1)\phi(e_2)\, de_1 de_2$$

$$= \int_{-\infty}^{+\infty} \int_{-\infty}^{\sqrt{m}(\bar{x}-\bar{y}+\frac{1}{\sqrt{n}}e_2)} \phi(e_1)\phi(e_2)\, de_1 de_2$$

$$= \int_{-\infty}^{+\infty} \Phi(\sqrt{m}(\bar{x}-\bar{y}+\frac{1}{\sqrt{n}}e_2))\phi(e_2)\, de_2$$

$$= \Phi\left(\frac{\sqrt{m}(\bar{x}-\bar{y})}{\sqrt{1+\frac{m}{n}}}\right) = \Phi\left(\frac{\bar{x}-\bar{y}}{\sqrt{\frac{1}{m}+\frac{1}{n}}}\right),$$

and from Lemma A.1 (b), we have

$$\iint\limits_{\bar{x}-\frac{1}{\sqrt{m}}e_1 \geq \bar{y}-\frac{1}{\sqrt{n}}e_2} e_1\phi(e_1)\phi(e_2)\, de_1 de_2$$

$$= \int_{-\infty}^{+\infty} \int_{-\infty}^{\sqrt{m}(\bar{x}-\bar{y}+\frac{1}{\sqrt{n}}e_2)} e_1\phi(e_1)\phi(e_2)\, de_1 de_2$$

$$= -\int_{-\infty}^{+\infty} \phi(\sqrt{m}(\bar{x}-\bar{y}+\frac{1}{\sqrt{n}}e_2))\phi(e_2)\, de_2$$

$$= -\phi\left(\frac{\sqrt{m}(\bar{x}-\bar{y})}{\sqrt{1+\frac{m}{n}}}\right) \Big/ \sqrt{1+\frac{m}{n}} = -\frac{\sqrt{n}}{\sqrt{m+n}}\,\phi\left(\frac{\bar{x}-\bar{y}}{\sqrt{\frac{1}{m}+\frac{1}{n}}}\right)$$

since $\int_{-\infty}^{a} e_1\phi(e_1)\, de_1 = -\int_{-\infty}^{a} d\phi(e_1) = -\phi(a)$.

Similarly, we have

$$\iint\limits_{\bar{x}-\frac{1}{\sqrt{m}}e_1 \geq \bar{y}-\frac{1}{\sqrt{n}}e_2} e_2\phi(e_1)\phi(e_2)\, de_1 de_2 = \frac{\sqrt{m}}{\sqrt{m+n}}\,\phi\left(\frac{\bar{x}-\bar{y}}{\sqrt{\frac{1}{m}+\frac{1}{n}}}\right).$$

Hence,

$$\begin{aligned} E\Theta_1 &= \bar{x} - \frac{\bar{x}}{2}\,\Phi\left(\frac{\bar{x}-\bar{y}}{\sqrt{\frac{1}{m}+\frac{1}{n}}}\right) + \frac{\bar{y}}{2}\,\Phi\left(\frac{\bar{x}-\bar{y}}{\sqrt{\frac{1}{m}+\frac{1}{n}}}\right) \\ &\qquad - \frac{1}{2\sqrt{m}}\frac{\sqrt{n}}{\sqrt{m+n}}\,\phi\left(\frac{\bar{x}-\bar{y}}{\sqrt{\frac{1}{m}+\frac{1}{n}}}\right) - \frac{1}{2\sqrt{n}}\frac{\sqrt{m}}{\sqrt{m+n}}\,\phi\left(\frac{\bar{x}-\bar{y}}{\sqrt{\frac{1}{m}+\frac{1}{n}}}\right) \\ &= \frac{(\bar{x}+\bar{y})}{2} - \frac{(\bar{y}-\bar{x})}{2}\,\Phi\left(\frac{\bar{y}-\bar{x}}{\sqrt{\frac{1}{m}+\frac{1}{n}}}\right) - \frac{1}{2}\sqrt{\frac{1}{m}+\frac{1}{n}}\,\phi\left(\frac{\bar{x}-\bar{y}}{\sqrt{\frac{1}{m}+\frac{1}{n}}}\right) \end{aligned}$$

and similarly

$$E\Theta_2 = \frac{(\bar{x}+\bar{y})}{2} + \frac{(\bar{y}-\bar{x})}{2}\,\Phi\left(\frac{\bar{y}-\bar{x}}{\sqrt{\frac{1}{m}+\frac{1}{n}}}\right) + \frac{1}{2}\sqrt{\frac{1}{m}+\frac{1}{n}}\,\phi\left(\frac{\bar{x}-\bar{y}}{\sqrt{\frac{1}{m}+\frac{1}{n}}}\right).$$

Moreover, by Lemma A.2, we have $E\Theta_1 \leq (\bar{x}+\bar{y})/2 \leq E\Theta_2$.
The proofs of Theorem 3.2 and Theorem 3.3 are similar to the proofs of Theorem 3.1 in Shi, Gao and Zhang (2001) and Theorem 3.1 in Shi (1998).

References

1. Barlow, E. E., Bartholomew, D. J., Bremner, J. M. & Brunk, H. D. (1972). *Statistical inference under order restrictions.* London: John Wiley.
2. Barnard, G. A. (1995). Pivotal models and the fiducial argument. *Inter. Statist. Rev.* **63**, 309–323.
3. Dawid, A. P. & Stone, M. (1982). The functional model basis of fiducial inference. *Ann Statist.* **10**, 1054–1067.
4. Fisher, R. A. (1930a). Inverse probability. *Proc. Cambridge Philos. Soc.* **26**, 528–535.
5. Fisher, R. A. (1930b). The fiducial argument in statistical inference. *Ann. Eugenics.* **6**, 391–398.
6. Fisher, R. A. (1950). *Contributions to mathematical statistics.* New York: John Wiley.
7. Fisher, R. A. (1973). *Statistical methods and scientific inference.* New York: Hafner Press.
8. Fraser, D. A. S. (1961a). On fiducial inference. *Ann. Math. Statist.* **32**, 661–671.

9. Fraser, D. A. S. (1961b). The fiducial method and invariance. *Biometrika.* **37**, 643–656.
10. Fraser, D. A. S. (1968). *The structure of inference.* New York: Wiley.
11. Horot, R., Pardalos, P. M. & Thoat, N. V. (1995). *Introduction to global optimization.* London: Kluwer Academic Publishers.
12. Pedersen, J. G. (1978). Fiducial inference. *Inter. Statist. Rev.* **46**, 147–170.
13. Robertson, T., Wright, F. T. & Dykstra, R. L. (1988). *Order Restricted Statistical Inference.* New York: John Wiley.
14. Shi, N. Z. (1994). Maximum likelihood estimation of means and variances from normal populations under simultaneous order restrictions. *J. Mult. Anal.* **50**, 282–293.
15. Shi, N. Z., Gao, W. & Zhang, B. X. (2001). One-sided estimating and testing problems for location models from grouped samples. *Comm. Statist. — Simul. Comput.* **30**, 885–898.
16. Shi, N. Z. & Jiang, H. (1998). Maximum likelihood estimation of isotonic normal means with unknown variances. *J. Mult. Anal.* **64**, 183–195.
17. Stoer, J. & Witzgall, C. (1970). Convexity and optimization in finite dimension I, Berlin: Springer.
18. Wright, F. T. (1978). Estimating strictly increasing regression functions. *J. Amer. Statist. Assoc.* **73**, 636–639.
19. Xu, X. Z. & Li, G. Y. (2003). The fiducial inference in pivotal family of distributions. *Technical Report, Department of Mathematics, Institute of Technology Beijing.*
20. Zabell, S. L. (1992). R. A. Fisher and fiducial argument. *Statist Sci.* **7**, 369–387.
21. Zhang, B. X., Gao, W. & Shi, N. Z. (2003). One-sided estimating and testing problems for scale models from grouped samples. *Comm. Statist. — Theory and Methods.* **32**.

PART D

Probabilistic Models in Economics and Finance

When Large Claims are Extremes

Roger Gay
Department of Accounting and Finance, Monash University, Clayton VIC 3800, Australia
E-mail: roger.gay@BusEco.monash.edu.au

New results for tail-ratios of distributions are derived and applied in an insurance context. They are used to demonstrate independence of consecutive ratios of the largest Pareto claims, and that the minimum-variance unbiased maximum likelihood estimator (MVUMLE) for the Pareto tail-index is equivalent to Hill's estimator. For sufficiently large Pareto claims, the scale factor can be ignored in tail-index estimation. An analogue of this idea is available for all distributions with a regularly varying tail, when the pool of claims is sufficiently large; in this case too, Hill's estimator is pivotal, i.e. is MVUMLE. By graphically comparing Hill's estimator in the Pareto case with Hill's estimator for a distribution with regularly varying tail, it is possible for any given distribution function to determine when claims are large enough to be classified as 'extremes'. Ratios of such extremes form the efficient set for tail-index estimation. Simulation and actual examples are provided.

Keywords: regularly varying tails; tail-index estimation; ratios of extremes; Hill estimator.

1. Introduction and motivation

In *The economics of liability losses — insuring a moving target*, Enz and Holzheu (2004), economists with the world's largest reinsurer, Swiss Re, expressed concern about the rate of growth of insured losses in commercial liability insurance in the 10 largest Western economies where rate of claims growth is increasing at 1.5 to 2 times as fast as GDP. In 2002 claims totalled about 84 billion USD of which 67 billion USD (0.64% of US GDP) arose in the US. The authors were particularly critical of inefficiencies in the US Tort system (under which less than 50% of claims awarded found its way back to the insured). With apparent increase in severity of large commercial liability claims, insurers were urged to monitor their premium determination processes, to ensure that these systems remain in place and affordable. For these classes of insurance involving heavy-tailed claims with infinite variance (see for instance Mikosch, 1997, for justification), a major part of an insurer's total claims cost is likely to arise from a few large claims. It makes sense to focus attention on these sets of largest claims.

In this paper a general theorem about the tail ratios of absolutely continuous distribution functions is established. It is applied to heavy-tailed distributions within the domain of attraction of the Frechet distribution (i.e. distributions with a regularly varying tail). For these distributions, the largest observations from large samples can be assumed to have Pareto or Frechet marginal distributions for values k up to some maximum which can be determined approximately using a version of a Hill's plot. Ratios of consecutive extremes from this qualifying set have distributions like powers of beta random variables. The maximum likelihood estimator of the tail-index based on this set is minimum variance unbiased and achieves the Cramer-Rao minimum variance bound.

2. Heavy-Tailed Distributions and General Insurance Claims

A classic insurance model (see for instance Bowers *et al.* (1986)) describes annual aggregate claims S_N which comprise non-negative claims X_i assumed independently identically distributed, and independent of N, the number of claims, i.e.

$$S_N = X_1 + X_2 + \cdots + X_N .$$

To model large claims the X_i are supposed to arise from a distribution with a regularly varying tail, now explicitly defined.

2.1. *Distributions with regularly varying tail of exponent δ*

Definition 2.1. The class **F** consists of distribution functions $F(.)$ for which

$$1 - F(x) = L(x)x^{-\delta} = L(x)x^{-1/\rho}, \quad (x > 0, \delta > 0)$$

where $L(x)$ is slowly varying; i.e. $(L(ax)/L(x)) \to 1$ as $x \to \infty$, for all $a > 0$; see for instance, Feller (1971, p.278). The parameter δ is the tail-index, while $\rho = \frac{1}{\delta}$ is the extreme value index.

For the sort of large claims realised in commercial liability insurance, commonly assumed is that $E[X]$ exists, but not $E[X^2]$. This means that $1 > \rho \geq \frac{1}{2}$ is the special interest class, and that S_N is not governed by conventional central limit theorems; see e.g. Mikosch (1997). For ρ increasingly closer to 1, the major expected cost of annual aggregate claims is attributable to a smaller proportion of the largest of them. The actuarial rule of thumb — the "20–80" rule; 20% of claims account for 80% of claims cost (e.g. Embrechts *et al.* (1997, Chapter 8)) — implies that ρ is about 0.86 for claims arising from Pareto distributions .

Theoretical support for the empirical rule is available from the following:

$$\text{For } F(.) \in \mathbf{F} \text{ and large } x, \ \Pr[S_N > x] \approx \Pr[\max X_i > x]$$

$$\approx E[N]L(x)x^{-\delta};$$

see Feller (1971, p.279 and p.288, Exercise 31).

Standard methods of estimation of tail-index ρ (e.g. Smith (1987), Embrechts *et al.* (1997, Chapter 6)) assume that the total number of claims n is large, as is k, with $n \to \infty$, $k \to \infty$ but $\frac{k}{n} \to 0$. However, the relation between the two sequences needed for convergence in law of tail-index estimators may be quite prescriptive (e.g. Drees *et al.* 2004). In this paper k may be quite small, 'qualifying' values of k being determined by results of Theorem 3.1. The Hill (1975) estimator based on these k values is the pivotal quantity for tail-index estimation.

2.2. *Two distributions important for heavy-tailed claims*

Two distribution families especially relevant for large claims modelling are:

Pareto, with distribution function

$$1 - F(x) = (1 + \frac{x}{\lambda})^{-\frac{1}{\rho}}, \qquad (x > 0, \rho > 0, \lambda > 0), \tag{1}$$

and

Frechet, with density

$$f_k(x) = \frac{\rho^{-1} x^{-(k/\rho)-1} e^{-x^{-1/\rho}}}{\Gamma(k)}, \qquad (x > 0, \rho > 0, k \geq 1). \tag{2}$$

Both distributions are themselves inclusive of broader classes of families under different (but heuristically similar) statistical conditions.

Pareto is relevant because of the peaks-over-thresholds methodology initiated by Balkema and De Haan (1974), Pickands (1975), Smith (1987) and Leadbetter (1991). Most relevant (for $\rho > 0$) is the following:

For sufficiently large t the conditional distribution

$$F_t(x) = \Pr[X \leq t + x \mid X > t] = \frac{F(t+x) - F(t)}{1 - F(t)}$$

where $1 - F(t) > 0$, $t < x^* = \sup(x : F(x) < 1) \leq \infty$, is well-approximated by Pareto (1) for suitably chosen $\lambda = \lambda(t)$. More precisely, the approximation arises from convergence to the generalised Pareto distribution (GPD) shown by Gnedenko (1943) who proved the existence of normalising $\sigma(t)$ such that

$$\lim_{t \to x^*} F_t(x\sigma(t)) \to H_\rho(x)$$

where $1 - H_\rho(x) = (1 + \rho x)^{1/\rho}$ if and only if $F(.)$ is in class **F**. Methods of estimating t and $\sigma(t)$ (equivalently t and $\lambda(t)$) are discussed in Embrechts *et al.* (1997, Chapter 6).

The Frechet distribution is more directly inclusive of a broader class of distributions, because the normalised order statistic X_{n-k+1}/v_n from any $F(.)$ in **F** converges weakly to the k-th extreme $X^*_{(k)}$ with density $f_k(x)$. The normalising constants v_n derive from the *tail-quantile function* satisfying for large n the equation $n[1 - F(v_n)] = 1$; class **F** is the *maximum domain of attraction* of the Frechet distribution (see Embrechts *et al.* (1997, p.131), Teugels and Vanroelen (2004)).

2.3. *Implications for large claims of membership of* F*: the* $k^{-\rho}$ *law*

Membership of class **F** means that *the largest claims* arising from any of its distributions can roughly speaking be equivalently described by the Pareto distribution (because for exceedances over a sufficiently high threshold t, the conditional distributions $F_t(x)$ are approximately Pareto for any $F(.)$ in **F**), or by the Frechet distribution (because for sufficiently large n, suitably normalised order statistics from any distribution $F(.)$ in **F** converge weakly to Frechet extremes).

For both distributions, expected relative claim size is governed by a $k^{-\rho}$ *law:*

$$E\left[\frac{X^*_{(1+k)}}{X^*_{(1)}}\right] = \frac{E[X^*_{(1+k)}]}{E[X^*_{(1)}]} \cong \Gamma(1+\rho)(k+1)^{-\rho}.$$

This result is established in the sequel. It is for this reason that precise determination of tail-index ρ is of critical concern to insurers.

2.4. *Extension of GPD*

Bierlant, Joossens and Segers (2004) proposed an extension of the GPD devised to approximate the conditional distribution of $X - t$ given $X > t$

for much lower thresholds t than the GPD is capable of. Proof of their result involves a refinement of Pickands' original (1975) workings. The new model is:

$$1 - F_t(x : \delta, \lambda, \gamma) = \left(1 + \frac{x}{\lambda} + \nu\left(1 - (xt^{-1} + 1)\right)^{1+\gamma}\right)^{-\delta} \tag{3}$$

where $\nu = \lambda t - 1$ of which Pareto (1) is a special case when $\gamma = -1$.

The extended model has the capacity to fit a much larger proportion of claims in large data sets. For the Society of Actuaries, Large Claims Database, the authors report good fits for 75,789 claims above \$25,000 compared with generalised Pareto which provided a good fit only for 7,860 claims above \$100,000. The principal deficiency of the new 3-parameter distribution is its inability to handle *the very largest claims.*

Such claims are precisely the focus of this paper (for which the GPD can be assumed to provide an adequate model, as does the Frechet distribution, with restrictions presently to be examined).

3. Main Results

The following result holds generally for any absolutely continuous distribution function $F(.)$. Denote by $b(u : j, k)$, the beta density

$$b(u : j, k) = B^{-1}(j, k)u^{j-1}(1 - u)^{k-1}, \quad u \in [0, 1], \quad (j \geq 1, k \geq 1)$$

where $B(j, k) = \Gamma(j)\Gamma(k)/\Gamma(j + k)$ and where $\Gamma(.)$ denotes the gamma function.

In the context of Theorem 3.1 below, the density $b(u_m : m, 1) = m(u_m)^{m-1}$; $u_m \in [0, 1]$ is a special case, with distribution function $B(u_m : m, 1) = (u_m)^m$, $u_m \in [0, 1]$.

Theorem 3.1. *Denote by $X_{(1)}, X_{(2)}, \ldots, X_{(n)}$ the ascending order statistics from any absolutely continuous distribution function $F(.)$. Then 'tail-ratios' (the random variables $\{U_m\}, m = 1, 2, \ldots, k$) have the following properties:*

(i)

$$U_k = \frac{1 - F(X_{(n-k+1)})}{1 - F(X_{(n-k)})}$$

$$U_{k-1} = \frac{1 - F(X_{(n-k+2)})}{1 - F(X_{(n-k+1)})}$$

$$\vdots$$

$$U_1 = \frac{1 - F(X_{(n)})}{1 - F(X_{(n-1)})}$$

are independently distributed beta random variables, with respective densities

$$b(u_k : k, 1),\ b(u_{k-1} : k-1, 1),\ \ldots,\ b(u_1 : 1, 1)$$

all being independent of $X_{(n-k)}$ *(and in particular of* $1 - F(X_{(n-k)})$*, which has density* $b(u : k+1, n-k)$*; c.f. Renyi (1953); Kendall and Stuart (1969, Exercise 14.17)).*

(ii) The ratio

$$V_k = \frac{1 - F(X_{(n)})}{1 - F(X_{(n-k)})}$$

has a beta distribution with density $b(v_k : 1, k)$ *which is independent of* $X_{(n-k)}$*.*

(iii)

$$E\left[\frac{1 - F(X_{(n-m+1)})}{1 - F(X_{(n-m)})}\right] = \frac{E\left[1 - F(X_{(n-m+1)})\right]}{E\left[1 - F(X_{(n-m)})\right]} = \frac{m}{m+1}, \qquad m = 1, 2, \ldots, n-1.$$

Proof.

The proof of Theorem 3.1(i) is quite long and is relegated to the Appendix. Theorem 3.1(ii) follows from Theorem 3.1(i) and Kendall and Stuart (1969, Exercise 11.8), because

$$V_k = U_k \times U_{k-1} \times \cdots \times U_1$$

is the product of independent beta variates. It is independent of $X_{(n-k)}$ because by Theorem 3.1(i), all its component factors are.

Theorem 3.1 (iii): Since the tail ratio U_m has density $b(u_m : m, 1) = mu_m^{m-1}$, $u_m \in [0, 1]$, $m = 1, 2, \ldots, n-1$, it follows that

$$E\left[\frac{1 - F(X_{(n-m+1)})}{1 - F(X_{(n-m)})}\right] = E[U_m] = \frac{m}{m+1} .$$

Since the density of $F(X_{(n-m+1)})$ is $b(u : n-m+1, m)$ it follows that $E[1 - F(X_{(n-m+1)})] = m/(n+1)$, $E[1 - F(X_{(n-m)})] = (m+1)/(n+1)$ so that we also have

$$\frac{E[1 - F(X_{(n-m+1)})]}{E[1 - F(X_{(n-m)})]} = \frac{m}{m+1}, \quad m = 1, 2, \ldots, n+1 . \qquad \square$$

4. Application of Main Results to Pareto Claims

Consider the largest $(k+1)$ claims $X_{(n-k)}, X_{(n-k+1)}, \ldots, X_{(n)}$ from Pareto distribution (1), and the corresponding standardised claims by $Z_{(m)} = 1 + X_{(n-m+1)}/\lambda$, $m = 1, 2, \ldots, (k+1)$, where λ is an unknown scale parameter. From Theorem 3.1 (i) the ratios

$$U_m = \left(\frac{Z_{(m)}}{Z_{(m+1)}}\right)^{-\delta}, \quad m = 1, 2, \ldots, k$$

are independently distributed and have densities $b(u_m : m, 1) = m(u_m)^{m-1}$. Their distribution functions (d.f.) are $B(u_m : m, 1) = (u_m)^m$, $0 \leq u_m \leq 1$, $m = 1, 2, \ldots, k$.

5. Estimation of the Tail-Index $\rho = 1/\delta$ in the Pareto Case

(Optimality of the Hill estimator for Pareto claims)

- Maximum likelihood estimation of the tail-index ρ

$U_m = (Z_{(m)}/Z_{(m+1)})^{-\delta}$ where U_m has d.f. $B(u_m : m, 1) = (u_m)^m$, $m = 1, 2, \ldots, k$.
$Y_m = Z_{(m)}/Z_{(m+1)}$ has density $f(y_m) = m\delta(y_m)^{-m\delta-1}$, $(y_m \geq 1)$, $m = 1, 2, \ldots, k$.

Since the $\{Y_m\}$ are independently distributed, the likelihood L based on k consecutive ratios of the largest Pareto claims is

$$L = \prod_{m=1}^{k} m\delta(ym)^{-m\delta-1}$$

and the log likelihood ℓ is given by

$$\ell = \text{const.} + k \log \delta - \delta \times \sum_{m+1}^{k} m \log y_m - \sum_{m=1}^{k} \log y_m .$$

The derivative of ℓ with respect to δ is zero when

$$\rho = \frac{1}{\delta} = k^{-1} \sum_{m=1}^{k} m \log y_m$$
$$= k^{-1} \sum_{m=1}^{k} m \log \left\{ \frac{Z_{(m)}}{Z_{(m+1)}} \right\}$$

which is the Hill estimator based on k consecutive ratios of the largest claims (hereafter denoted by $\hat{\rho}_k$).

Noting that:

(i)

$$E \left[m \log \left\{ \frac{Z_{(m)}}{Z_{(m+1)}} \right\} \right] = E[-\rho m \log(U_m)]$$

where U_m is distributed with beta density $b(u_m : m, 1) = m(u_m)^{m-1}$, and that for integer j

(ii)

$$E \left[\left(-\rho m \log(U_m)^j \right) \right] = \Gamma(j+1) l^j, \qquad j = 0, 1, 2, \ldots,$$

it follows that

(a) $E[\hat{\rho}_k] = \rho$ (i.e. $\hat{\rho}_k$ is unbiased for ρ),

(b) $\text{Var}[\hat{\rho}_k] = \rho^2/k$. It is easy to check that the variance ρ^2/k represents *the Cramer-Rao minimum variance bound for* ρ, and we now do so. Differentiating ℓ twice and taking expectations gives

$$E \left[-\frac{d^2}{d\delta^2} \ell \right] = \frac{k}{\delta^2} \,.$$

The Cramer-Rao bound for any function $g(\delta)$ of δ is

$$g'(\delta)^2 / E \left[-\frac{d^2}{d\delta^2} \ell \right]$$

(see for instance, Rao (1973, p.324)).
Here $g(\delta) = 1/\delta$, and the bound is $\delta^{-4} \times \delta^2/k = \rho^2/k$
Thus in the case of Pareto claims, Hill's estimator is the minimum variance unbiased maximum likelihood estimator.

(c) Furthermore $\hat{\rho}_k$ has moments of all orders and moment generating function

$$M_\rho(\theta) = (1 - \frac{\rho\theta}{k})^{-k}, \quad (0 < \frac{\rho\theta}{k} < 1)$$

so that small sample distribution of $\hat{\rho}_k$ is gamma with density

$$f_\rho(x,k) = \exp\left(\frac{-kx}{\rho}\right) x^{k-1} \left(\frac{k}{\rho}\right)^k / \Gamma(k), \quad x \geq 0, k \geq 1, \rho > 0.$$

Proof. Recalling the independence of the ratios $\{Z_{(m)}/Z_{(m+1)}\}$, $m = 1, 2, \ldots, k$, the moment generating function of $\hat{\rho}_k$ is

$$\begin{aligned} E[\exp(\theta\hat{\rho}_k)] &= E\left[\prod_{m=1}^{k} \left\{\frac{Z_{(m)}}{Z_{(m+1)}}\right\}^{m\theta/k}\right] \\ &= \left(1 - \frac{\rho\theta}{k}\right)^{-k}. \end{aligned}$$

This follows since $E[\{Z_{(m)}/Z_{(m+1)}\}^{m\theta/k}] = E[\{U(m,1)\}^{-\rho m\theta/k}]$ where $U(m,1)$ has density $B(u : m, 1) = mu^{m-1}$, $(0 \leq u \leq 1)$, and its expectation is $(1 - \rho\theta/k)^{-1}$. The moment generating function is recognisable as that of the gamma distribution with density

$$f_\rho(x,k) = \exp\left(\frac{-kx}{\rho}\right) x^{k-1} \left(\frac{k}{\rho}\right)^k / \Gamma(k), \quad x \geq 0, k \geq 1, \rho > 0. \square$$

6. The Pareto Scale Parameter λ

For estimation of the tail-index, is knowledge of the scale parameter λ necessary?

Heuristically, the ratio of large standardised Pareto claims

$$\begin{aligned} \frac{Z_{(m)}}{Z_{(m+1)}} &= \frac{(1 + X_{(n-m+1)}/\lambda)}{(1 + X_{(n-m)}/\lambda)} \\ &\approx \frac{X_{(x-m+1)}/\lambda}{X_{(n-m)}/\lambda} \\ &= \frac{X^*_{(m)}}{X^*_{(m+1)}}. \end{aligned}$$

The diagram below (Figure 1) shows the progress of Hill's estimator for a pool of 2000 Pareto claims (generated with $\lambda = 1500$, $\rho = 0.75$) when up to 200 ratios are used ($k = 200$). For the lower plot the value of the scale parameter is assumed known and used in calculation of the Hill estimator ; in the upper plot its value is ignored. Hill plots for the two series progressively diverge as further ratios are added to the estimator.

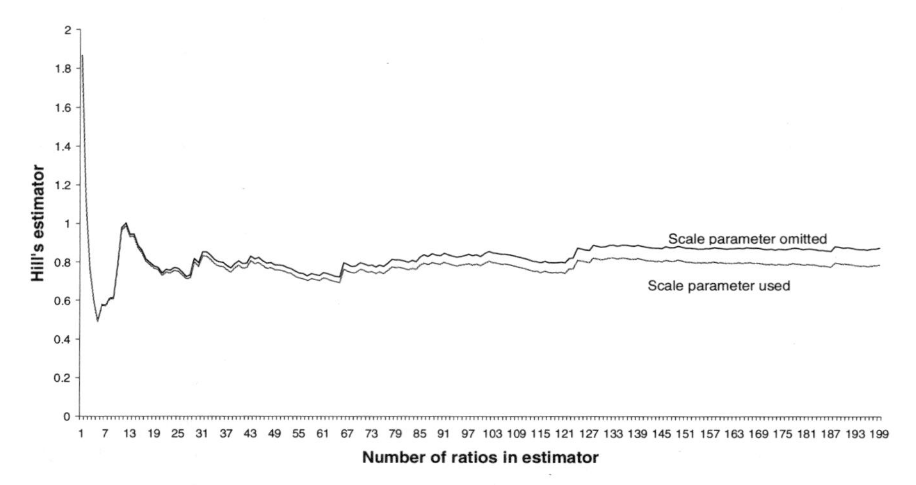

Fig. 1. Hill's plot showing progress of the estimator over 200 Pareto ratios with and without knowledge of λ. Pool is 2000 claims.

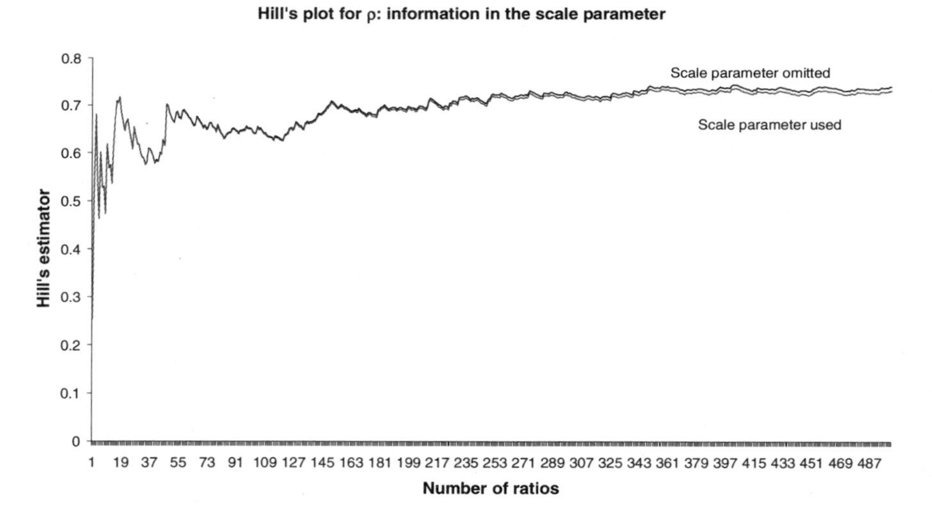

Fig. 2. Hill's plot for 500 ratios from a pool of 100,000 claims. The penalty for ignoring the scale factor is negligible for quite large k.

However, not surprisingly in view of the foregoing discussion, as the sample size is increased, the size of the largest claims increases in probability, and the difference between the plots can be ignored for quite large k values. In Figure 2 the pool of claims is 100,000.

7. Application of Main Results to Claims from Distributions with a Regularly Varying Tail

To what extent can this reasoning be applied to a general distribution with a regularly varying tail where $1 - F(x) = L(x)x^{-\delta}(x > 0, \delta > 0)$ and where $L(x)$ is slowly varying?

Consider the ratios

$$U_k = \frac{1 - F(X_{(n-k+1)})}{1 - F(X_{(n-k)})} \approx \frac{1 - F(v_n X^*_{(k)})}{1 - F(v_n X^*_{(k+1)})} = \frac{L(v_n X^*_{(k)}(X^*_{(k)})^{-\delta})}{L(v_n X^*_{(k+1)}(X^*_{(k+1)})^{-\delta})}$$

$$U_{k-1} = \frac{1 - F(X_{(n-k+2)})}{1 - F(X_{(n-k+1)})} \approx \frac{1 - F(v_n X^*_{(k-1)})}{1 - F(v_n X^*_{(k)})} = \frac{L(v_n X^*_{(k-1)}(X^*_{(k-1)})^{-\delta})}{L(v_n X^*_{(k)}(X^*_{(k)})^{-\delta})}$$

$$\vdots$$

$$U_1 = \frac{1 - F(X_{(n)})}{1 - F(X_{(n-1)})} \approx \frac{1 - F(v_n X^*_{(1)})}{1 - F(v_n X^*_{(2)})} = \frac{L(v_n X^*_{(1)}(X^*_{(1)})^{-\delta})}{L(v_n X^*_{(2)}(X^*_{(2)})^{-\delta})}$$

If the pool of claims generating the order statistics $X_{(n-k)}$, $X_{(n-k+1)}$, ..., $X_{(n)}$ (appropriately normalised) is sufficiently large to justify the assumption of their joint weak convergence to Frechet marginal distributions, then the variables U_k, U_{k-1}, ..., U_1 provide the distributions of ratios of extremes

$$\left\{\frac{X^*_{(k)}}{X^*_{(k+1)}}\right\}^{-\delta}, \left\{\frac{X^*_{(k-1)}}{X^*_{(k)}}\right\}^{-\delta}, \ldots, \left\{\frac{X^*_{(1)}}{X^*_{(2)}}\right\}^{-\delta}$$

respectively.

This means that the ratios of the slowly varying functions

$$\frac{L(v_n X^*_{(m)})}{L(v_n X^*_{(m+1)})}$$

can be assumed to be 1 in each of the expressions for $\{U_m\}$, $m = 1, 2, \ldots, k$, above.

7.1. *Discussion: Heuristics*

The key question to be answered is the following:

On the basis of observed claims values, for what values of m (if any) can the ratio

$$\frac{L(v_n X^*_{(m)})}{L(v_n X^*_{(m+1)})}$$

be taken as 1? And if it is true for some values, what is the largest value k which the data supports?

Heuristically for low values of m (involving the largest extremes) the ratio is based on claims realisations well out in the tail of the distribution. As m increases, the ratio involves extremes which are retreating back into the body of the distribution. Based on observed values, just how far can one retreat without compromising optimal estimation properties?

7.2. *Deciding on largest k (how many large claims are 'extremes'?)*

Example 7.1. Figure 3 shows the Hill's plot for 100,000 claims and the corresponding $k = 1000$ ratios generated from the distribution function

$$1 - F(x) = \frac{\lambda}{x^{2h} - 2bx^h}, \quad \left(2h = \frac{8}{3}, \ \rho = 0.75\right), \ x \geq a^* = b + (b^2 + \lambda)^{\frac{1}{2}}.$$

The plot classically exemplifies the dilemma faced by insurers and statisticians attempting to estimate the tail-index of distributions with a regularly varying tail.

- If a small number of the largest claims is used (e.g. about 30), the plot exhibits high volatility.
- If a large number is used (more than 100), there is material bias.

The only practical way round the dilemma (since the functional form of $F(x)$ is an imponderable) is to increase sample size, i.e. the pool of claims. For achievement of this end it behoves insurers to:

(1) pool experience across national boundaries,
(2) aggregate year-on-year claims in current dollar terms.

Increasing in total the total pool of claims reduces the bias to manageable proportions (Figure 4). Roughly speaking, this means that a test of hypothesis along the lines that the ratio of slowly varying functions involving any of these 500 order statistics is unity would be accepted. Moreover, with a large pool of claims, initial volatility of the series is reduced; a relatively small number of ratios may lead to a reasonable tail-index estimate (e.g. Figure 5).

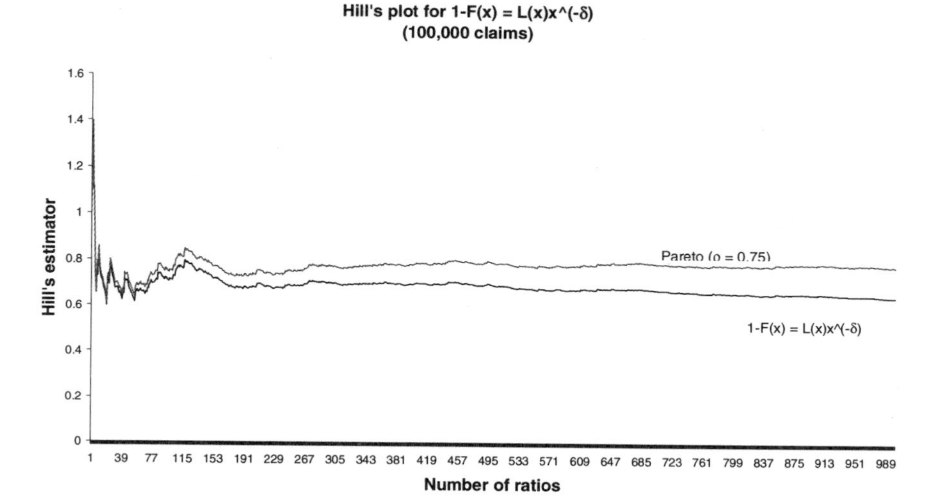

Fig. 3. Hill's plot for distribution above: $\lambda = 1\,500$, $b = 100$ showing divergence from Pareto as number of ratios used increases.

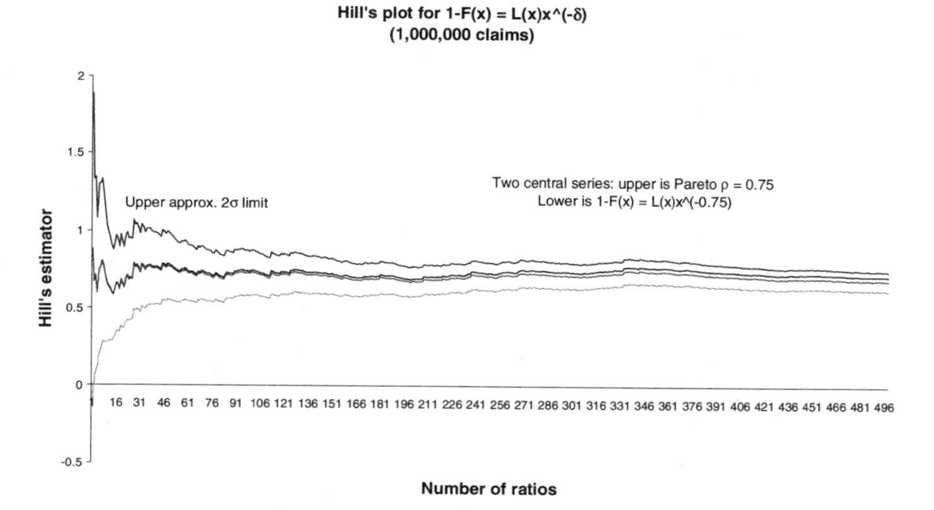

Fig. 4. Hill's plot for increased pool claim size, showing approximate 2σ confidence bands. The bias, while material, is nowhere near significant.

Example 7.2. (Largest man-made disasters)

In this example, the size n of the total pool of claims, while unknown,

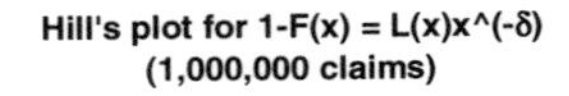

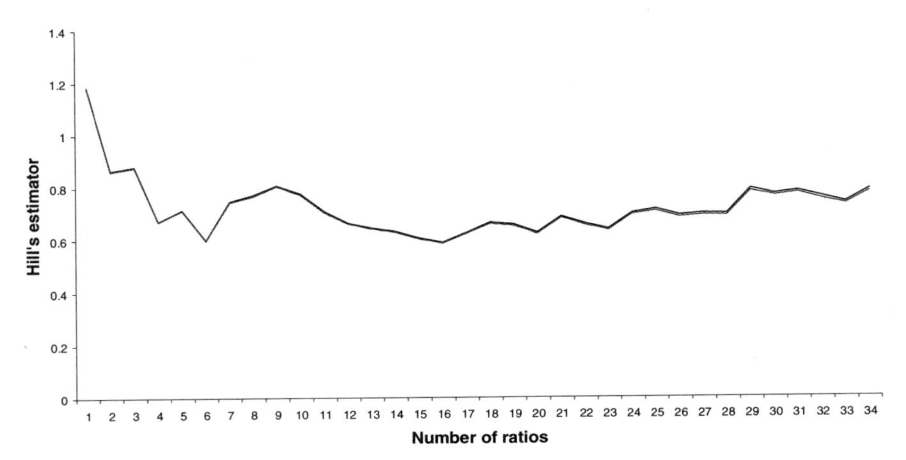

Fig. 5. Hill's plot for the 35 largest claims; the two series are indistinguishable, and volatility is low.

can be assumed to be very large, large enough to justify the assumption that the claims in Table 1 are 'extremes'.

Table 1. Large claims data: Insured losses, (1999) Man-made disasters.

Place	Event	Insured Loss USD m.
US, Dearborn, MI	Explosion & fire at power station	650
US, Gramercy, LA	Explosion, Aluminium plant	275
US, Richmond CA	Oil refinery explosion	247
US, Kansas City	Power plant explosion	196
UK, Edinburgh	Explosion at transformer factory	137
Germany, Wuppertal-Eberfield	Chemicals plant explosion	102.5
Germany, Gendorf	Polymer plant explosion	92.2
Germany, Vahdorf	Turkey slaughter-house fire	82
Germany, Darmstadt-Arheilgen	Fire liquid crystal plant	71.7
US, Martinez, CA	Oil refinery explosion	71

Source: Swiss Re, Sigma No. 2 (2000)

The relative size of claims is shown in Figure 6 below.

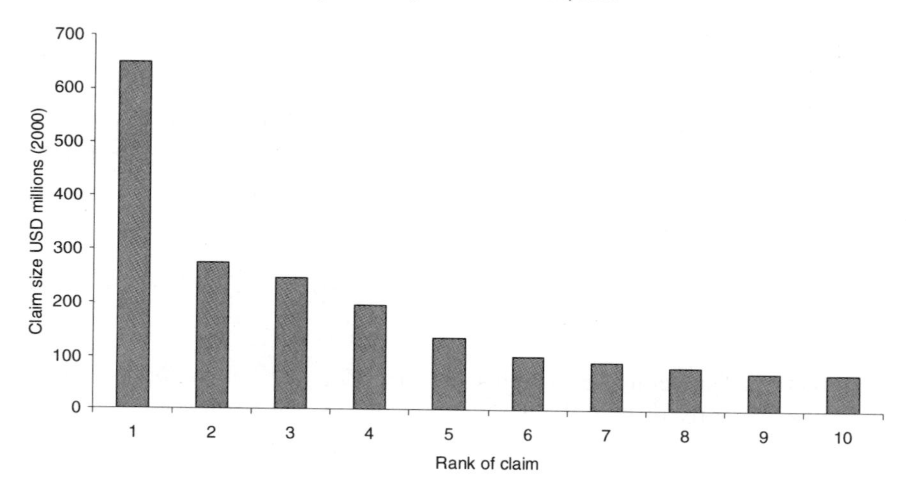

Fig. 6. Ten largest man-made disaster insurance claims of 1999.

7.3. *Estimates of ρ from the empirical structure function*

The pattern of the largest claims is very suggestive. We use it to provide another estimator of the tail-index to compare with the Hill estimator. Estimates of the tail-index ρ can be 'backed out' of the expected structure of the extremes, by equating the expected structure with observed structure for the large claims of Table 1.

Assuming weak convergence of the largest claims to Frechet extremes, and since V_m has density $b(v_m : 1, m)$ from Theorem 3.1(ii):

$$E\left[\frac{X^*_{(m+1)}}{X^*_{(1)}}\right] = E\left[V_m^\rho\right].$$

That the expected value of the ratio of the weak limits is the same as the expected value of that limit follows from Chung (1974, Exercise 8, p.100 and Theorem 4.5.4 p.97). Thus

$$\begin{aligned} E\left[\frac{X^*_{(m+1)}}{X^*_{(1)}}\right] &= mB(1+\rho, m) \\ &= (1+\rho)^{-1}\left(1+\frac{\rho}{2}\right)^{-1}\cdots\left(1+\frac{\rho}{m}\right)^{-1} \\ &\approx \Gamma(1+\rho)(m+1)^{-\rho} \end{aligned}$$

using Feller (1968, exercise 12.24), the approximation being excellent for m greater than 5, and $\rho \in [0.5, 1]$ with exact equality at $\rho = 1$.

This is the power law of expected relative claim size for distributions with a regularly varying tail of index δ.

Furthermore, since $E[X^*_{(1)}/X^*_{(m+1)}] = E[X^*_{(1)}]/E[X^*_{(m+1)}]$ is implied by Theorem 3.1(i) and Theorem 3.1(ii), the ratio $X^*_{(1)}/X^*_{(m+1)}$ *is a consistent as well as an unbiased estimator* of $mB(1-\rho, m)$ for all $m = 1, 2, \ldots, k$ (it is easy to check that

$$E\left[\frac{X^*_{(1)}}{X^*(m+1)}\right] = \frac{E\left[X^*_{(1)}\right]}{E\left[X^*_{(m+1)}\right]}$$

by direct computation from Frechet density (2)).

Thus estimates of ρ can be derived by equating the *estimating functions* $\left\{G_m(\rho, X^*_{(1)}/X^*_{(m+1)})\right\}, m = 1, 2, \ldots, k$ to *empirical structure functions*

$$\frac{X^*_{(1)}}{X^*_{(2)}}, \quad \frac{X^*_{(1)}}{X^*_{(3)}}, \quad \ldots, \quad \frac{X^*_{(1)}}{X^*_{(1+k)}}.$$

Here $k = 9$ and the estimating equations are:

$$G_1\left(\rho, \frac{X^*_{(1)}}{X^*_{(2)}}\right) = (1-\rho)^{-1} - \frac{650}{275}$$

$$G_2\left(\rho, \frac{X^*_{(1)}}{X^*_{(3)}}\right) = (1-\rho)^{-1}\left(1 - \frac{\rho}{2}\right)^{-1} - \frac{650}{247}$$

et c., and the quasi-likelihood estimation theory of Heyde (1997) is invoked. In this context the methodology is akin to the L-moment estimations procedure of Pandey *et al.* (2001). The full set of estimates deriving from the 9 observed ratios is provided in Table 2.

Table 2. Estimates of the tail-index deriving from the empirical structure function. (The largest insured losses of 1999; man-made disaster)

	$j=1$	2	3	4	5	6	7	8	9
$\hat{\rho}$	0.5769	0.495	0.5115	0.5735	0.6144	0.6139	0.6195	0.6306	0.6178

Plots of these quasi-likelihood estimates are compared with the corresponding Hill estimates in Figure 7.

Are estimates derived from the empirical structure function likely to unlock the true value of ρ?

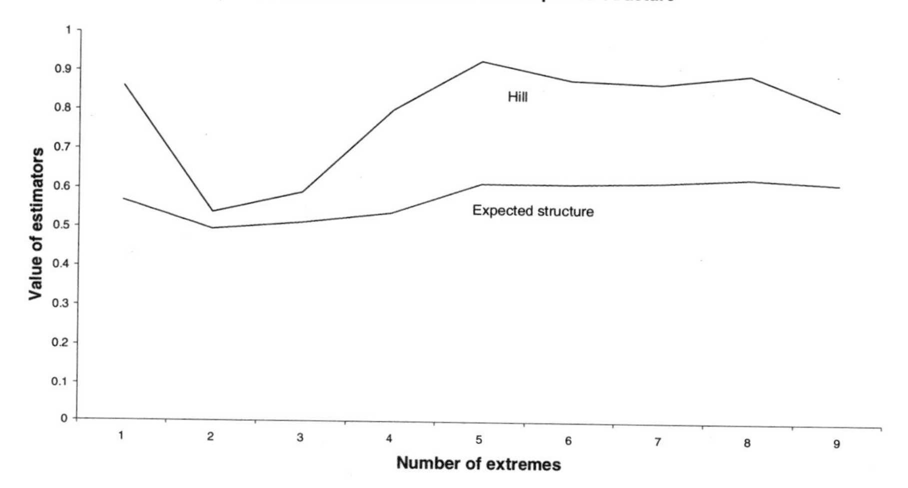

Fig. 7. Comparison of Hill's estimator and tail-index estimates deriving from expected structure of extremes (the latter likely to underestimate ρ).

Given that for $F \in \mathbf{F}$, Theorem 3.1(ii) asserts that the distribution of $\left\{X^*_{(1)}/X^*_{(m+1)}\right\}$ for a sufficiently large pool is that of $(V_m)^{-\rho}$ where V_m has density $b(v_m; 1, m)$ for each $m = 1, 2, \ldots, 9$, it is possible to examine the probability that this ratio exceeds its expected value.

Thus

$$\begin{aligned}
&\Pr\left[\frac{X^*_{(1)}}{X^*_{(m+1)}} > \{mB(1-\rho, m)\}\right] \\
&= \Pr\left[\frac{X^*_{(1)}}{X^*_{(m+1)}} > (1-\rho)^{-1}\left(1-\frac{\rho}{2}\right)^{-1}\cdots\left(1-\frac{\rho}{m}\right)^{-1}\right] \\
&= 1 - \left\{1 - \{mB(1-\rho, m)\}^{-1/\rho}\right\}^m
\end{aligned}$$

for $m = 1, 2, \ldots, 9$.

For instance:

(i) if $\rho = 0.5$, $E[X^*_{(1)}/X^*_{(2)}] = (1-0.5)^{-1} = 2$, but $\Pr[E[X^*_{(1)}/X^*_{(2)}] > 2]$ is only 0.25;

(ii) if $\rho = 0.9$, $E[X^*_{(1)}/X^*_{(2)}] = (1-0.9)^{-1} = 10$, but $Pr[E[X^*_{(1)}/X^*_{(2)}] > 10]$ is only 0.0774.

Table 3 shows some values of these probabilities for $\rho = 0.5, 0.6, 0.7, 0.8$ and 0.9, and $m = 1, 2, \ldots, 9$.

Table 3. The probability that the observed value of $X^*_{(1)}/X^*_{(1+m)}$ exceeds its expected value $\{mB(1-\rho, m)\}$, $m = 1, 2, \ldots, 9$.

	p-values				
m-values	0.5	0.6	0.7	0.8	0.9
1	0.2500	0.2172	0.1791	0.1337	0.0774
2	0.2615	0.2253	0.1842	0.1363	0.0781
3	0.2653	0.2279	0.1858	0.1370	0.0783
4	0.2672	0.2292	0.1865	0.1374	0.0784
5	0.2683	0.2300	0.1870	0.1375	0.0784
6	0.2690	0.2305	0.1872	0.1377	0.0784
7	0.2695	0.2308	0.1874	0.1378	0.0785
8	0.2699	0.2311	0.1876	0.1378	0.0785
9	0.2702	0.2313	0.1877	0.1379	0.0785

These low probabilities stem in part from the fact that if ρ is anywhere near $1, X^*_{(1)}$ is a rare event, unlikely to exceed its expected value $\Gamma(1-\rho)$. However they also imply that *estimates of ρ backed out from the empirical structure function are likely to underestimate its true value.* From an insurer's perspective the news is not good: 'the big claims are still out there'.

8. Summary and Conclusions

New results for tail-ratios of distributions are applied to the Pareto distribution and other distributions within the domain of attraction of the Frechet (extreme value) distribution.

For Pareto claims, the maximum likelihood estimator is equivalent to Hill's estimator, is minimum variance achieving the Cramer-Rao minimum variance bound and has a gamma distribution.

For claims from distributions with regularly varying tails, the same results hold true if the sample sizes available are large enough.

Distributional results for ratios of jointly distributed extremes show that the empirical structure of claims is likely to lead to underestimation of the tail-index.

For sufficiently large pools of claims it is possible to monitor the tail-index using inference based on a relatively small number of the largest claims.

To avail themselves of the methodology insurers require access to large pools of heavy-tailed claims, probably necessitating aggregation over time and across national boundaries. By this route, adherence to the exhortations of Enz and Holzheu (2004) as described in the introduction, would seem to be within reach of general insurers.

Appendix

Proof of Theorem 3.1 (i)

Theorem 3.1. (Transformation of the k largest order statistics to independently distributed random variables)

The joint distribution of the k largest order statistics $X_{(n-k+1)}$, $X_{(n-k+2)}, \ldots, X_{(n-1)}, X_{(n)}$ from an absolutely continuous distribution $F(.)$ factorises into the product of k independent random variables under the following transformations.

$$u_1(x_{(n-k+1)}, x_{(n-k+2)}, \ldots, x_{(n-1)}, x_{(n)}) = \frac{1 - F(x_{(n)})}{1 - F(x_{(n-1)})}$$

$$u_2(x_{(n-k+1)}, x_{(n-k+2)}, \ldots, x_{(n-1)}, x_{(n)}) = \frac{1 - F(x_{(n-1)})}{1 - F(x_{(n-2)})}$$

$$\vdots$$

$$u_j(x_{(n-k+1)}, x_{(n-k+2)}, \ldots, x_{(n-1)}, x_{(n)}) = \frac{1 - F(x_{(n-j+1)})}{1 - F(x_{(n-j)})}$$

$$\vdots$$

$$u_{k-1}(x_{(n-k+1)}, x_{(n-k+2)}, \ldots, x_{(n-1)}, x_{(n)}) = \frac{1 - F(x_{(n-k+2)})}{1 - F(x_{(n-k+1)})}$$

$$u_k(x_{(n-k+1)}, x_{(n-k+2)}, \ldots, x_{(n-1)}, x_{(n)}) = x_{(n-k+1)}.$$

The transformations

$$u_j(x_{(n-k+1)}, x_{(n-k+2)}, \ldots, x_{(n-1)}, x_{(n)}) = \frac{1 - F(x_{(n-j+1)})}{1 - F(x_{(n-j)})}$$

imply transformations to mutually independent random variables $\{U_j\}$ having beta densities

$$\boldsymbol{B}(u_j : j, 1) = B^{-1}(j, 1) u_j^{j-1}, \quad j = 1, 2, \ldots, k-1,$$

all being independent of $X_{(n-k+1)}$.

Here $B(a,b) = \Gamma(a)\Gamma(b)/\Gamma(a+b)$ $(a > 0, b > 0)$, where $\Gamma(.)$ denotes the gamma function.

Proof. The joint density of the k largest order statistics (denoted by $f^{\#}(.)$) is:

$$
\begin{aligned}
&f^{\#}(x_{(n-k+1)}, x_{(n-k+2)}, \ldots, x_{(n-1)}, x_{(n)}) \\
&= \frac{\Gamma(n+1)}{\Gamma(n-k+1)\Gamma(k)} \{F(x_{(n-k+1)})\}^{n-k} f(x_{(n-k+1)}) \\
&\quad \times f(x_{(n-k+2)}) \times f(x_{(n-k+3)}) \times \cdots \times f(x_{(n)})
\end{aligned}
$$

It can be re-written as:

$$
\begin{aligned}
&f^{\#}(x_{(n-k+1)}, x_{(n-k+2)}, \ldots x_{(n-1)}, x_{(n)}) \\
&= \frac{\Gamma(n+1)}{\Gamma(n-k+1)\Gamma(k)} \{F(x_{(n-k+1)})\}^{n-k} \\
&\times \{1 - F(x_{(n-k+1)})\}^{k-1} f(x_{(n-k+1)}) \\
&\times (k-1) \times \left[\frac{1 - F(x_{(n-k+2)})}{1 - F(x_{(n-k+1)})}\right]^{k-2} \times \frac{f(x_{(n-k+2)})}{\{1 - F(x_{(n-k+1)})\}} \\
&\times (k-2) \times \left[\frac{1 - F(x_{(n-k+3)})}{1 - F(x_{(n-k+2)})}\right]^{k-3} \times \frac{f(x_{(n-k+3)})}{\{1 - F(x_{(n-k+2)})\}} \\
&\vdots \\
&\times 2 \times \left[\frac{1 - F(x_{(n-1)})}{1 - F(x_{(n-2)})}\right]^{1} \times \frac{f(x_{(n-1)})}{\{1 - F(x_{(n-2)})\}} \\
&\times 1 \times \left[\frac{1 - F(x_{(n)})}{1 - F(x_{(n-1)})}\right]^{0} \times \frac{f(x_{(n)})}{\{1 - F(x_{(n-1)})\}} \,.
\end{aligned}
$$

The result of Theorem 1 follows by observing that the form of the density of the k largest order statistics can also be written as:

$$
\begin{aligned}
&f^{\#}(x_{(n-k+1)}, x_{(n-k+2)}, \ldots x_{(n-1)}, x_{(n)}) \\
&= \frac{\Gamma(n+1)}{\Gamma(n-k+1)\Gamma(k)} \{F(x_{(n-k+1)})\}^{n-k} \\
&\quad \times \{1 - F(x_{(n-k+1)})\}^{k-1} f(x_{(n-k+1)}) \\
&\quad \times (k-1) \times (u_{k-1})^{k-2} \times \frac{f(x_{(n-k+2)})}{\{1 - F(x_{(n-k+1)})\}} \quad \cdots
\end{aligned}
$$

$$\times (k-2) \times (u_{k-2})^{k-3} \times \frac{f(x_{(n-k+3)})}{\{1 - F(x_{(n-k+2)})\}}$$

$$\vdots$$

$$\times 2 \times (u_2)^{2-1} \times \frac{f(x_{(n-1)})}{\{1 - F(x_{(n-2)})\}}$$

$$\times 1 \times (u_1)^{1-1} \times \frac{f(x_{(n)})}{\{1 - F(x_{(n-1)})\}}$$

and then as

$$\begin{aligned}
&f^{\#}(x_{(n-k+1)}, x_{(n-k+2)}, \ldots, x_{(n-1)}, x_{(n)}) \\
&= \frac{\Gamma(n+1)}{\Gamma(n-k+1)\Gamma(k)} \{F(x_{(n-k+1)})\}^{n-k} \\
&\quad \times \{1 - F(x_{(n-k+1)})\}^{k-1} f(x_{(n-k+1)}) \\
&\quad \times (k-1) \times (u_{k-1})^{k-2} \\
&\quad \times (k-2) \times (u_{k-2})^{k-3} \\
&\quad \vdots \\
&\quad \times 2 \times (u_2)^{2-1} \\
&\quad \times 1 \times (u_1)^{1-1} \times \frac{\partial(u_1, u_2, \ldots, u_k)}{\partial(x_{(n-k+1)}, x_{(n-k+2)}, \ldots, x_{(n)})}
\end{aligned}$$

where the Jacobian

$$\frac{\partial(u_1, u_2, \ldots, u_k)}{\partial(x_{(n-k+1)}, x_{(n-k+2)}, \ldots, x_{(n)})} = \left| \frac{\partial u_i}{\partial x_{(n-j+1)}} \right|, \quad i, j = 1, 2, \ldots, k$$

has the value

$$\left| \frac{\partial u_i}{\partial x_{(n-j+1)}} \right| = \left[\frac{f(x_{(n)})}{\{1 - F(x_{(n-1)})\}} \right] \times \left[\frac{f(x_{(n-1)})}{\{1 - F(x_{(n-2)})\}} \right] \times \cdots$$

$$\cdots \times \left[\frac{f(x_{(n-k+2)})}{\{1 - F(x_{(n-k+1)})\}} \right] .$$

Notice that the determinant $\left| \frac{\partial u_i}{\partial x_{(n-j+1)}} \right|$ $(i, j = 1, 2, \ldots, k)$ has zero entries in all positions in the first $k-1$ rows except for:

(i) $j = i$, in which case

$$\frac{\partial u_i}{\partial x_{(n-i+1)}} = \frac{-f(x_{(n-i+1)})}{\{1 - F(x_{(n-i)})\}};$$

(ii) $j = i + 1$, in which case

$$\frac{\partial u_i}{\partial x_{(n-i)}} = u_i \times \frac{f(x_{(n-i)})}{\{1 - F(x_{n-i})\}}.$$

In the k-th row all entries are zero except that $\frac{\partial u_k}{\partial x_{(n-k+1)}} = 1$.

Thus the transformations

$$u_j(x_{(n-k+1)}, x_{(n-k+2)}, \ldots, x_{(n-1)}, x_{(n)}) = \frac{1 - F(x_{(n-j+1)})}{1 - F(x_{(n-j)})}$$

imply transformations to mutually independent random variables $\{U_j\}$ having beta densities

$$\mathbf{B}(u_j : j, 1) = B^{-1}(j, 1)u_j^{j-1}, \quad j = 1, 2, \ldots, k-1,$$

all being independent of $X_{(n-k+1)}$; see, for instance, Kendall and Stuart (1969, pp. 23–24).

This completes the proof. □

References

1. Balkema, A. A. and De Haan, L. (1974), Residual lifetime at great age, *Annals of Probability*, 2, pp. 792–804.
2. Bierlant, J., Joossens, E. and Segers, J. (2004), Discussion of "Generalized Pareto fit to the Society of Actuaries' Large Claims Database" by A. Cebtrain, M. Denuit and P. Lambert, *North American Actuarial Journal*, 8, pp. 108–111.
3. Bowers, N. L. Gerber, H. U., Hickman, J. C., Jones, D. A. and Nesbitt, C. J. (1986), *Risk Theory*, Society of Actuaries, USA.
4. Chung, K.L. (1974), *A Course in Probability Theory*, Academic Press, New York.
5. Drees, H., Ferreira, A. and De Haan, L. (2004), On maximum likelihood estimation of the extreme value index, *Annals of Applied Probability*, 14, pp. 1179–1201.
6. Embrechts, P., Kluppelberg, C. and Mikosch, T. (1997), *Modelling Extremal Events for Insurance and Finance*, Springer, New York.
7. Enz, R. and Holzheu, T. (2004), The economics of liability losses — insuring a moving target, *Sigma* No. 6, Swiss Re. (Available online at www.swissre.com/sigma.)
8. Feller, W. (1968), *An Introduction to Probability Theory and Its Applications*, Vol. I, Wiley, New York.

9. Feller, W. (1971), *An Introduction to Probability Theory and Its Applications*, Vol. II, Wiley, New York.
10. Gnedenko, B. V. (1943), Sur la distribution limite du terme maximum d'une serie aleatoire, *Annals of Mathematics*, 44, pp. 423–433.
11. Heyde, C. C. (1997), *Quasi-likelihood and Its Applications; A General Approach to Optimal Parameter Estimation*, Springer, New York.
12. Hill, B. M. (1975), A simple general approach to inference about the tail of a distribution, *Annals of Statistics*, 5, pp. 1163–1174.
13. Kendall, M. G. and Stuart, A. (1969), *The Advanced Theory of Statistics*, Vol. 1. Griffin, London.
14. Leadbetter, M. R. (1991), On a basis for 'peaks over threshold' modelling, *Statistics and Probability Letters*, 12, pp. 357–362.
15. Mikosch, T. (1997), Heavy-tailed modelling in insurance, *Communications in Statistics - Stochastic Models*, 13, pp. 799–815.
16. Pandey, M. D., Van Gelder, P.H.A.J.M. and Vrijling, J. K. (2001), The estimation of extreme quantiles using L-moments in the peaks-over-thresholds approach, *Structural Safety*, 23, pp. 179–192.
17. Pickands, J. (1975), Statistical inference using extreme order statistics, *Annals of Statistics*, 3, pp. 119–131.
18. Rao, C.R. (1973), *Linear Statistical Inference and Its Applications*, Wiley, New York.
19. Renyi, A. (1953), On the theory of order statistics, *Acta Math. Acad. Sci. Hungar.* 4, pp. 191–232
20. Smith, R. L. (1987), Estimating the tails of probability distributions, *Annals of Statistics*, 15, pp. 1174–1207.
21. Teugels, J. L. and Vanroelen, G. (2004), Box-Cox transformations and heavy-tailed distributions, *Journal of Applied Probability (Special Volume) 43a-Festschrift for C.C. Heyde, 2004*, Applied Probability Trust.

Shrinkage Estimation of Gini Index

R. Ghori, S. E. Ahmed and A. A. Hussein

Department of Mathematics and Statistics, University of Windsor, Windsor, Canada N9B 3P4

E-mail (1): rghori@uwindsor.ca

E-mail (2): seahmed@uwindsor.ca

E-mail (3): ahussein@uwindsor.ca

The Gini index is an indicator of the degree of income inequality in a society. In recent decades, the Gini index has become a popular means of describing the inequality measure. In this article, we propose shrinkage estimation strategies for estimating the Gini indices for multiple populations. The classical estimator is investigated as a competitor of the proposed estimators. In light of a quadratic loss function, the asymptotic risks of the estimators are derived. A central theme of this paper is that the shrinkage method provides a powerful extension of its classical counterpart for the estimation of Gini indices in a multi-population situation.

Keywords: Gini index; shrinkage estimation; L-moments; Ł-statistics; asymptotic risk analysis; retrospective sampling.

1. Introduction

There are several ways to express the degree of income inequality in a society. The Gini index is perhaps one of the most used indicators of economic and social condition. The Gini coefficient is a measure of inequality developed by the Italian scientist Corrado Gini and published in his 1912 paper "variabilità e mutabilità". Algebraically, it is defined as "expected value of the ratio of the difference of two arbitrary specimens to the mean value of all specimens". However, it is understood generally by the geometric definition "area enclosed by the Lorenz curve and the diagonal", which has a meaning equal to the algebraic definition. A Lorenz curve plots the cumulative percentages of total income received against the cumulative number of recipients, starting with the poorest individual or household. The gap between the actual lines and the mythical line is a function of the degree of inequality. The Gini index measures the area between the Lorenz curve and a mythical or hypothetical line of absolute equality, expressed as a percentage of the maximum area under the line. In the egalitarian society, the Gini index would be 0, since the Lorenz curve would match the 45° line perfectly; the higher the Gini index, then the greater the distance and the more unequal the distribution of income. In practice, the Gini index usually falls between 0.20 and 0.45. It is an interesting fact that while the most developed European nations tend to have values between 0.24 and 0.36, the United States has been above 0.4 for several decades. This index can be viewed as an approach to quantify the perceived differences in welfare and compensation policies and philosophies. Income or resource distribution could be found to have direct impact on the poverty rate of a country or region.

The Gini index is usually used to measure income inequality, but can be used to measure any form of uneven distribution. For example, it can also be used to measure wealth inequality. This use requires that no one has a negative net wealth. The Gini index has been used to study several aspects of health inequities. As in the often cited studies of income distributions, comparisons can be made between countries. For example, the Pan American Health Organization (`www.paho.org`) published a short survey applying the Gini index to infant mortality rates across a group of South American countries ([10] and [7]). Ref. [15] evaluated the predictive power of different branch prediction features using the metric Gini index, which is used as a featured selection measure in the construction of decision trees. That study showed that the Gini index is a good metric for comparing branch prediction features. The Gini index has also been used to study spatial patterns of care and health care access in Canada [9], with the goal of understanding the relationship between competitive conditions and practitioner location. The researches on the Gini index have been developed remarkably and extended into various directions as evidenced by the bibliographies of [27], [28] and [16]. Further, a substantial literature has been devoted to the construction of indices of economic inequality that are consistent with axiomatic systems of fairness. We refer to [3], [11], [22], [12], [23] and [4] for comprehensive surveys on measures of inequality including the Gini index.

In summary, the Gini index of income or resource inequality is a measure of the degree to which a population shares that resource unequally. It is based on the statistical notion known in literature as the "mean difference" of a population. The index is scaled to vary from a minimum of zero to a maximum of one, zero representing no inequality and one representing the maximum possible degree of inequality. A major statistical limitation of the Gini index is the absence and the intractability of an appropriate sampling distribution ([18], [21]). In this article, we give the large sampling distribution of the Gini index when m independent retrospective samples of different sizes are available. Thus, this paper generalises the classical one-sample set-up into a multi-sample situation and develops inference techniques for the indices. Further, the shrinkage estimation of the Gini index is proposed and its statistical properties investigated. The requirements of a large sample size may not be stringent. There are many situations where taking a large sample is more economical than taking frequent samples. The overall cost of sampling may be reduced by judiciously collecting a relatively large amount of inexpensive data to increase the accuracy of the estimator. Thus, the goal of finding large sample methods for such problems seems well worth achieving.

2. Preliminaries and Statement of the Problem

In order to define Gini's mean difference, suppose that $X_1, \ldots, X_n$ are independent and identically distributed (i.i.d.) random variables with nonzero mean and cumulative distribution function (c.d.f.) F. We shall assume that F, instead of being restricted to a parametric family, is completely unknown, subject only to some very general conditions such as continuity and existence of second moment. The "parameter" $\theta = \theta(F)$ to be estimated is a real valued function defined over this nonparametric class $\mathcal{F}$. The classical Gini index is defined as

$$\gamma = \frac{1}{2\mu} E|X_j - X_i| = \frac{\Delta}{2\mu}, \tag{1}$$

where X_i and X_j are independent copies of the random variable X with c.d.f. F and $E(X) = \mu$. The quantity $\Delta = E|X_j - X_i|$ is known as the *Gini mean difference.* The classical Gini index estimator is defined (see [13], [14] and references therein) by,

$$\hat{\gamma} = \frac{1}{2n^2\bar{X}} \sum_{i=1}^{n} \sum_{j=1}^{n} |X_j - X_i|, \tag{2}$$

Suppose that there are m independent retrospective samples of size $n_1, n_2, \ldots, n_m$ acquired from $k = 1, 2, \ldots, m$ populations, respectively. Denote the observed data by X_{ki}, $i = 1, 2, \ldots, n_k$. In this article, we are interested in estimating the m-dimensional parameter vector of Gini indices

$\boldsymbol{\gamma} = (\gamma_1, \gamma_2, \ldots, \gamma_m)'$ based on the random samples of size $n_1, n_2, \ldots, n_m$, respectively. The classical Gini index of each component is

$$\gamma_k = \frac{\Delta_k}{2\mu_k} = \frac{E|X_{kj} - X_{ki}|}{2\mu_k}, \tag{3}$$

and its *classical Gini index estimator (CGI)* is

$$\hat{\gamma}_k = \frac{1}{2\bar{X}_k} \frac{1}{\binom{n_k}{2}} \sum\sum_{i<j} |X_{kj} - X_{ki}|, \tag{4}$$

where $\bar{X}_k = \frac{1}{n_k} \sum_{i=1}^{n_k} X_{ki}$. One of the main objectives of this paper is to provide estimators when it is suspected that $\boldsymbol{\gamma} = \boldsymbol{\gamma}^o$ may hold, where $\boldsymbol{\gamma}^o = (\gamma_1^o, \gamma_2^o, \ldots, \gamma_m^o)'$. The $\boldsymbol{\gamma}^o$ is a prior guessed vector of Gini indices obtained by previous census or recent enumeration.

2.1. *Asymptotic normality of Gini index*

Lemma 2.1: Assume as above that for each $k = 1, 2, \ldots, m$, X_{ki}, $i = 1, 2, \ldots, n_k$ are independent and identically distributed random variables with CDF $F_k(.)$. If the X_{ki} have nonzero mean and finite second moments, then for each $k = 1, 2, \ldots, m$ and for $n_k \to \infty$,

$$n_k^{1/2}\{\hat{\gamma}_k - \gamma_k\} \xrightarrow{D} \mathcal{N}\{0, \tau_k^2\},$$

where $\hat{\gamma}_k$ are the classical Gini index estimators, defined above, and

$$\tau_k^2 = \frac{1}{\mu_k^2}\left(\sigma_{1k}^2 \gamma_k^2 - 2\sigma_{2k}^2 \gamma_k + \sigma_{3k}^2\right), \tag{5}$$

$$\sigma_{1k}^2 = \iint_{x<y} 2F_k(x)[1 - F_k(y)]\, dx\, dy,$$

$$\sigma_{2k}^2 = \iint_{x<y} [(2F_k(x) - 1) + (2F_k(y) - 1)]F_k(x)[1 - F_k(y)]\, dx\, dy,$$

$$\sigma_{3k}^2 = \iint_{x<y} 2(2F_k(x) - 1)(2F_k(y) - 1)F_k(x)[1 - F_k(y)]\, dx\, dy. \tag{6}$$

Proof. As in [19], we note that $\hat{\gamma}_k$ is just the ratio of two L-statistics or two sample L-moments, namely the first two sample L-moments, l_{2k}/l_{1k} with $l_{1k} = \bar{X}_k$ and $l_{2k} = \frac{1}{2n^2}\sum_{i=1}^{n}\sum_{j=1}^{n}|X_j - X_i|$. As such, and following the method of [19], it easy to see that under the assumptions of the Lemma,

$$n_k^{1/2}\begin{pmatrix} l_{1k} - \mu_k \\ l_{2k} - \Delta_k/2 \end{pmatrix} \xrightarrow{D} N(\mathbf{0}, \Sigma)$$

where

$$\Sigma = \begin{pmatrix} \sigma_{1k}^2 & \sigma_{3k}^2 \\ \sigma_{3k}^2 & \sigma_{2k}^2 \end{pmatrix}$$

with entries defined in (6) and $\mathbf{0} = (0,0)'$. The result now follows by simple application of the multivariate delta method. □

3. The Proposed Improved Estimation Strategy

We now turn to the main objective of this investigation, the parameter vector $\boldsymbol{\gamma} = (\gamma_1, \gamma_2, \dots, \gamma_m)'$ which is estimated by $\hat{\boldsymbol{\gamma}} = (\hat{\gamma}_1, \hat{\gamma}_2, \dots, \hat{\gamma}_m)'$. We are primarily interested in the estimation of $\boldsymbol{\gamma}$ when it is plausible that $\boldsymbol{\gamma}$ is "close" to some specified $\boldsymbol{\gamma}^o$. As one basis for identifying model-estimator uncertainty, ref. [24] demonstrated the inadmissibility of the maximum likelihood estimator (MLE) when estimating the m-variate normal mean vector $\boldsymbol{\theta}$ under quadratic loss. Following this result, refs. [20], [25] and [6] combined the m-variate MLE $\hat{\boldsymbol{\theta}}$ and m-dimensional fixed null vector, under the normality assumption, as

$$\hat{\boldsymbol{\theta}}^S = (1 - c\lambda/||\hat{\boldsymbol{\theta}} - \mathbf{0}||^2)(\hat{\boldsymbol{\theta}} - \mathbf{0}), \quad \text{when} \quad 0 < \lambda < 2(m-2),$$

and demonstrated that for $m > 2$ this estimator dominates the MLE.

We therefore propose Stein-type methodology for the vector $\boldsymbol{\gamma}$. In this case, the construction of an estimator rests on the choice of a special value $\boldsymbol{\gamma}^o$ of $\boldsymbol{\gamma}$, called the pivot, whose plausibility plays a basic role in the motivation. It should be noted however that the Stein-type estimator is primarily used for location parameter, and in this paper, we have extended this method for the estimation of Gini index parameters for arbitrary populations.

Define $\hat{\boldsymbol{\gamma}}_n^o = \sqrt{n}(\hat{\boldsymbol{\gamma}} - \boldsymbol{\gamma}^o)$ with $n = \sum_{k=1}^m n_k$ and $T = (\hat{\boldsymbol{\gamma}}_n^o)'\hat{\boldsymbol{\Omega}}_n\hat{\boldsymbol{\gamma}}_n^o$, where

$$\hat{\boldsymbol{\Omega}}_n = \operatorname{diag}\left(\frac{\omega_{1,n}}{\hat{\tau}_1^2}, \dots, \frac{\omega_{m,n}}{\hat{\tau}_m^2}\right)$$

with $\omega_{i,n} = \frac{n_i}{n}$. The *Stein-type shrinkage estimator (SE)* is then defined by

$$\hat{\boldsymbol{\gamma}}^S = \boldsymbol{\gamma}^o + \left[1 - \lambda T^{-1}\right](\hat{\boldsymbol{\gamma}} - \boldsymbol{\gamma}^o), \quad m \geq 3,$$

where λ is the shrinkage constant such that $0 \leq \lambda \leq 2(m-2)$. The choice $\lambda = m - 2$ which minimises the risk will be used throughout this paper. We will demonstrate that the proposed Stein-type estimator provides uniform improvement over $\hat{\boldsymbol{\gamma}}$. The estimator $\hat{\boldsymbol{\gamma}}^S$ is generally called a shrinkage estimator since it shrinks the classical estimator towards the $\boldsymbol{\gamma}^o$. For some insights into shrinkage estimation, we refer to [8] and [26], among others.

In cases where the estimated parameter is positive, shrinking the classical estimator towards the specified vector $\boldsymbol{\gamma}^o$ may result in an over-shrinkage and as a consequence, the proposed estimator may become negative. To avoid this nuisance, we truncate $\hat{\boldsymbol{\gamma}}^S$ to obtain a convex combination of $\hat{\boldsymbol{\gamma}}$ and $\boldsymbol{\gamma}^o$ which is called the *positive-rule shrinkage estimator (PSE)*. The *PSE* may be defined as follows:

$$\hat{\boldsymbol{\gamma}}^{S+} = \boldsymbol{\gamma}^o + \left[1 - (m-2)T^{-1}\right]^+ (\hat{\boldsymbol{\gamma}} - \boldsymbol{\gamma}^o),$$

where $z^+ = \max(0, z)$. The positive-part estimator is particularly important to control the over-shrinking inherent in $\hat{\boldsymbol{\gamma}}^S$.

Clearly, in practice, parameter spaces are always restricted. Although restrictions may be difficult to specify, the gain in performance that may be realised from incorporating the restrictions can be tremendous, as this paper will demonstrate. Reductions in risk (under quadratic loss function) of 50% are attainable over the restricted parameter space if the restrictions are judiciously exploited.

An unusual and novel feature of this paper is that we do not assume the shape of the income/wealth populations. More importantly, we consider the asymptotic shrinkage estimation of Gini index parameters in a retrospective sampling. Based on the review literature, this kind of study is not available for practitioners.

In passing we remark that the application of shrinkage estimators is subject to the condition that $m \geq 3$. Therefore, we recommend the use of pretest estimators when such a constraint holds. The *pretest estimators (PTE)* are widely used by researchers, notably in economic research. For details on the subject, we refer to the extensive bibliographies of [5] and [17].

In the present investigation, emphasis is on a situation where all sample sizes are large while the parameter vector is believed to be close to $\boldsymbol{\gamma}^o$. In this context, the notion of asymptotic distributional quadratic risk will enable us to study the large sample properties of the proposed estimators. To study the statistical properties of the estimators, we introduce a quadratic loss function:

$$L(\boldsymbol{\gamma}^*) = n(\boldsymbol{\gamma}^* - \boldsymbol{\gamma})'\boldsymbol{\Lambda}(\boldsymbol{\gamma}^* - \boldsymbol{\gamma}),$$

where $\boldsymbol{\gamma}^*$ is any estimator of $\boldsymbol{\gamma}$, and $\boldsymbol{\Lambda}$ is a nonnegative definite (n.n.d.) weighting matrix. Then, the *quadratic risk* for $\boldsymbol{\gamma}^*$ is given by

$$R(\boldsymbol{\gamma}^*) = nE\{(\boldsymbol{\gamma}^* - \boldsymbol{\gamma})'\boldsymbol{\Lambda}(\boldsymbol{\gamma}^* - \boldsymbol{\gamma})\}.$$

We plan to investigate the risk performance of the proposed estimators in the entire parameter space induced by the information that γ is close to γ^o. In the case $\gamma = \gamma^o$ the asymptotic risk can be calculated without technical difficulty; however, when $\gamma \neq \gamma^o$, then $\hat{\gamma}, \hat{\gamma}^S$ and $\hat{\gamma}^{S+}$ will be risk equivalent. To avoid this technical problem, we define a sequence which may be specified as $\gamma_{(n)} = \gamma^o + \boldsymbol{\xi}/n^{1/2}$, where $\boldsymbol{\xi}$ is a vector of fixed real numbers. Evidently, $\gamma_{(n)}$ approaches γ^o at a rate proportional to $n^{-1/2}$. Now, we can compute the *asymptotic distributional quadratic risk (ADQR)*. To do so, let the *asymptotic distribution function* of $\{\sqrt{n}(\gamma^* - \gamma)\}$ be given by

$$G(\mathbf{z}) = \lim_{n\to\infty} \Pr\{\sqrt{n}(\gamma^* - \gamma) \leq \mathbf{z}\}.$$

Further, let

$$\mathbf{Q} = \int\int \cdots \int \mathbf{z}\mathbf{z}' dG(\mathbf{z}).$$

Finally, the ADQR is defined by $R(\gamma^*) = \mathrm{tr}(\boldsymbol{\Lambda}\mathbf{Q})$. The asymptotic risk under the local alternatives is defined as $R(\gamma^*) = \lim_{n\to\infty} R_n(\gamma^*)$, where $R_n(\gamma^*) = nE\{(\gamma^* - \gamma_{(n)})'\boldsymbol{\Lambda}(\gamma^* - \gamma_{(n)})\}$. Assuming that the limit exists, one may need extra regularity conditions to evaluate this risk function. This point has been explained in detail in various other contexts, by a host of researchers. In obtaining our result for the shrinkage estimators, we use the technique that was developed by [2] and [1].

First, we give expressions for the *asymptotic distributional biases (ADB)* of the estimators as follows:

$$ADB(\hat{\gamma}^S) = -(m-2)\boldsymbol{\xi}E(\chi^{-2}_{m+2}(\Delta)),\ \Delta = \boldsymbol{\xi}'\boldsymbol{\Omega}^{-1}\boldsymbol{\xi},$$

$$ADB(\hat{\gamma}^{S+})$$
$$= -\boldsymbol{\xi}[\psi_{m+2}(m-2;\Delta) + (m-2)E\{\chi^{-2}_{m+2}(\Delta)I(\chi^2_{m+2}(\Delta) > (m-2))\}],$$

where $\boldsymbol{\Omega} = \lim_{n\to\infty} \boldsymbol{\Omega}_n$; $\psi_m(x\,;\Delta)$ means the noncentral chi-squared cumulative distribution function with noncentrality parameter Δ and m degrees of freedom;

$$E\left(\chi_m^{-2k}(\Delta)\right) = \int_0^\infty x^{-2k} d\psi_m(x\,;\Delta),$$

and

$$I(\mathcal{A}) = \begin{cases} 1 & \text{if } \mathcal{A} \text{ is true,} \\ 0 & \text{otherwise,} \end{cases}$$

where $\mathcal{A}$ is any expression or condition.

Further, define

$$B(\gamma^*) = [ADB(\gamma^*)]'\boldsymbol{\Omega}^{-1}[ADB(\gamma^*)]$$

as the quadratic bias of an estimator γ^* of parameter vector γ. Then,

$$B(\hat{\gamma}^S) = (m-2)^2\Delta[E(\chi_{m+2}^{-2}(\Delta))]^2,$$

$$B(\hat{\gamma}^{S+})$$
$$= \Delta\left[\psi_{m+2}(m-2;\Delta) + (m-2)E\{\chi_{m+2}^{-2}(\Delta)I(\chi_{m+2}^2(\Delta) > (m-2))\}\right]^2.$$

Note that, since $E(\chi_m^{-2}(\Delta))$ is a decreasing log-convex function of Δ, the quadratic bias of $\hat{\gamma}^S$ starts from 0 at $\Delta = 0$, increases to a maximum, and then decreases towards 0. The behaviour of $\hat{\gamma}^{S+}$ is similar to $\hat{\gamma}^S$; however, the quadratic bias curve of $\hat{\gamma}^{S+}$ remains below the curve of $\hat{\gamma}^S$ for all values of Δ.

The ADQR functions of the estimators are given in the following theorem.

Theorem 3.1

$$\begin{aligned} R(\hat{\gamma}^{S+}) = R(\hat{\gamma}^S) - R(\hat{\gamma}) & \\ & \times E[\{1-(m-2)\chi_{m+2}^{-2}(\Delta)\}^2 I(\chi_{m+2}^2(\Delta) \le (m-2))] \\ & + \boldsymbol{\xi}'\boldsymbol{\Lambda}\boldsymbol{\xi}\big[E[2\{1-(m-2)\chi_{m+2}^{-2}(\Delta)\}I(\chi_{m+2}^2(\Delta) \le (m-2))] \\ & - E[\{1-(m-2)\chi_{m+4}^{-2}(\Delta)\}^2 I(\chi_{m+4}^2(\Delta) \le (m-2))]\big], \end{aligned}$$

where

$$\begin{aligned} R(\hat{\gamma}^S) &= R(\hat{\gamma}) + \boldsymbol{\xi}'\boldsymbol{\Lambda}\boldsymbol{\xi}(m-2)(m+2)E(\chi_{m+4}^{-4}(\Delta)) \\ &\quad - (m-2)\operatorname{tr}(\boldsymbol{\Lambda}\boldsymbol{\Omega}^{-1})\{2E(\chi_{m+2}^{-2}(\Delta)) - (m-2)E(\chi_{m+2}^{-4}(\Delta))\}, \end{aligned}$$

and $R(\hat{\gamma}) = \operatorname{tr}(\boldsymbol{\Lambda}\boldsymbol{\Omega}^{-1})$.

Proof. This theorem can be proved using arguments similar to those in [2]. □

Using the results of the above theorem, we investigate the comparative statistical properties of the proposed estimators. First, note that for $\boldsymbol{\xi} = \mathbf{0}$,

$$R(\hat{\gamma}) - R(\hat{\gamma}^S) = \operatorname{tr}(\boldsymbol{\Lambda}\boldsymbol{\Omega}^{-1})(m-2)E\{2\chi_{m+2}^{-2} - (m-2)\chi_{m+2}^{-4}\} > 0.$$

Therefore, $\hat{\gamma}^S$ is a superior alternative to the classical estimator of Gini indices vector. For the general case, that is $\boldsymbol{\xi} \ne \mathbf{0}$, consider the class of n.n.d. matrices,

$$\boldsymbol{\Lambda}^D = \left\{\boldsymbol{\Lambda}_{n.n.d.} : \frac{\operatorname{tr}(\boldsymbol{\Lambda}\boldsymbol{\Omega}^{-1})}{\lambda_{\max}(\boldsymbol{\Lambda}\boldsymbol{\Omega}^{-1})} \ge \frac{m+2}{2}\right\},$$

where $\lambda_{\max}(A)$ means the largest eigenvalue of A. Moreover, assume that

$$\lambda_{\min}(\mathbf{\Lambda\Omega}^{-1}) \leq \frac{\boldsymbol{\xi}'\mathbf{\Lambda}\boldsymbol{\xi}}{\boldsymbol{\xi}'\mathbf{\Omega}\boldsymbol{\xi}} \leq \lambda_{\max}(\mathbf{\Lambda\Omega}^{-1}), \quad \text{for } \boldsymbol{\xi} \neq \mathbf{0} \text{ and } \mathbf{\Lambda} \in \mathbf{\Lambda}^{\mathbf{D}}.$$

Then, under the class of matrices defined above, we have established that $R(\hat{\gamma}^S) \leq R(\hat{\gamma})$ for all $\boldsymbol{\xi}$, with strict inequality holding for some $\boldsymbol{\xi}$. It follows that $\hat{\gamma}$ is asymptotically inferior to $\hat{\gamma}^S$ in the entire parameter space. The risk of $\hat{\gamma}^S$ begins with an initial value less than the risk of $\hat{\gamma}$ and then increases monotonically towards $\text{tr}(\mathbf{\Lambda\Omega}^{-1})$. Therefore, the risk of $\hat{\gamma}^S$ is uniformly smaller than that of $\hat{\gamma}$, and the upper limit is attained when $||\boldsymbol{\xi}|| \to \infty$.

Finally, we investigate the relative risk performance of $\hat{\gamma}^{S+}$ to $\hat{\gamma}^S$ and $\hat{\gamma}$, respectively. Using the simple fact that each indicator function in the risk expression of $\hat{\gamma}^{S+}$ has a value of 1 if $\chi^2_{m+l}(\Delta) \leq (m-3)$ and a value of 0 otherwise, it follows that $1-(m-2)/(\chi^2_{m+l}(\Delta)) \leq 0$, for $l = 2, 4$. Hence $R(\hat{\gamma}^{S+}) - R(\hat{\gamma}^S) \leq 0$ for all $\boldsymbol{\xi}$, where strict inequality holds for some $\boldsymbol{\xi}$. In other words, $\hat{\gamma}^{S+}$ asymptotically dominates $\hat{\gamma}^S$ under local alternatives and therefore $\hat{\gamma}^{S+}$ is also superior to $\hat{\gamma}$.

It is noted that the asymptotic risks of the proposed estimators depend on the matrices $\mathbf{\Lambda}$ and $\mathbf{\Omega}^{-1}$ and may not be useful for numerical work. However, the numerical computation of the risk functions can be simplified by considering the particular choice $\mathbf{\Lambda} = \mathbf{\Omega}^{-1}$; then $\text{tr}(\mathbf{\Lambda\Omega}^{-1}) = m$, and $\boldsymbol{\xi}'\mathbf{\Lambda}\boldsymbol{\xi} = \Delta$. The numerical values of the risk functions were obtained by writing a computer program in Fortran, and using subroutines from the IMSL library. The percentage of improvement in risk is calculated by using the formula

$$PI = \frac{100\{R(\hat{\gamma}) - R(\gamma^*)\}}{R(\hat{\gamma})}.$$

Table 1 provides the percentage improvements in risk of $\hat{\gamma}^{S+}$ and $\hat{\gamma}^S$ over $\hat{\gamma}$.

Table 1. Percentage risk improvement of PSE and SE over UE for $m = 8$

Δ	0	1.0	2.0	3.0	4.0	5.0	6.0	7.0	9.0	10	11
PSE	71	60	51	44	39	35	31	28	26	24	21
SE	63	52	45	40	35	32	29	27	25	23	20

Table 1 reveals that both proposed shrinkage estimators have maximum risk gain over $\hat{\gamma}$ at $\Delta = 0$, and the value of the improvement is a decreasing function of Δ. As an example, the improvement of $\hat{\gamma}^{S+}$ over $\hat{\gamma}$ at $\Delta = 0$ is 71%, which is an unprecedented improvement. The risk performance of $\hat{\gamma}^{S+}$ is superior to both $\hat{\gamma}^{S}$ and $\hat{\gamma}$ in the entire parameter space. For large values of Δ both $\hat{\gamma}^{S+}$ and $\hat{\gamma}^{S}$ behave equally. Hence, the numerical findings support the analytical results.

Similarly, Table 2 provides the percentage of risk improvement of $\hat{\gamma}^{S+}$ over $\hat{\gamma}^{S}$ for various choices of m with $\Delta = 0$, 0.5, 1.0, 1.5 and 3.0.

Table 2. Percentage Risk Improvement of PSE over SE.

m	$\Delta = 0$	$\Delta = 0.5$	$\Delta = 1.0$	$\Delta = 1.5$	$\Delta = 3.0$
4	13	11	9	7	4
5	18	14	12	10	6
10	24	20	16	14	9
15	26	21	18	15	10
20	27	22	19	16	11
25	28	23	19	16	11
30	29	23	19	17	12

Again, we notice the supremacy of $\hat{\gamma}^{S+}$ over $\hat{\gamma}^{S}$. At $m = 4$, the improvement at $\Delta = 0$ is 13% and this increases as the value of m increases. Moreover, the improvement decreases with an increase in Δ. However, the important issue here is not the improvement in the sense of lowering the risk by using the positive part of the $\hat{\gamma}^{S}$. The focal point in this estimation strategy is to preserve the sign of $\hat{\gamma}$. By considering only the positive part of $\hat{\gamma}^{S}$ the resulting estimator $\hat{\gamma}^{S+}$ removes the absurd behaviour of $\hat{\gamma}^{S}$ and does not change the sign of the estimators. Thus, components of $\hat{\gamma}^{S+}$ will have the same sign as that of components of $\hat{\gamma}$. For this pragmatic reason, we strongly recommend that the usual shrinkage estimator should be used as a tool for developing the positive part shrinkage estimator.

4. Final Comments

In this paper, asymptotic statistical estimation procedures are established for the Gini index parameters for nonnormal distributions, a commonly used income index. We proposed shrinkage and positive part shrinkage estimator when the true Gini indices are suspected to be close to a guessed value. We illustrated how the classical large sample theory for the usual estimators can be extended to the shrinkage estimators for the index parameters. We provided the risk analysis of the estimators both theoretically and numerically.

Perhaps, the most important message in this paper is that very large gains in precision may be achieved by judiciously exploiting the guessed values of the parameters which in practice will be available in any realistic problem. Our numerical findings indicate that a reduction of 50% or more in the risk seems quite realistic depending on the values of Δ and m, standardised distance between the true and the guessed values and number of populations in consideration. Thus, it seems inconceivable that so little attention has been paid to these situations in the development of statistical theory. Like the statistical models underlying the statistical inferences to be made, the guessed values will be susceptible to uncertainty and practitioners may be reluctant to impose the additional information regarding parameters in the estimation process. More importantly, the estimation strategy combining the sample and parameter information is superior to a strategy based on sample information only. Research on statistical implications of these and other estimators for a range of statistical models is ongoing.

References

1. Ahmed, S.E. (2001). Shrinkage estimation of regression coefficients from censored data with multiple observations. *Empirical Bayes and Likelihood Inference, Lecture Notes in Statistics*, Editors: S.E. Ahmed and N. Reid. Springer-Verlag.
2. Ahmed, S.E. and E. Saleh (1999). Improved nonparametric estimation of location vector in a multivariate regression model. *Nonparametric Statistics* **11**, 51–78.
3. Anand, S. (1983). *Inequality and poverty in Malaysia: Measurement and Decomposition.* Oxford University Press.
4. Atkinson, A.B. and F. Bourguignon (2000). *Introduction: Income distribution and Economics, Handbook of Income Distribution.* Elsevier.
5. Bancroft, T.A. and C. Han (1977). Inference based on conditional specification: A note and a bibliography. *International Statistical Review* **45**, 117–127.
6. Baranchik, A.M. (1964). *Multiple regression and estimation of the mean of a multivariate normal distribution.* Technical Report 51, Stanford University, Dept. of Statistics.
7. Berndt, D.J., J.W. Fisher and R.V. Rajendrababu (2003). Measuring healthcare inequities using the Gini index. In: *Proceedings of the 36th Hawaii International Conference on System Sciences (HICSS 36).*
8. Brandwein, A. and W.E. Strawderman (1990). Stein estimation: the spherically symmetric case. *Statistical Science* **3**, 356–369.
9. Brown, M. (1994). Using Gini-style indicies to evaluate the spatial patterns of health practitioners: Theoretical considerations and an application based on Alberta data. *Social Science and Medicine* **38(9)**, 1243–1256.
10. Castillo-Salgado, C., C. Schneider, E. Loyola, O. Mujica, A. Roca and T. Yerg

(2001). Measuring health inequalities: Gini coefficient and concentration index. *Epidemiological Bulletin, PAHO.*
11. Chakravarty, S.R. (1990). *Ethical Social Index Numbers.* Springer-Verlag, New York.
12. Cowell, F. (1999). *"Measurement of Inequality" in Atkinson, A.B. and F. Bourguignon (eds) Handbook of Income Distribution.* North Holland, Amsterdam.
13. David, H.A. (1968). Gini's mean difference rediscovered. *Biometrika* **55**, 573–575.
14. David, H.A. (1970). *Order Statistics.* Wiley, New York.
15. Desmet, V., L. Eeckhout and K. De Bosschere (2004). Evaluation of the Gini index for studying branch prediction features. In: *Proceedings of the 6th International Conference on Computing Anticipatory Systems (CASYS). American Institute of Physics. AIP Conference Proceedings.* Vol. 718. pp. 376–384.
16. Giorgi, G.M. (1990). Bibliographic portrait of the Gini concentration ratio. *Metron* **XLVIII, n 1-4**, 183–221.
17. Han, Chien-Pai, C.V. Rao and J. Ravichandran (1988). Inference based on conditional specification: A second bibliography. *Communications in Statistics — Theory and Methods* **17**, 1945–1964.
18. Hart, P.E. (1971). Entropy and other measures of concentration. *Journal of the Royal Statistical Society, Series A,* **134**, 73–85.
19. Hosking, J.R.M. (1990). *L*-moments: Analysis and estimation of distributions using linear combinations of order statistics. *Journal of the Royal Statistical Society, Series B,* **52**, 105–124.
20. James, W. and C. Stein (1961). Estimation with quadratic loss. In: *Proceedings of the Fourth Berkeley Symposium on Mathematical Statistics and Probability.* University of California Press, Berkeley, CA.
21. Nygard, F. and A. Sandstrom (1981). Measuring income inequality. *Almqvist and Wiksell International, Sweden.*
22. Sen, A.K. (1997). *On Economic Inequality.* Clarendon Press, Oxford.
23. Silber, J. (1999). *Handbook on Income Inequality Measurement. Recent Economic Thought.* Vol. 71. Kluwer Academic Publishers, Boston.
24. Stein, C. (1956). Inadmissibility of the usual estimator for the mean of a multivariate normal distribution. In: *Proceedings of the Fourth Berkeley Symposium on Mathematical Statistics and Probability.* University of California Press, Berkeley, CA. 197–206.
25. Stein, C. (1962). Confidence sets for the mean of a multivariate normal distribution. *Journal of the Royal Statistical Society, Series B,* **24**, 265–285.
26. Stigler, S.M. (1990). A Galtonian perspective on shrinkage estimators. *Statistical Science* **5**, 147–155.
27. Xu, K. (2004). *How has the literature on Gini's index evolved in the past 80 years.* Technical Report, Dalhousie University.
28. Yitzhaki, S. (1998). More than a dozen alternative ways of spelling Gini. *Research in economic inequality* **8**, 13–30.

On the Problem of Discriminating between the Tails of Distributions

Chris C. Heyde

Department of Statistics, Columbia University,
New York NY 10027, USA
and
CMA, Mathematical Sciences Institute, Australian National University,
Canberra ACT 0200, Australia
E-mail: chris@maths.anu.edu.au

Khanhav Au

CMA, Mathematical Sciences Institute, Australian National University,
Canberra ACT 0200, Australia
and
Department of Mathematics and Statistics, University of Melbourne,
Parkville VIC 3010, Australia
E-mail: l.au@ms.unimelb.edu.au

In areas such as financial and insurance risk and communication network design the heaviness of the tail of the underlying distribution is crucial for the calculations. However, although it seems straightforward theoretically to distinguish between (say) exponential tails and power tails, this requires unexpectedly large samples in practice. Here we will use quantiles to compare the tails of distributions which are standardised to unit interquartile range to allow for possible infinite variance. We present some chi-squared tests of goodness-of-fit focussed on the tails. We also provide methods of quick comparison of distributions using counts over high thresholds and using extreme values.

Keywords: tailweight discrimination; quantiles; goodness-of-fit tests; counts over thresholds; extreme values.

1. Introduction

The behaviour of probabilistic models in many areas, and particularly in financial and insurance risk and communication network design, is critically influenced by the tails of the underlying distribution(s) which drive the models. The literature, however, reveals considerable uncertainty about the choice of tailweight in the model. This is despite the fact that it seems straightforward theoretically to distinguish between the main classes of tail-weights, such as exponential or power tails, using simple methods such as Q–Q plots. Indeed, a whole collection of widely applied methods of discrimination are available and described in books such as Embrechts, Klüppelberg and Mikosch (1997), Reiss and Thomas (2001) and Rolski, Schmidli, Schmidt and Teugels (1999).

In this paper we will discuss why the distinctions are impossible to make without particularly large samples, at least in the tens of thousands. The paper is a sequel to Heyde and Kou (2004) in which the fundamental shortcomings in the standard methods of discrimination were emphasised. Here we will broaden the comparisons to include distributions with infinite variance, for which we use quantiles and unit interquartile range to standardise the distributions for comparison instead of the usual unit standard deviation. This is necessary for dealing with risks of rare events, such as for property insurance claims, where even the mean may not be finite, and also allowing for the school of thought that favours risky asset returns distributions with infinite variance (e.g. Rachev and Mittnik (2000)). We will first use quantiles to compare the tails of distributions, and to assess the precision of comparisons on real data. Then we shall present chi-squared tests of goodness-of-fit which focus on the tails. The paper concludes with a discussion of quick comparison of distribution methods using counts over high thresholds and using extreme values.

Suppose random variable X, with zero mean, has distribution function $F_X(x)$. Define the p-th quantile ξ_p by $F_X(\xi_p) = p$, where $0 < p < 1$. Then the random variable

$$Y = X(\xi_{0.75} - \xi_{0.25})^{-1}$$

has distribution function given by $H_Y(y) = F_X((\xi_{0.75} - \xi_{0.25})y)$ and unit interquartile range, and direct comparison between finite and infinite variance distributions is possible.

We shall address the question of how large a sample is needed to distinguish between the tails of particular target distributions at a fixed quantile p. Specifically, we focus on the normal and Laplace distributions as exemplars of light and exponential tailed distributions, and the t distribution with low degrees of freedom and the stable distribution with index $\alpha < 2$ as exemplars of heavy power tailed distributions, the last having infinite variance.

2. Quantile Comparisons

The class of distributions that will be looked at closely is the Student t_ν, Laplace l_β and symmetric-stable s_α distributions with the Gaussian g_σ distribution being a benchmark for comparison. The densities for the normal, Laplace and t_ν are, respectively,

$$g_\sigma(x) = \frac{1}{\sqrt{2\pi}\,\sigma} \exp\left(-\frac{x^2}{2\sigma^2}\right), \qquad \sigma > 0,\ x \in (-\infty, \infty),$$

$$l_\beta(x) = \frac{1}{2\beta} \exp\left(-\frac{|x|}{\beta}\right), \qquad \beta > 0,\ x \in (-\infty, \infty),$$

$$t_\nu(x) = \frac{1}{\sqrt{\pi\nu}} \cdot \frac{\Gamma(\frac{\nu+1}{2})}{\Gamma(\frac{\nu}{2})\left(1 + \frac{x^2}{\nu}\right)^{(\nu+1)/2}}, \qquad \nu > 0,\ x \in (-\infty, \infty).$$

The density for the symmetric-stable s_α with tail-index α $(0 < \alpha \leq 2)$ when $\alpha \neq 1$ is given in Nolan (1997, Theorem 1) as

$$s_\alpha(x) = \begin{cases} \dfrac{\alpha x^{1/(\alpha-1)}}{\pi|\alpha-1|} \displaystyle\int_0^{\frac{\pi}{2}} v_\alpha(\theta) \exp\left(-x^{\frac{\alpha}{\alpha-1}} v_\alpha(\theta)\right) d\theta, & x > 0, \\ \frac{1}{\pi}\Gamma\left(1 + \frac{1}{\alpha}\right), & x = 0, \end{cases}$$

where

$$v_\alpha(\theta) = \left(\frac{\cos(\theta)}{\sin(\alpha\theta)}\right)^{\alpha/(\alpha-1)} \frac{\cos((\alpha-1)\theta)}{\cos(\theta)}$$

and when $x < 0$, $s_\alpha(x) = s_\alpha(-x)$. On the other hand, for $\alpha = 1$,

$$s_1(x) = s_1(-x) = t_1(x) = \frac{1}{\pi(1 + x^2)}.$$

No analytical expression for the stable densities is available in general, but there are useful power series representations (e.g. Feller (1971, Lemma 1, p. 583)).

The computer programs STABLE written by J. P. Nolan for computing the stable density, distribution function and quantile are available for free from
`http://academic2.american.edu//~jpnolan/stable/stable.html`.
This product is also commercially available in an SPLUS version from
`http://www.robustanalysis.com`.
Table 1 gives various (absolute) quantile values η_p of the (interquartile

range) standardised Gaussian, Laplace, t and stable distributions for some p values. It should be noted that the quantile value η_p is independent of the parameters σ and β in the Gaussian and Laplace cases, respectively.

It is immediately evident from Table 1 that little discrimination between distributions is available at the $p = 0.01$ level and that for useful discrimination sample sizes need to be at least sufficient to use $p = 0.001$. Note that even at the $p = 0.001$ level it is not possible to distinguish between the t_3 and $s_{1.8}$ distributions.

Table 1. Quantile value η_p (absolute) of the (interquartile range) standardised distribution for different p values.

p	g_σ	l_β	t_3	t_4	t_5	t_6	$s_{1.5}$	$s_{1.6}$	$s_{1.7}$	$s_{1.8}$
0.01	1.72	2.82	2.97	2.53	2.32	2.19	3.99	3.25	2.68	2.23
0.001	2.29	4.48	6.68	4.84	4.05	3.63	17.71	12.80	9.27	6.56
0.0001	2.76	6.14	14.51	8.80	6.66	5.59	81.81	53.57	35.51	23.09
0.00001	3.16	7.80	31.33	15.75	10.70	8.38	379.56	225.74	137.42	82.79

Distinguishing between quantiles in practice depends on the precision with which the quantiles can be estimated. This is elucidated by a result of Walker (1968) which gives that, if Y has a density $h_Y(y)$, which is continuous and positive at $y = \eta_p$, then $\hat{\eta}_p = Y_{([np])}$ (the $[np]$ order statistic of a sample of size n from the distribution of Y) is approximately normal with mean η_p and variance $p(1-p)/nh_Y^2(\eta_p)$, i.e.

$$\hat{\eta}_p = Y_{([np])} \sim N\left(\eta_p\,, \frac{p(1-p)}{nh_Y^2(\eta_p)}\right) \qquad (0 < p < 1), \qquad (1)$$

provided that n is sufficiently large. It is interesting to note from this result that the $[np]$ order statistics asymptotically have a finite variance, even in the case when Y has a stable distribution which has infinite variance.

To compute the standard deviation of $\hat{\eta}_p$, the density $h_Y(y)$ is required. Explicit formulae for $h_Y(y)$ in the case of the (interquartile range standardised) Gaussian and Laplace distributions are:

Gaussian

$$h_Y(y) = \frac{1.349}{\sqrt{2\pi}}\, e^{-\frac{1}{2}1.349^2 y^2}, \qquad y \in (-\infty, \infty),$$

Laplace

$$h_Y(y) = \frac{\ln 4}{2}\, e^{-|y|\ln 4}, \qquad y \in (-\infty, \infty).$$

The corresponding densities $h_Y(y)$ for t_ν and s_α are *dt.iqr* and *dstable.iqr*, respectively[a].

Tables 2, 3 and 4 use (1) and the densities given above to provide various standard deviation values of $\hat{\eta}_p$ when $n = 5\,000$, $10\,000$ and $15\,000$ respectively and for different values of p.

Table 2. Standard deviation of $\hat{\eta}_p$ for $n = 5\,000$.

p	g_σ	l_β	t_3	t_4	t_5	t_6	$s_{1.5}$	$s_{1.6}$	$s_{1.7}$	$s_{1.8}$
0.01	0.04	0.10	0.16	0.11	0.09	0.08	0.35	0.25	0.18	0.11
0.001	0.10	0.32	1.02	0.57	0.40	0.32	5.24	3.53	2.38	1.56

Table 3. Standard deviation of $\hat{\eta}_p$ for $n = 10\,000$.

p	g_σ	l_β	t_3	t_4	t_5	t_6	$s_{1.5}$	$s_{1.6}$	$s_{1.7}$	$s_{1.8}$
0.01	0.03	0.07	0.11	0.08	0.06	0.05	0.25	0.18	0.13	0.08
0.001	0.07	0.23	0.72	0.41	0.29	0.23	3.70	2.50	1.69	1.10
0.0001	0.19	0.72	4.86	2.24	1.39	1.01	54.50	33.44	20.84	12.77

Table 4. Standard deviation of $\hat{\eta}_p$ for $n = 15\,000$.

p	g_σ	l_β	t_3	t_4	t_5	t_6	$s_{1.5}$	$s_{1.6}$	$s_{1.7}$	$s_{1.8}$
0.01	0.02	0.06	0.09	0.06	0.05	0.04	0.20	0.15	0.10	0.07
0.001	0.06	0.19	0.59	0.33	0.23	0.19	3.02	2.04	1.38	0.90
0.0001	0.16	0.59	3.96	1.83	1.14	0.82	44.50	27.30	17.02	10.43

Although our focus is on comparing tailweights it is worth pointing out that the density at the origin for the distribution of Y in the Laplace case is much higher than for all the other distributions. Details are given in Table 5. Note that the results for the t distribution are very similar to those for the stable distribution.

[a] These are functions written in SPLUS codes which are available from the authors.

Table 5. The density at zero for the distribution of Y.

g_σ	l_β	t_3	t_4	t_5	t_6	$s_{1.5}$	$s_{1.6}$	$s_{1.7}$	$s_{1.8}$
0.538	0.693	0.562	0.556	0.552	0.549	0.557	0.551	0.547	0.543

3. Required Sample Size

In equation (1), the variance of $\hat{\eta}_p$ is a function of η_p and this can be replaced by $\hat{\eta}_p$, as $\hat{\eta}_p$ is a consistent estimator. So, the length of an approximate $100(1-\phi)\%$ confidence interval for η_p is

$$2z_{\phi/2}\sqrt{\frac{p(1-p)}{nh_Y^2(\hat{\eta}_p)}} \quad (0<\phi<1)$$

where $z_{\phi/2}$ is such that $P(Z > z_{\phi/2}) = \frac{\phi}{2}$. Hence, the required sample size n to ensure that η_p is estimated with precision ϵ is given by

$$n > \frac{p(1-p)\,z_{\phi/2}^2}{\epsilon^2 h_Y^2(\hat{\eta}_p)}\,. \tag{2}$$

From this result we can see how large a sample size n is needed; for example, to distinguish the t distribution from stable.

To illustrate we use data that consist of the S&P 500 index for the period July 1962 to December 2004[b] (returns standardised to have unit interquartile range) which has an estimated value of $\hat{\eta}_{0.01}$ of 2.58. Assume $Y = t_4$ is the true distribution for the data. Then the required sample size n at the 95% confidence interval (i.e. $\phi = 0.05$) for $\eta_{0.01}$ to be estimated with precision ϵ is given by

$$n > \frac{268.47}{\epsilon^2}.$$

To distinguish t_4 from the stable distribution with tail index $\alpha = 1.7$, say, the value for ϵ should be the difference between $\hat{\eta}_{0.01} = 2.58$ and the value of $\eta_{0.01} = 2.68$ under the stable distribution with $\alpha = 1.7$ obtained in Table 1. This requires that the the sample size n should be greater than 26 847. This corresponds to having 100 years of daily data! Sample sizes clearly need to be at least sufficient to use $p = 0.001$.

---

[b]data from `http://finance.yahoo.com`.

4. Chi-squared Tests

The limitations in distinguishing between distributions on the basis of their tail behaviour is clear from Sections 2 and 3, but it is useful to make quantitative comparisons of goodness-of-fit in the tails. This can be done using chi-squared tests.

We must begin by deciding on a percentage $p\,\%$ of the tail of the distribution to use. The choice of p may be motivated by the application, such as 1% for a value-at-risk calculation. Then consider partitioning the tail into m bins, where the probability for a value in bin i, p_i, is such that $\sum_{i=1}^{m} p_i = p/100$. Also, the expected number in bin i, $E_i = np_i$, n being the size of the sample, is chosen to be larger than 5 for each i. Then, the chi-squared test statistic is given by

$$T = \sum_{i=1}^{m} \frac{(O_i - E_i)^2}{E_i} \sim \chi^2_{m-1},$$

where O_i represents the observed number in the i-th bin.

We shall illustrate with two applications, one involving foreign exchange rates in Section 4.1, and the other the S & P 500 index in Section 4.2. We begin with $m = 8$ bins, $p_i = 0.00125$ and $p = 1$.

4.1. *Foreign exchange rates*

The data consist of foreign exchange rates involving the Australian dollar against the currencies of Australia's major trading partners (i.e. United States (USD), Japan (JPY), United Kingdom (GBP) and New Zealand (NZD))[c]. There are 5055 daily returns each of the countries for the period 12th December 1983 to 5th December 2003, and the data are standardised to have unit interquartile range.

Table 6. Chi-squared test statistics for foreign exchange rate returns at the 1% tail.

	g_σ	l_β	t_3	t_4	t_5	t_6	$s_{1.5}$	$s_{1.6}$	$s_{1.7}$	$s_{1.8}$
USD	1623.18	6.83*	9.86*	10.31*	46.56	91.94	28.62	20.25	18.37	76.40
JPY	1343.57	4.39*	14.11	16.99	31.83	68.26	36.25	24.65	25.62	86.33
GBP	643.33	17.28	23.53	12.53*	13.85*	21.56	44.13	31.15	21.17	33.36
NZD	1380.61	11.95*	10.17*	16.01	58.07	124.08	28.42	20.27	25.30	68.26

Table 6 gives value from a chi-squared goodness-of-fit test with results

[c]data from http://www.rba.gov.au/statistics/historical/index.html.

which are acceptable at the 5% significance level indicated by an asterisk. They suggest the Laplace or t distribution as a suitable model for the foreign exchange rate returns in the 1% tail.

Based on the results of Table 6, the Laplace distribution may seem to be a more suitable general purpose model than the t distribution for the 1% tail. It should be noted, however, in case it is relevant, that the empirical densities at zero are much better fitted by t-distributions.

Although the above chi-squared goodness-of-fit test rejects the stable distribution, it is interesting to note that going further out into the tail, to $p = 0.5$, and performing another chi-squared goodness-of-fit test (with $m = 4$ and $p_i = 0.00125$ and at the 5% significance level) shows a reasonable fit for the stable distribution with $\alpha = 1.7$ in the case of USD and NZD. The results are presented in Table 7.

Table 7. Chi-squared test statistics for foreign exchange rate returns at the 0.5% tail.

	g_σ	l_β	t_3	t_4	t_5	t_6	$s_{1.5}$	$s_{1.6}$	$s_{1.7}$	$s_{1.8}$
USD	1608.38	3.72*	5.56*	4.44*	30.47	73.73	21.91	14.66	7.14*	23.25
JPY	1272.79	2.63*	9.82	4.47*	21.32	49.06	25.28	21.59	12.09	42.71
GBP	639.16	8.04	18.86	7.52*	5.04*	20.58	25.28	23.43	18.86	8.75
NZD	1356.70	7.86	3.83*	9.67	54.76	117.74	18.22	14.86	7.75*	24.89

4.2. *Standard & Poors 500 index*

In this example, we look at a much larger data set than that of Section 4.1. The data set is of the S & P 500 index for the period July 1962 to December 2004. This provides $n = 10\,691$, enough to distinguish the tail of the distribution.

Table 8 gives values from a chi-squared goodness-of-fit test and clearly indicates t_4 as a suitable model for the 1% tail distribution (where, as above, an acceptable result at the 5% significance level is indicated by an asterisk).

Table 8. Chi-squared test statistics of S &P 500 index returns at the 1% tail.

	g_σ	l_β	t_3	t_4	t_5	t_6	$s_{1.5}$	$s_{1.6}$	$s_{1.7}$	$s_{1.8}$
S&P 500	2235.69	15.95	27.02	11.76*	36	79	74.52	48.20	37.93	115.37

Examining the chi-squared goodness-of-fit test (now with $m = 4$ and $p_i = 0.00125$) further out in the tail at $p = 0.5$ leads to the same result. This is presented in Table 9.

Table 9. Chi-squared test statistics of S&P 500 index returns at the 0.5% tail.

	g_σ	l_β	t_3	t_4	t_5	t_6	$s_{1.5}$	$s_{1.6}$	$s_{1.7}$	$s_{1.8}$
S&P 500	2201.11	11.50	17.58	7.67*	21.07	59.98	42.80	37.48	19.37	30.20

5. Counts over Threshold Indicators

If we go further into the tails of the distribution than the 1% or 0.5% level discussed in Section 4, then relative frequencies are low but counts over threshold indicators can still yield quite useful results. We shall illustrate with an assessment of the returns for the Dow Jones 30 Industrials for the period January 2, 1991 to January 2, 2001[d]. The returns have been standardised to variance one since infinite variance models do not fit at all well.

Table 11 gives the sample maxima and minima and the counts over 5 and 6 standard deviations for the absolute values of the returns for the 10 years of data. Table 10 gives the expected number of exceedances of the 5 and 6 standard deviation thresholds for the absolute values of the variance-standardised Laplace, t_5, t_4 and t_3 distributions.

Table 11 supplies a rating for each stock of *,** or -. One (two) star(s) suggest (strongly suggest) the use of the t-distribution with 3 or 4 degrees of freedom to model the tails while - suggests the use of, say, a t_6-distribution or a Laplace distribution.

Table 10. Expected exceedances for 2500 observations with variance-standardised distributions

distribution	> 5	> 6
$\|L\|$	2.12	0.53
$\|t_5\|$	2.25	0.75
$\|t_4\|$	5.25	2.75
$\|t_3\|$	8.00	4.75

6. Extreme Value Indicators

It may be tempting to use extreme values alone to suggest appropriate distributions to use, but, as we shall see, this gives much weaker information than that supplied by counts over threshold methods.

[d]data from http://finance.yahoo.com.

Table 11. Dow Jones 30 Industrials variance-standardised returns: Jan. 2, 1991 to Jan. 2, 2001.

Lookup	max	min	$\sum(\lvert obs\rvert > 5)$	$\sum(\lvert obs\rvert > 6)$	rating
AA	6.46	-5.24	5	1	*
AXP	5.68	-5.52	2	0	-
ATT	6.16	-14.70	10	4	**
BA	5.71	-9.45	5	3	**
CAT	4.97	-6.33	2	2	*
C	7.5	-5.23	3	1	*
KO	5.54	-6.55	4	1	*
DD	4.58	-6.04	3	1	*
EK	6.03	-15.2	8	5	**
XOM	6.59	-5.54	2	1	*
GE	4.90	-5.97	1	0	-
GM	3.69	-5.46	1	0	-
HWP	5.84	-7.54	7	2	**
HD	4.11	-15.6	1	1	**
HON	12.4	-9.10	6	3	**
INTC	4.68	-9.21	2	1	**
IBM	5.87	-8.07	8	2	**
IP	5.77	-5.70	2	0	-
JMP	5.77	-4.93	3	0	-
JNJ	4.58	-6.26	1	0	-
MCD	6.03	-6.37	4	3	**
MRK	5.13	-5.36	4	0	*
MSFT	7.67	-7.38	5	3	**
MMM	6.91	-6.71	4	3	**
MO	7.25	-12.8	10	5	**
PG	5.03	-20.1	6	1	**
SBC	6.11	-7.84	5	2	**
UTX	4.80	-4.67	0	0	-
WMT	4.25	-5.08	1	0	-
DIS	7.41	-8.90	6	3	**

The qualitatively different asymptotic behaviour engendered by different tailweights certainly suggests the use of sample extremes as a useful indicator. To see this we first examine the record behaviour of a sample $|X_1|, |X_2|, \ldots, |X_n|$ of i.i.d. random variables with distribution function F. Clearly, $\Pr(\max_{1\le k\le n} |X_k| > x) = 1 - F^n(x)$.

If

$$1 - F(x) \sim cx^{-\alpha} \tag{3}$$

as $x \to \infty$ for some $\alpha > 0$, it is easily checked that

$$\Pr(\max_{1\le k\le n} |X_k| > n^{1/\alpha} y) \to 1 - e^{-c/y^\alpha}.$$

Thus, for a power tail, $n^{-\frac{1}{\alpha}} \max_{1\le k\le n} |X_k|$ converges in distribution and the records increase at a power rate.

If, on the other hand, the tail is exponential,

$$1 - F(x) \sim ce^{-x\beta} \tag{4}$$

as $x \to \infty$ for some $\beta > 0$, it is easily checked that

$$\Pr(\max_{1\le k\le n} |X_k| > \frac{(\ln ny)}{\beta}) \to 1 - e^{-c/y},$$

that is, $\beta(\ln n)^{-1} \max_{1\le k\le n} |X_k|$ converges in probability to 1; the records increase at a logarithmic rate.

In the case of (3) we have a random limit and it is useful to note that, after an easy calculation,

$$E(\max_{1\le k\le n} |X_k|) \approx c^{1/\alpha}\Gamma(1 - \frac{1}{\alpha})n^{1/\alpha},$$

while for (4) the corresponding result is

$$E(\max_{1\le k\le n} |X_k|) \approx \beta^{-1} \ln n.$$

Despite the evident differences in growth rates, these two results give quite similar values in many cases of practical interest. For example, a 20-year sample of daily returns ($n = 5\,000$) involves comparison of $\ln(5\,000) = 8.51$ with $(5\,000)^{1/4} = 8.41$, $\alpha = 4$ being a favoured model parameter (see e.g. Heyde and Liu (2001)). The constants of proportionality also make comparisons difficult.

If we have finite variance and we standardise the variables to have unit standard deviation, we find for the t_ν distribution, $\nu > 2$, that the constant c in (3) is given by

$$c = c_\nu = \frac{2(\nu - 2)^{\nu/2}\Gamma(\frac{\nu+1}{2})}{\sqrt{\pi}\,\Gamma(\frac{\nu}{2})}$$

and $\alpha = \nu$. Thus, in particular, for $\nu = 3$ we have

$$E(\max_{1\le k\le n} |X_k|) \approx 1.468n^{1/3} \tag{5}$$

and for $\nu = 4$,

$$E(\max_{1\le k\le n} |X_k|) \approx 1.918n^{1/4}. \tag{6}$$

For the standardised Laplace distribution the corresponding result is

$$E(\max_{1\le k\le n} |X_k|) \approx 0.707 \ln n. \tag{7}$$

The result for the Laplace distribution, which of course has a pure exponential form, is an accurate reflection of what can be expected from simulations, but this is not the case with the results for the t-distributions, where (5) and (6) give substantially larger results for n around 5 000 than are observed from simulations. For example, 50 sets of 5 000 simulated variance-standardised t_4 variables gave a 95% confidence interval of 10.81 $\pm$ 0.84 for the mean maximum (and a range of 6.95 to 19.71) compared with a value of 16.13 for the approximation (5) in this case. On the other hand, 50 sets of 5 000 simulated variance-standardised Laplace variables gave a 95% confidence interval of 6.53 $\pm$ 0.25 for the mean maximum (and a range of 5.38 to 8.22) compared with a value of 6.02 for the approximation (7).

The poor comparisons for the t-distribution are the reflection of the intrinsic differences, for moderate tail values, between the t and Pareto distributions. There is slow convergence of the distribution of the normalised maximum to the corresponding extreme value distribution.

One must conclude that extreme value comparisons, if made, require very careful interpretation.

7. Conclusions

(1) Distinguishing between exponentially tailed and power tailed distributions, or within the family of power tailed distributions, requires large samples, perhaps 50 000 observations.
(2) Chi-squared tests of goodness-of-fit in the tails are useful but are sensitive to the part of the tails that is chosen.
(3) Counts over high thresholds provide useful quick general indications of tail behaviour.
(4) Extreme values are rather too heavily dependent on the exact form of the parent distribution to be of much help in discriminating between tail behaviours unless sample sizes are very large.

References

1. Embrechts, P., Klüppelberg, C. and Mikosch, T. (1997). *Modelling Extremal Events for Insurance and Finance.* Springer, Germany.
2. Feller, W. (1971). *An Introduction to Probability Theory and its Applications.* Vol. II, 2nd Ed. Wiley, New York.
3. Heyde C.C. and Kou, S.G. (2004). On the controversy over tailweight of distributions. *Operations Research Letters*, **32**, 399–408.

4. Heyde, C.C. and Liu, S. (2001). Empirical realities for a minimal description risky asset model: the need for fractal features. *J. Korean Math. Soc.*, **38**, 1047–1059.
5. Nolan, J.P. (1997). Numerical calculation of stable densities and distribution functions. *Commun. Statist. — Stochastic Models*, **13**, 759–774.
6. Rachev, S. and Mittnik, S. (2000). *Stable Paretian Models in Finance.* Wiley, New York.
7. Reiss, R.-D. and Thomas, M. (2001). *Statistical Analysis of Extreme Values.* 2nd Edition, Birkhauser, Basel.
8. Rolski, T., Schmidli, H., Schmidt, V. and Teugels, J. (1999). *Stochastic Processes in Insurance and Finance.* Wiley, New York.
9. Walker, A.M. (1968). A note on the asymptotic distribution of sample quantiles. *J. Roy. Statist. Soc. Series B*, **30**, 570–575.

PART E

Numerical Methods

An Approximate Maximum *a Posteriori* Method with Gaussian Process Priors

Markus Hegland

Statistical Machine Learning Group, National ICT Australia,
Canberra ACT 2601, Australia
and
CMA, Mathematical Sciences Institute, Australian National University,
Canberra ACT 0200, Australia
E-mail: markus.hegland@anu.edu.au

The maximum *a posteriori* method is generalised for infinite dimensional problems and it is shown that in this case the problem can be reduced to a nonlinear variational problem. This is not a trivial generalisation as the probability density used for the finite dimensional case does not exist. It is shown how the logarithmic gradient can be used to characterise stationary points for the Gaussian process prior case. A nonconforming finite element method is suggested using sparse grids to solve the resulting variational problem.

Keywords: maximum *a posteriori* method; machine learning.

1. Introduction

Computational learning theory deals with the problem of constructing computational systems which "learn" their behaviour from examples. A particular learning problem aims to find a function $f : X \to Y$ from a given sequence of training or data points

$$\mathcal{D} = \{(x_1, y_1), \ldots, (x_n, y_n)\}.$$

A slightly more general problem considered here assumes that these points are drawn from a probability distribution with density $p(x, y)$ and the aim is to reconstruct the conditional distribution $p(y|x) = p(x, y)/ \int_Y p(x, y)\, dy$ from the data $\mathcal{D}$. An important branch of computational learning theory considers the problem of how well $f(x)$ or $p(y|x)$ can be recovered from $\mathcal{D}$ with increasing data sizes n. This discussion is based on a rich mathematical framework and draws insights from several branches of mathematics and statistics [1–9].

For (small) finite sets X and Y one may estimate $p(y|x)$ by counting the occurrence of each combination of values (x, y) in $\mathcal{D}$. When either one or both of the sets X and Y are infinite, the learning problem as stated above is ill-posed and requires further information about $p(y|x)$ for the solution. If Y is finite, the learning problem is referred to as the *classification problem*; the case $Y = \mathbb{R}$ is the *regression problem*; and if X contains only one element, one gets a *density estimation* problem.

In the case where p is everywhere nonzero one can introduce a function $u(x, y)$ such that

$$p(y|x) = \exp(u(x, y) - g(u, x)).$$

For example, if $p(y|x)$ is a member of the *exponential family* it will have this form. The introduction of the *log partition function* $g(u, x)$, which at first sight appears to be redundant, allows the specification of additional information about the shape of $p(y|x)$ independent of the normalisation $\int p(y|x)\, dy = 1$ (which is taken care of by $g(u, x)$) and it follows that

$$g(u, x) = \log \int_Y \exp(u(x, y))\, dy.$$

The required additional information often relates to the fact that $u(x, y)$ is chosen from a given function class. This includes the normal distribution where $u(x, y)$ is a second order polynomial and other members of the exponential family. Alternatively, u may satisfy smoothness conditions in the form of bounds on the derivatives or a Sobolev norm. Here, this prior information is given in the form of a *prior probability distribution* for u over a function space. If this function space is finite-dimensional, one can then use the *maximum a posteriori* method to get an estimate for u and thus p. This method estimates u by the maxima of the *a posteriori* density of the u. In the following a method is proposed which generalises this maximum *a posteriori* method to the case of infinite dimensional function spaces. Note that one cannot in general assume the existence of a density for the posterior probability. This is because the density is the Radon-Nikodym derivative of the probability distribution with respect to the Lebesgue measure which does not exist for general function spaces as there are no translationally invariant measures in this case [10] (This fact is well known to theoretical physicists and has to be dealt with in the definition of Feynman path integrals.).

If u is a function $u : X \times Y \to \mathbb{R}$ then it is an element of the *locally convex* space $\mathbb{R}^{X\times Y}$. The particular prior considered here is a *Gaussian measure* on $\mathbb{R}^{X\times Y}$, or, equivalently, a Gaussian stochastic process or random field with index set $X \times Y$. While the prior or posterior probability distributions cannot be characterised by a probability density, the *likelihood function* is a density relating to the probability distribution of the data $\mathcal{D}$ given the function u and is defined as $L(u, \mathcal{D}) = p(\mathcal{D}|u)$ and one gets

$$L(u, \mathcal{D}) = \prod_{i=1}^{n} \exp(u(x_i, y_i) - g(u, x_i)).$$

In section 2 the maximum *a posteriori* method will be generalised for $u \in \mathbb{R}^{X \times Y}$ and the maximum *a posteriori* method will be reduced to the variational problem of minimising the functional

$$J(u) = \|u\|_{H(\gamma)}^2 + l(u),$$

where $\|\cdot\|_H$ is the Cameron-Martin or reproducing kernel Hilbert space norm and where $l(u) = -\log L(u, \mathcal{D})$, i.e.,

$$l(u) = -\sum_{i=1}^{n} (u(x_i, y_i) - g(u, x_i)).$$

In section 3 the numerical solution of this problem is discussed. In particular, a nonconforming Galerkin method using sparse grids is suggested and some preliminary approximation results derived. A more thorough discussion including numerical examples will be given in a follow-up paper.

This discussion is an alternative to the common approach found in the literature which is based on *finite dimensional distributions* of the form $p(u(x_1, y_1), \ldots, u(x_n, y_n))$ [11–13]. The characterisation by a variational problem used here opens the path to the application of standard techniques used for the solution of nonlinear partial differential equations.

2. Gaussian Priors and the Maximum *a Posteriori* Method

2.1. *Gaussian Priors*

In contrast to the finite dimensional case, a Gaussian probability measure γ on $\mathbb{R}^{X \times Y}$ does not have a density in general. However, as in the finite dimensional case it is still characterised by the mean and a covariance operator.

The domain of the probability measure is the locally convex space $\mathbb{R}^{X \times Y}$. The topology of this space relates to the pointwise continuity of the functions and is the product topology with respect to the index set $X \times Y$. With respect to this topology the point evaluation functionals

$\delta_{(x,y)} : u \to u(x,y)$ are continuous. In fact, the (topological) dual of $\mathbb{R}^{X\times Y}$ is the set of all functionals of the form

$$\phi(u) = \sum_{j=1}^{n} c_j u(x_j, y_j),$$

for some $x_j \in X$, $y_j \in Y$ and $c_j \in \mathbb{R}$. An in-depth treatment of locally convex spaces can be found in [14].

The continuous linear functionals are used to establish the sigma-algebra $\mathcal{E}(\mathbb{R}^{X\times Y})$ for the probability measure. This sigma algebra is generated by sets of the form $\{u \in \mathbb{R}^{X\times Y} \mid \phi(u) < c\}$ where the ϕ are continuous functionals on $\mathbb{R}^{X\times Y}$. The prior is then a probability measure γ on $\mathbb{R}^{X\times Y}$ with respect to this sigma algebra. Here it is assumed that the measure is a Radon Gaussian measure so the measurable sets can be approximated by compact sets and the values of any linear functionals $\phi(u)$ are normally distributed random variables. An excellent discussion of Gaussian measures can be found in [15].

If u is distributed according to this Gaussian measure γ, the values $u(x,y)$ are a family of normally distributed random variables and so u is a *Gaussian random field*. The values of a continuous functional $\phi = \sum_{i=1}^{n} c_i \delta_{(x_i,y_i)}$ are then normally distributed with expectation

$$\int_{\mathbb{R}^{X\times Y}} \sum_{i=1}^{n} c_i u(x_i, y_i) \gamma(du) = \sum_{i=1}^{n} c_i \overline{u}(x_i, y_i)$$

and variance

$$\int_X \sum_{i,j=1}^{n} c_i c_j (u(x_i,y_i) - \overline{u}(x_i,y_i))(u(x_j,y_j) - \overline{u}(x_j,y_j))\gamma(du)$$
$$= \sum_{i,j=1}^{n} c_i c_j K(x_i, y_i, x_j, y_j)$$

where $\overline{u}(x,y)$ is the mean value of the random field and $K(x,y,x',y')$ the kernel or covariance matrix. These two functions, which correspond to linear and bilinear operators on $\mathbb{R}^{X\times Y}$ respectively, uniquely determine the Gaussian measure γ. These two functions define the prior and encode the "most likely function" u when no data is available and the covariance between the function values of u at different locations (x,y). In this way one can introduce areas with high correlation between the function values so that the function values are very similar in these areas.

The prior γ defines a norm

$$\|u\|_{H(\gamma)} = \sup_{\phi}\{\phi(u) \mid R(\phi, \phi) \leq 1\}$$

where the supremum is taken over all continuous linear functionals ϕ with

$$\phi(u) = \sum_{i=1}^{n} c_i u(x_i, y_i), \quad u \in \mathbb{R}^{X \times Y},$$

and

$$R(\phi, \phi) = \sum_{i,j=1}^{n} c_i c_j K(x_i, y_i, x_j, y_j).$$

The domain of this seminorm is the *Cameron-Martin space*

$$H(\gamma) = \{u \mid \|u\|_{H(\gamma)} < \infty\},$$

which is a reproducing kernel Hilbert space (RKHS). It can be seen that this space consists of all $u \in \mathbb{R}^{X \times Y}$ which can be represented by (pointwise) limits of sequences $u_1, u_2, \ldots$ where

$$u_n(x, y) = \sum_{i=1}^{n} c_{i,n} K(x_{i,n}, y_{i,n}, x, y)$$

and the sequences $c_{i,n}$, $x_{i,n}$ and $y_{i,n}$ are such that the corresponding functionals ϕ_n with

$$\phi_n(u) = \sum_{i=1}^{n} c_{i,n}(u(x_{i,n}, y_{i,n}) - \overline{u}(x_{i,n}, y_{i,n}))$$

converge in $L_2(\gamma)$. Note that in general the limit $\tilde{\phi}$ of the ϕ_n is not a continuous linear functional on $X \times Y$ but it can be viewed as a continuous linear functional on the Hilbert space $H(\gamma)$ and one has

$$\tilde{\phi}(u) = \|u\|^2_{H(\gamma)}.$$

The set of $\tilde{\phi}$ defined in this way contains all the continuous linear functionals ϕ on $\mathbb{R}^{X \times Y}$.

2.2. *Logarithmic derivatives and stationary points*

For the following, consider (nonlinear) functionals of the form

$$\psi(u) := \theta(\phi_1(u), \ldots, \phi_m(u))$$

where $\theta \in C_b^\infty(\mathbb{R}^m)$ and $\phi_i \in X^*$. These are the *smooth cylindrical functionals* and they have Fréchet derivatives of the form

$$\partial_v \psi(u) = \sum_{i=1}^m \frac{\partial \theta}{\partial \phi_i}(u)\phi_i(v).$$

One now says that a Radon measure μ on $\mathbb{R}^X$ is *differentiable along* $v \in \mathbb{R}^X$ (in the sense of Fomin) if there exists a functional $\beta_v^\mu \in L^1(\mu)$ such that, for all smooth cylindrical functions ψ, one has

$$\int_X \partial_v \psi(u)\mu(du) = -\int_X \psi(u)\beta_v^\mu(u)\mu(du).$$

The functional β_v^μ is called the *logarithmic derivative of* μ *along* v. One can now further show that, for a Radon Gaussian measure γ with expectation zero, one has

$$\beta_v^\gamma(u) = -(v,u)_{H(\gamma)}$$

and γ is differentiable for all $u \in H(\gamma)$ along $v \in H(\gamma)$.

For any $L \in L^1(\gamma)$ let $\mu = L \cdot \gamma$ be a measure such that $L = d\mu/d\gamma$ is the Radon-Nikodym derivative. It can now be seen that if $L \in L^1(\gamma)$ is differentiable, then $\mu = L \cdot \gamma$ is differentiable and the logarithmic derivative along v is in this case

$$\beta_v^\mu(u) = -(v,u)_{H(\gamma)} + \partial_v L(u)/L(u).$$

For the following discussion, only measures of the form $\mu = L \cdot \gamma$ will be considered.

The logarithmic derivative provides the means to define stationary points of probability distributions which generalise the concept of stationary points of the probability density functions. Note first that regular variable transformations, while leading to transformed densities, do preserve the stationary points. The notion of a stationary point turns out not to depend on the existence of a density at all and one has:

Definition 2.1. A point $u \in \mathbb{R}^{X \times Y}$ is called a *stationary* point for the measure μ if the logarithmic derivative

$$\beta_v^\mu(u) = 0$$

for all v for which μ is differentiable along v.

One can see that this characterises the stationary points of the density for the finite dimensional case. Furthermore, a Gaussian measure has only one stationary point which coincides with the mean of the distribution. In the case of the measure $\mu = L \cdot \gamma$ one obtains the following variational characterisation of stationary points u:

$$-(v,u)_{H(\gamma)} + \partial_v L(u)/L(u) = 0$$

for $v \in H(\gamma)$. A particular class of stationary points consists of the modes of the distribution and thus these equations lead to a (partial) characterisation of the modes. Distinguishing modes from other stationary points, however, does require additional conditions which will be discussed elsewhere.

2.3. *Maximum a posteriori method with exponential families*

In order to apply the concepts of the previous sections to the maximum *a posteriori* method one requires sufficient regularity of the likelihood function. This mainly translates into conditions for the log partition function $g(u,x)$ and the prior. First it will be assumed that the prior is centred, i.e., that the mean is zero. A major ingredient of the log partition function is the integral operator

$$u \in H \mapsto \int_Y u(x,y)\, dy.$$

Denote by $(\mathbb{R}^{X\times Y})^*_\gamma$ the closure of the dual of $\mathbb{R}^{X\times Y}$ with respect to $L^2(\gamma)$. The main condition suggested here is that the integral operator is well defined and an element of $(\mathbb{R}^{X\times Y})^*_\gamma$. It follows that the operator is a continuous linear functional on $H(\gamma)$ [15]. In the following, let $k_{x,y}$ denote the reproducing kernel of $H(\gamma)$ which is defined by

$$(k_{x,y}, u)_{H(\gamma)} = u(x,y).$$

Using the Riesz representation theorem in $H(\gamma)$, there also is a $k_x \in H(\gamma)$ such that

$$\int_Y u(x,y)\, dy = (k_x, u)_{H(\gamma)}.$$

Formally, we will write

$$k_x = \int_Y k_{x,y} dy.$$

We will assume that for every $u \in H(\gamma)$ one has $\exp(u) \in H(\gamma)$ and so the log partition function is

$$g(u,x) = \log((k_x, \exp(u))_{H(\gamma)}),$$

where $(\cdot,\cdot)_{H(\gamma)}$ is the scalar product associated with $\|\cdot\|_{H(\gamma)}$.

By theorem 2.10.9 and lemma 2.4.1 in [15] the linear functional $u \to \int_Y u(x,y)\,dy$ is measurable and it follows that $g(\cdot, x)$ is measurable as it is a composition of measurable functionals. Furthermore, $g(u,\cdot)$ is defined for any $u \in \mathbb{R}^{X\times Y}$ γ-almost-everywhere.

For the following it will also be assumed that $H(\gamma)$ is continuously embedded in $C(X \times Y)$. For a $h \in H(\gamma)$ one now has

$$\left|\frac{dg(u+th,x)}{dt}\right| = \left|\frac{\int_Y \exp(u(x,y)+th(x,y))h(x,y)\,dy}{\int_Y \exp(u(x,y)+th(x,y))\,dy}\right| \le \|h\|_\infty \le C\|h\|_{H(\gamma)}$$

where C is the *embedding constant* of the reproducing kernel Hilbert space $H(\gamma)$. Consequently, $g(u+th,x)$ is an absolutely continuous function of t. Consider now the likelihood function L defined by

$$L(u,\mathcal{D}) = \prod_{i=1}^{n} \exp(u(x_i,y_i) - g(u,x_i))$$

where the data n-tuple is $\mathcal{D} = \{(x_1,y_1),\ldots,(x_n,y_n)\}$. From the definition and the properties of g it follows that L is measurable with respect to the product measure $\gamma(du) \times dy_1 \times \cdots \times dy_n$ and

$$\int_{\mathbb{R}^n} L(u,\mathcal{D})\,dy_1\cdots dy_n = 1.$$

By *Tonelli's theorem*, $L(\cdot,\mathcal{D})$ is in $L^1(\gamma)$ for almost every $\mathcal{D}$ and one has

$$\int_Y \int_{\mathbb{R}^{X\times Y}} L(u,\mathcal{D})\gamma(du)\,dy_1\cdots dy_n = \int_{\mathbb{R}^{X\times Y}} \int_Y L(u,\mathcal{D})\,dy_1\cdots dy_n\,\gamma(du) = 1.$$

A direct application of corollary 5.1.10 in [15] provides:

Proposition 2.1. *Let γ be a Radon Gaussian measure on $\mathbb{R}^{X\times Y}$ such that $H(\gamma) \subset C(X\times Y)$ is continuously embedded and $h \in H(\gamma)$. Furthermore, let the functional $u \to \int_Y u(x,y)\,dy$ be in $L_2(\gamma)$ for all $x \in X$. Then the logarithmic derivative (defined in Section 2.2) of the measure $\mu = L(\cdot,\mathcal{D})\cdot\gamma$ satisfies*

$$\beta_h^\mu = \beta_h^\gamma + \partial_h L(\cdot,\mathcal{D})/L(\cdot,\mathcal{D}).$$

The *a posteriori* measure is the conditional measure defined by

$$\mu(A|\mathcal{D}) = \int_A L(u,\mathcal{D})\gamma(du) \Big/ \int_{\mathbb{R}^{X\times Y}} L(u,\mathcal{D})\gamma(du)$$

for any measurable $A \subset \mathbb{R}^{X\times Y}$. This is just a scaled version of the measure μ from the previous proposition. It follows that the two distributions share their extremal and stationary points. Now set

$$l(u) = -\log L(u,\mathcal{D})$$

and one has for the exponential family case

$$l(u) = -\sum_{i=1}^{n}(u(x_i, y_i) - g(u, x_i)). \tag{1}$$

A consequence of the proposition is that any stationary point must satisfy

$$(h, u)_H + \partial_h l(u) = 0$$

for all $h \in H(\gamma)$ as $\beta_h^\gamma(u) = -(h, u)_H$. Furthermore, it will be shown in an upcoming paper that a maximal point of $\mu(A|\mathcal{D})$ minimises

$$J(u) = \|u\|_H^2 + l(u).$$

In summary, the maximum *a posteriori* method is well defined for the infinite dimensional case as well and leads to a variational problem in a Hilbert space. The next section covers the numerical solution of this problem using a nonconforming Galerkin approach.

3. Numerical Solution

The maximum *a posteriori* method was reduced to the problem of finding a minimum of the functional

$$J(u) = \|u\|_{H(\gamma)}^2 + l(u)$$

where $\|u\|_{H(\gamma)}$ is the Cameron-Martin or RKHS norm defined by the prior and where

$$l(u) = -\sum_{i=1}^{n}(u(x_i, y_i) - g(u, x_i)).$$

It has been assumed that the prior is such that $H(\gamma) \subset C(X \times Y)$ (the set of continuous functions on $X \times Y$) and that $\int_Y dy = 1$. By the mean value theorem for every x_i there is an $\eta_i \in Y$ such that

$$g(u, x_i) = \log \int_Y \exp(u(x_i, y))\, dy = \log \exp(u(x_i, \eta_i)) = u(x_i, \eta_i).$$

It follows then that $l(u) = -\sum_{i=1}^{n}(u(x_i, y_i) - u(\xi_i, \eta_i))$ or, with the Riesz representation theorem:

$$l(u) = -\sum_{i=1}^{n}\left(k_{x_i, y_i} - k_{\xi_i, \eta_i}, u\right)_{H(\gamma)}.$$

With $w = -\sum_{i=1}^{n}(k_{x_i, y_i} - k_{\xi_i, \eta_i})/2$, one gets $J(u) = \|u\|_{H(\gamma)}^2 + 2(w, u)_{H(\gamma)} = \|u + w\|_{H(\gamma)}^2 - \|w\|_{H(\gamma)}^2$. While w depends on u, the prior probability distribution has been chosen such that $\|k_{x,y}\|_{H(\gamma)} \leq C$ (where C is the embedding constant introduced previously), and, by the triangle inequality, one then gets

$$\|w\|^2_{H(\gamma)} \leq 4n^2C^2.$$

Thus $J(u) \geq -4n^2C^2$ and it follows that $J(u)$ is a *proper* functional, i.e., bounded from below and not identically ∞. One also gets that J is *coercive* in the sense that, if for some set $S \subset H(\gamma)$ one has $\sup_{u\in S} \|u\|_{H(\gamma)} = \infty$, then one also has $\sup_{u\in S} J(u) = \infty$. It follows from the differentiability of l that J is *weakly lower semicontinuous*, i.e., that every weakly converging sequence $v_n \rightharpoonup v$ satisfies

$$J(v) \leq \liminf_{n\to\infty} J(v_n).$$

One then has

Proposition 3.1. *Let $J(u) = \|u\|^2_{H(\gamma)} + l(u)$ where $\|\cdot\|_{H(\gamma)}$ is the Cameron-Martin norm for a Gaussian measure γ and l is the negative log likelihood from equation (1). If γ is centred with uniformly bounded kernel $\|k_{x,y}\|_{H(\gamma)} \leq C$, then J has at least one (global) minimum $u^* \in H(\gamma)$.*

Proof. Consider a minimising sequence u_n such that $J(u_n) \to \inf_{u\in H(\gamma)} J(u)$. One can choose a subsequence which is bounded as $J(u)$ is proper. Thus there is a weakly converging subsequence u_{n_i}; let the limit be u^*. As J is weakly semicontinuous one has

$$J(u^*) \leq \liminf J(u_{n_i}) = \inf J(u)$$

and so u^* is a minimum of J. □

While in general J can have many minima, $J(u)$ is strictly convex in some important cases. This includes the case of regression where $p(y|x)$ are all normal and where $u(x,y)$ is a quadratic function of y.

For regression and classification the representer theorem leads to an explicit representation for u of the form

$$u = \sum_{i=1}^{n} c_i k(\cdot, x_i)$$

for the minimiser. However, the determination of u using this approach requires computing the coefficients c_i which leads to the (repeated) solution of large dense linear systems which makes this approach very expensive for large data sets. A feasible, more generally applicable alternative is based on approximations of u and shall be further developed in the following.

Consider first conforming methods where one has a family of finite dimensional subspaces $V_\alpha \subset H(\gamma)$ which approximate elements of $H(\gamma)$ in the $\|\cdot\|_\infty$ norm such that

$$\inf_{v \in V_\alpha} \|u - v\|_\infty \leq \epsilon_\alpha \|u\|_{H(\gamma)}, \quad \text{for } u \in H(\gamma).$$

One can now get a bound on how close the global minimum of $J(u)$ in $H(\gamma)$ is to the minimum in V_α:

Proposition 3.2. *If $V_\alpha \subset H(\gamma)$ is such that $\operatorname{dist}_\infty(u, V_\alpha) \leq \epsilon_\alpha \|u\|_{H(\gamma)}$ for all $u \in H(\gamma)$ and if J is Lipschitz continuous with Lipschitz constant L_J with respect to $\|\cdot\|_\infty$, then*

$$J(u^*) \leq J(u_\alpha^*) \leq J(u^*) + L_J \epsilon_\alpha \|u^*\|_{H(\gamma)}.$$

Proof. As $J(u_\alpha^*) \leq J(P_\alpha u^*)$ one has

$$J(u_\alpha^*) \leq J(u^*) + |J(P_\alpha u^*) - J(u^*)| \leq J(u^*) + L_J \epsilon_\alpha \|u^*\|_{H(\gamma)}.$$

The lower bound follows from the definition of u^*. □

Consequently, for small ϵ_α the functional is "close to minimised" and, if the optimisation problem is locally well posed, the set of u_α^* has limit points at the minima of J.

In most of the practical choices of $H(\gamma)$ the conforming approach is not feasible. For a *nonconforming method* one can first derive a generalisation of the representer theorem. For this one needs the Fréchet derivative of $J(u)$ which shall be denoted by ∇J. By definition, $\nabla J \in H(\gamma)^*$ where $H(\gamma)^*$ is the dual of the Hilbert space $H(\gamma)$. Furthermore, let $K^* : H(\gamma)^* \to H(\gamma)$ be the bijection given by the Riesz representation theorem such that

$$\phi(v) = (K^* \phi, v)_{H(\gamma)}, \quad \text{for } \phi \in H(\gamma)^* \text{ and } v \in H(\gamma).$$

The operator K^* is essentially the *reproducing kernel* of the Hilbert space; more specifically, one has

$$v(x, y) = \delta_{x,y}(v) = (K^* \delta_{x,y}, v)_{H(\gamma)}, \quad (x, y) \in X \times Y$$

and K^* will be called the *reproducing kernel operator*. From the stationarity of the functional one gets the following variational characterisation:

Proposition 3.3. *Let u be a minimiser of the nonlinear functional $J = \|\cdot\|^2_{H(\gamma)} + l$. Then*

$$u + \frac{1}{2}K^*\nabla l(u) = 0$$

where K denotes the reproducing kernel operator of $H(\gamma)$.

Proof. The minimum is a stationary point and so the values of the linear functional $\nabla J(u)$ are

$$\langle \nabla J(u), v\rangle = 0$$

for all $v \in H(\gamma)$. Inserting the derivative of $\|u\|^2_{H(\gamma)}$ and the definition of K^* one gets

$$\langle \nabla J(u), v\rangle = 2(u, v)_{H(\gamma)} + (K^*\nabla l(u), v)_{H(\gamma)}$$

and the equations follow as this is zero for all $v \in H(\gamma)$. □

Note that this characterises any stationary point including local minima, maxima and saddle points.

The approximation proceeds by first selecting a collection of approximating spaces $V_\alpha \subset C(X \times Y)$ and a projection operator $P_\alpha : C(X \times Y) \to V_\alpha$ such that, for any $u \in H(\gamma)$, one has

$$\|u - P_\alpha u\|_\infty \le \epsilon_\alpha \|u\|_{H(\gamma)}.$$

Particular examples depend on the type of problem solved. For the classification problem where $X = [0,1]^d$ and $Y = \{0, \ldots, m-1\}$ one may choose $V_\alpha = U_\alpha \otimes \mathbb{R}^Y$ where U_α is a space of piecewise multilinear functions on $[0,1]^d$ and α is an index describing the particular space chosen, e.g., the grid size or level. For the regression problem and $X = [0,1]^{d_1}$ and $Y = [0,1]^{d_2}$ one may choose for V_α a space of piecewise multilinear functions on $[0,1]^{d_1+d_2}$. For the density estimation problem where $X = [0,1]^d$ and $Y = \{0\}$ one may choose V_α to be a space of piecewise multilinear functions on $[0,1]^d$. In all these cases the operator P_α is chosen to be the interpolation on the grid points so that for all grid points (x_i, y_j) one has

$$(P_\alpha u)(x_i, y_j) = u(x_i, y_j).$$

The Galerkin approximation $u_\alpha \in V_\alpha$ is defined as the solution of the equations

$$u_\alpha + \frac{1}{2}P_\alpha K^*\nabla l(u_\alpha) = 0.$$

Error bounds for u_α will be discussed elsewhere. In many practical cases it turns out that one gets $\|u - u_\alpha\|_\infty \leq C(u)\epsilon_\alpha$ for some constant $C > 0$ depending on the solution u. The nonlinear equations for u_α are solved using a Newton method combined with a conjugate gradient based solver for the linear system. The details of the solver and its performance will be discussed elsewhere.

For large d in the previous examples the spaces of multilinear functions suffer under the curse of dimensionality as the number of grid points (and thus the dimension of the space) grows exponentially with the dimension. In practice, only up to around $d = 5$ is computationally feasible. For higher dimensions one considers sparse grid approximations [16]. Sparse grid spaces are function spaces defined as sums of the ordinary piecewise multilinear function spaces V_α:

$$V_I^{\text{sg}} = \sum_{\alpha \in I} V_\alpha$$

where I is the set of grids considered. If the set of spaces V_α is complete under intersections the interpolation onto the sparse grid space satisfies

$$P_I^{\text{sg}} = \sum_{\alpha \in I} c_\alpha P_\alpha.$$

The Galerkin approximation requires the solution of

$$u_I^G + \frac{1}{2} \sum_{\alpha \in I} P_\alpha K^* \nabla l(u_I^G) = 0$$

in a sparse grid space. The terms $P_\alpha K^* \nabla l(u_I^G)$ lead to large dense linear systems in the approximation. An alternative is the application of combination technique approximations [17] which take the form

$$u_I^C = \sum_{\alpha \in I} c_\alpha u_\alpha,$$

where the combination coefficients c_α are chosen as above and where the u_α are the solutions of the variational problem in the (smaller) partial spaces. It has been seen that these combination approximations can be slightly improved by adapting the coefficients c_α to the data as well; see [18,19]. A comprehensive discussion of the specific numerical scheme is the subject of current work and will be published in due course.

4. Conclusion

The maximum *a posteriori* method, while very popular in finite dimensional settings, suffers under the nonexistence of a prior probability density in the infinite dimensional case. It is shown here that the Cameron-Martin theory provides a framework for these cases. Generalisations of the representer theorem have been derived for this case and some initial discussion of the variational problems carried out. The current results suggest that computational learning problems can be efficiently solved using numerical techniques like the finite element method and sparse grid approximations. Current research is underways investigating particular implementations and application examples.

Acknowledgments

The author would like to thank the members of the NICTA Statistical Machine Learning Program, Prof. Michael Griebel (University Bonn), Prof. Ian Sloan (UNSW) and Dr Jochen Garcke (ANU) for many stimulating discussions and my former colleague Dr Boris Buchmann for introducing me to Gaussian measures and, in particular to [15]. The comments of the anonymous referees and the contributions by the editor Dr Shuangzhe Liu have greatly improved the presentation.

References

1. G. Kimeldorf and G. Wahba. Some results on Tchebycheffian spline functions. *J. Math. Anal. Appl.* **33**, 82–95 (1971).
2. G. Wahba. *Spline models for observational data.* CBMS-NSF Regional Conference Series in Applied Mathematics, Vol. 59 (Society for Industrial and Applied Mathematics (SIAM), Philadelphia, PA, 1990).
3. B. Schölkopf and A. Smola. *Learning with Kernels, Support Vector Machines, Regularization, Optimization and Beyond* (MIT Press, Cambridge, MA, 2002).
4. T. Poggio and S. Smale. The mathematics of learning: dealing with data. *Notices Amer. Math. Soc.* **50**, 537–544 (2003).
5. F. Cucker and S. Smale. On the mathematical foundations of learning. *Bull. Amer. Math. Soc. (N.S.)* **39**, 1–49 (2002).
6. B. D. Ripley. *Pattern Recognition and Neural Networks* (Cambridge University Press, Cambridge, 1996).
7. M. Anthony and P. Bartlett. *Neural Network Learning: Theoretical Foundations* (Cambridge University Press, 1999).
8. L. Devroye, L. Györfi and G. Lugosi. *A Probabilistic Theory of Pattern Recognition.* Applications of Mathematics, No. 31 (Springer, New York, 1996).
9. V. Vapnik. *The Nature of Statistical Learning Theory* (Springer Verlag, New York, 1995).
10. Y. Yamasaki. *Measures on infinite-dimensional spaces.* Series in Pure Mathematics, Vol. 5 (World Scientific Publishing Co., Singapore, 1985).

11. C. K. I. Williams and D. Barber. *IEEE Trans. Pattern Anal. Mach. Intell.* **20**, 1342–1351, (1998).
12. M. Seeger. Gaussian Processes for machine learning. *International Journal of Neural Systems* **14**, 69–106, (2004).
13. Y. Altun, A. J. Smola and T. Hofmann. Exponential families for conditional random fields. *AUAI '04: Proceedings of the 20th conference on Uncertainty in artificial intelligence*, 2–9, (AUAI Press, Arlington, Virginia, United States, 2004).
14. H. Jarchow. *Locally convex spaces.* (B. G. Teubner, Stuttgart, 1981). Mathematische Leitfäden. [Mathematical Textbooks].
15. V. I. Bogachev. *Gaussian measures.* Mathematical Surveys and Monographs, Vol. 62 (American Mathematical Society, Providence, RI, 1998).
16. C. Zenger. Sparse grids, in *Parallel algorithms for partial differential equations (Kiel, 1990)*, Notes Numer. Fluid Mech. Vol. 31, 241–251 (Vieweg, Braunschweig, 1991).
17. M. Griebel, M. Schneider and C. Zenger. A combination technique for the solution of sparse grid problems, in *Iterative methods in linear algebra (Brussels, 1991)*, 263–281 (North-Holland, Amsterdam, 1992).
18. M. Hegland. Adaptive sparse grids, in *Proc. of 10th Computational Techniques and Applications Conference CTAC-2001*, eds. K. Burrage and R. B. Sidje, *ANZIAM J.* **44**, C335–C353, April 2003.
19. M. Hegland. Additive sparse grid fitting, in *Curve and surface fitting (Saint-Malo, 2002)*, Mod. Methods Math. 209–218 (Nashboro Press, Brentwood, TN, 2003).

Data Extraction for Improved Prediction Outcomes in Organ Transplantation — A Hybrid Approach

Fariba Shadabi

Medical Informatics Centre, University of Canberra,
Canberra ACT 2601, Australia
E-mail: fariba.shadabi@canberra.edu.au

Dharmendra Sharma and Robert Cox

School of Information Sciences and Engineering, University of Canberra,
Canberra ACT 2601, Australia
E-mail (1): dharmendra.sharma@canberra.edu.au
E-mail (2): robert.cox@canberra.edu.au

Nikolai Petrovsky

Medical Informatics Centre, University of Canberra,
Canberra ACT 2601, Australia
E-mail: nikolai.petrovsky@canberra.edu.au

Neural network ensembles have made an impressive contribution in a number of different medical domains. Like simple neural network models, the neural network ensembles are known as 'black boxes' since how the outputs are produced is not obvious. Due to this limitation these techniques are not widely used by medical professionals. This paper first provides a short review of the different neural network rule extraction techniques. Then it describes a novel approach, namely "RDC-ANNE" that is designed to extract useful explanations from several combined neural network classifiers. The methodology employed utilises a dataset made available to us from a kidney transplant database. The dataset embodies a number of important properties, which make it a good starting point for the purpose of this research. Results reveal that this approach can be used to identify and extract the regions in the data space that have positive impact on the system performance, provide useful explanations from several combined neural networks and enhance the overall utility of current neural network models.

Keywords: neural network ensemble; rule extraction; hybrid intelligent method; kidney transplant.

1. Introduction

It has been shown that clinical databases that contain considerable noise often significantly deteriorate the generalisation ability of ordinary prediction models. In recent years there has been a rapid growth in the successful use of Artificial Intelligent (AI) systems with high power of generalisation ability in many diverse areas such as science, medicine and commerce. AI-based techniques such as Artificial Neural Networks (ANN) and Decision Trees (DT) can be applied in healthcare environments where an automated process must adapt to changing conditions. These techniques have provided great opportunities for researchers to enhance the information processing and retrieval capabilities of current knowledge-based systems.

The inspiration for Artificial Neural Networks (ANNs) came from the desire to simulate features of biological neural networks and learning systems which show high power in pattern recognition tasks and adaptability. Neural network models are examples of sub-symbolic methods. Unlike symbolic methods (such as Decision Trees), the sub-symbolic methods do not usually have the ability to induce symbolic representation from data and generate a set of rules that can be understood by humans. However, on the positive side, they are known to provide very good solutions to many difficult medical problems (Baxt, 1991; Ashutosh, Lee and Mohan, 1992; Tu and Guerriere, 1993; Baxt, 1995; Mobley, Leasure and Davidson, 1995; Ortiz, Ghefter and Silva, 1995; Itchhaporia, Snow and Almassy, 1996; La-Puerta, L'Italien and Paul, 1998; Pantazopoulos *et al.*, 1998; Dayhoff and DeLeo, 2001). Neural networks have the ability to provide good solutions in situations where a large number of variables contribute to an outcome but their individual influence is not well understood. Clinical data gathered from patients who underwent graft transplant surgery have this characteristic and are known to be complex (Doyle *et al.*, 1994; Liberati and Setti, 1994; Matis *et al.*, 1995; Sheppard *et al.*, 1999).

Over time, a wide variety of hybrid intelligent methods have been proposed to generate rules or extract knowledge from sub-symbolic and complex classifiers such as ANNs using concepts drawn from fuzzy logic (Masuoka *et al.*, 1990; Mitra, 1994; Castro, Mantas and Benitez, 2003) and from rule-based methods (Fu, 1991; Thrun *et al.*, 1991; Craven and Shavlik, 1996; Krishnan, 1997; Setiono, 1997; Zhou, Chen and Chen, 2000). This paper provides a brief overview of the variety of hybrid intelligent techniques that might be utilised for extracting rules from ANNs in clinical knowledge discovery and the decision making process. The study also describes a novel neural networks ensemble technology, known as RDC-ANNE (Rules

Driven by Consistency in Artificial Neural Networks Ensemble) for extracting rules from ANNs. This technique is applied and tested for predicting medical outcomes, using a trial data set from a kidney transplant database as a prototypical medical application.

2. Previous Work

2.1. *Extracting rules from neural networks*

Over time, neural networks have proven to be very powerful tools at hand for pattern recognition and classification tasks. The power of Artificial Neural Networks in comparison to other symbolic machine learning techniques such as decision trees has been well documented by numerous researchers (Shavlik, Mooney and Towell, 1991; Thrun *et al.*, 1991; Diederich, Hild and Bakiri, 1995). Their study revealed that for most problem domains Artificial Neural Networks are considered to be a very good choice. Artificial Neural Networks models are able to accept numerous input variables and adapt their criteria to better match the data they analyse. Also, given enough training time, appropriate numbers of hidden units and layers, ANN models are able to produce a high level of predictive accuracy, solve difficult problems and learn interesting linear or nonlinear relationships. These interesting features of ANN models provide a powerful and compact knowledge representation tool at hand with an efficient storage and individual patterns recall system. The learned patterns and relationships are stored as a set of values across all weights and thresholds. These values are meaningless and incomprehensible for human users.

Coupling Artificial Neural Networks and rule extraction algorithms can significantly enhance the overall utility of ANNs. Hybrid rule extraction-ANN algorithms can be grouped into four categories, namely the decompositional, pedagogical, eclectic and compositional algorithms (Andrews, Diederich and Tickle, 1995; Tickle *et al.*, 1998). These classification schemes are mainly based on the approach used to study and analyse the underlying ANN architecture or/and the classification given by the network for the processed input vectors.

The decompositional (also known as local) methods usually start by extracting rules from each unit (hidden and output) in a trained neural network. The rules extracted at the individual unit level are then combined to form a global relationship and the final rule base for the ANN architecture as a whole. Some examples of this style of algorithm are KT (Fu, 1994), Subset (Towell and Shavlik, 1993), COMBO (Krishnan, 1997), RX

(Setiono, 1997) and RULEX (Andrews and Geva, 1994; Andrews and Geva, 1995).

The second category of rule extraction algorithms is "pedagogical". The core idea in the pedagogical (also known as global) approach is to treat the trained neural network as a black box. It aims to extract rules that map inputs directly into outputs (Tickle *et al.*, 1998). The pedagogical algorithm uses the trained neural network only to generate test data for the rule generation algorithm. In this strategy the target concept is computed by the network and the input vectors are the actual network's input vectors (Craven and Shavlik, 1994). VIA (Thrun, 1994), TREPAN (Craven, 1996; Craven and Shavlik, 1996) and STARE (Zhou *et al.*, 2000) are other examples of this style of algorithm.

The third rule extraction category is eclectic algorithms. The methods from this category, like those in the decompositional category, carefully examine the ANNs at the level of individual units; they also extract rules at the global relationship level within trained neural networks. DEDEC (Tickle, Orlowski and Diederich, 1996) is an example of this style of algorithm.

The final category is compositional algorithms. The compositional algorithms neither focus on local models that mirror the behaviour of individual units, nor treat the network as a "black box", like pedagogical approaches. Representatives of this category include algorithms proposed by Giles *et al.*, 1992; Omlin, Giles and Miller, 1992; and Giles and Omlin, 1993.

2.2. *Extracting rules from a neural network ensemble*

Neural network ensembles are known to be good predictive models. However, a neural network ensemble is composed of several independently trained neural network models; therefore, naturally its comprehensibility is considerably more difficult than are those of its component classifiers.

Over the years, many authors have tackled the problem of rule extraction from individual networks; however, much less work has been done in the explanation of combined neural networks (Wall and Cunningham, 2000). This is of significance for acceptance of this technique in medicine.

One example of research on improving comprehensibility of artificial neural networks ensembles can be found in recent work by Craven (1996). This work uses the TREPAN algorithm (Craven, 1996; Craven and Shavlik, 1996) and the Addemup algorithm (Opitz and Shavlik, 1996) for generating rules from the ensembles in a telephone domain. TREPAN seems a good choice for this task because it does not try to translate all the individual units (hidden and output) of networks into rules and also more importantly it can be applied to a wide class of networks (also see Figure 1).

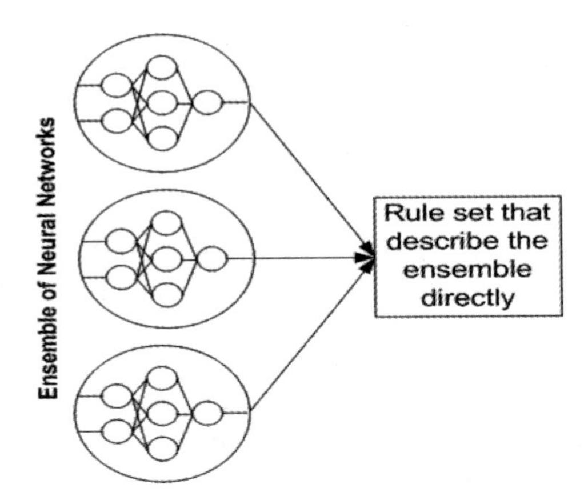

Fig. 1. A black box (global) explanation approach.

Recently, Wall *et al.* (2003) proposed a local explanation strategy as an alternative to black-box or global approaches such as TREPAN. In this study they tried to explain the ensemble on a case-by-case basis (see Figure 2). This strategy has been pursued successfully by other researchers such as Sima, 1995 and Das *et al.*, 1998. Local explanation strategy can be used to identify the ensemble members that are relevant in explaining the prediction associated with a particular case. In this strategy, once the ensembles of rule sets have been produced, a rule selection strategy (mainly based on majority voting) is applied and a number of coverage statistics is calculated in order to study the fitness of rules to any unseen example and provide a case-by-case explanation. It should be noted that this approach tries to discover how the elements of the ensemble contributed to the prediction. This is done by first extracting sets of rules from each of the member networks in the ensemble and producing ensembles of rule sets. The authors tested this strategy on relatively small ensembles of networks. Although there is no reason to believe that this approach will not perform well on larger ensembles, its suitability and computation time complexity remain to be verified.

More recently, a series of a black box approaches, namely REFNE (Rule Extraction from Neural Network Ensemble) , C4.5 Rule-PANE (C4.5 Rule

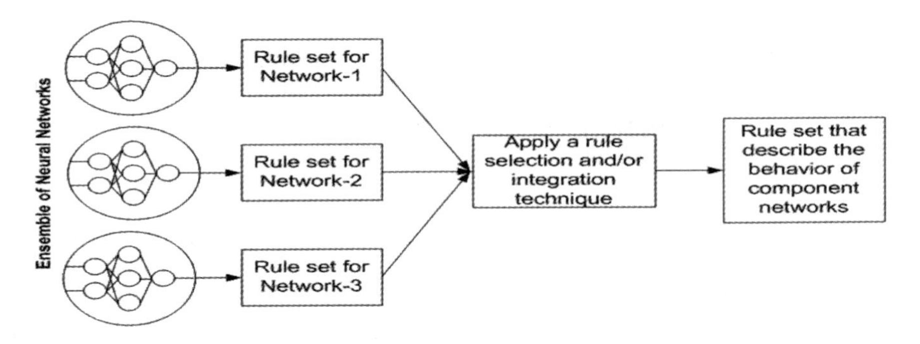

Fig. 2. A component-based (local) explanation approach.

Proceeded by Artificial Neural Ensemble), and NeC4.5 (Neural ensemble based on C4.5) have been suggested by Zhou and Jiang, 2003, Zhou, Jiang and Chen, 2003 and Zhou and Jiang, 2004 for extracting meaningful knowledge from neural network ensembles. Basically, these approaches utilise the trained neural network classifiers to generate instances and then extract rules from them. Their results revealed that, given enough training data, the C4.5 Rule-PANE approach can indeed provide comprehensible solutions with strong generalisation ability that are significantly more accurate than the standard decision trees.

3. The Proposed Method

The RDC-ANNE method is motivated by the Rule Extraction approach of Zhou and Jiang, 2003, Zhou and Jiang, 2004 and Wall *et al.*, 2003. In this approach first we identified the patterns that were consistently used across the classifier series with positive impacts (higher predictive sources). Then for the rule extraction stage we incorporated elements of both the local and global explanation strategies.

Like other black box or global rule extraction approaches (Zhou and Jiang, 2003; Zhou and Jiang, 2004), in this strategy the target concept is computed by the ensemble of networks and the input vectors are the actual network's input vectors. It also tries to consider the diversity and expertise of the component networks in the rule generation process. However the RDC-ANNE method is not purely a local strategy (Wall *et al.*, 2003) because it does not focus on identifying the ensemble members that are relevant in explaining the prediction (output) associated with a particular case. Instead it tries to explain the output of the ensemble based on a cluster of cases that consistently generate agreement across the classifiers with similar expertise. This form of regularisation could offer a significant improvement in prediction accuracy and comprehensibility (for a detailed discussion and description, see Quinlan, 1996; Shadabi *et al.*, 2004; Shadabi *et al.*, 2005).

4. A Case Study

For the purpose of this study a series ($n = 500$) of Multilayer Perceptron (MLP) networks were trained independently, to differentiate between successful and unsuccessful transplants. The data used in the project was obtained from a kidney transplant and dialysis database (ANZDATA, 2000). Some variables were removed because they were actually an indication of the outcome of the transplant. The variables that were retained are AGE (Recipient age at transplant), MISA (Number of mismatches A), MISB (Number of mismatches B), MISDR (Number of mismatches DR), MISDQ (Number of mismatches DQ), REFHOSP (Referring hospital), REFSTAT (Referring state), DONHOSP (Donor hospital), DONSTAT(Donor state), TRANHOS (Transplant hospital), TRANSTA (Transplant state), DONSOUR (Donor source), DONAGE (Donor age), DONSEX (Donor sex), ISCHEMIA (Total ischemia to nearest hour) and KIDPRESI (Initial kidney preservation). For ANN training, we divided the data set into three equal sized sets, the training set, the overtraining prevention (tuning) set and the test set. We also pre-processed the data further by performing normalisation. It should be noted that for classification using PART decision list in the data mining tool WEKA (Witten and Frank, 2000), the data was converted to ARFF format.

In this study, rather than reporting predictive accuracy alone to show best model choice, we modified the program in order to show the input patterns (examples) that were included across the ANN series in the final results. Patterns that consistently were in agreement across the classifiers can be considered as examples with positive impacts or higher predictive sources.

4.1. *Methodology*

For the purpose of this study, the following methodology was employed:

(1) Pre-process the data set. This includes: extracting the data from different tables, cleaning the data, transforming the nominal attributes

into numeric attributes, and choosing the appropriate parameters to be included in the dataset with the help of a domain expert.

(2) Split the dataset for training and testing (with balanced distribution of success and failure cases).
(3) Perform classification using the ANN ensemble approach.
(4) Generate new training sets by extracting the patterns (examples) that were consistently causing agreement across the ANN classifiers, in the testing phase.
(5) From the new training sets generated in step 4, choose a reasonably big training set that has provided both a good level of accuracy (based on its corresponding classification table) and a reasonable amount of model agreement.
(6) Extract rules from the patterns in the chosen training set, using the WEKA PART decision list.
(7) Analyse the results.

It should be noted that prior domain knowledge can also be applied to choose the significant rules that could be useful for predicting the outcomes of a medical event (e.g. failure or success of graft transplants) or to form new concepts.

4.2. *Results*

In this experiment, the balanced test set reached 70% accuracy rate with 87% agreement among the networks (435 of 500 networks), based on 19% of data points (i.e. 84 cases). The results indicate that the model was able to classify about 87% of successful transplants and 54% of unsuccessful cases.

For rule generation using the PART decision list, at first, we applied a more conventional strategy and generated a new training data set by feeding the entire examples in the test set to a trained ensemble and replacing the true class labels of the original test instances with the class labels assigned to them by the ensemble. The rule set generated by PART decision list produced 26 rules.

The rule set can be made substantially easier to understand by enforcing the model to consider mainly the examples whose class labels consistently caused agreement across the ANN classifiers. In effect, this strategy tries to safely remove the branches of rules that were generated by the presence of noise in the data set. The following rule set was produced by applying RDC-ANNE approach based on the 84 examples whose class labels (outputs from the ensemble) were in agreement across 87% of classifiers:

(1) donstate $\leq$ 4 AND donage $\leq$ 43: Success (38 cases)
(2) donhosp $\leq$ 109 AND misa $\leq$ 1 AND refstate $>$ 3 AND misb $>$ 0 AND refhosp $>$ 94: Failure (11 cases)
(3) misa $\leq$ 1 AND misb $\leq$ 1: Success (16 cases /1 false positive)
(4) donsex $>$ 0 (i.e. Female): Failure (9 cases)
(5) misa $>$ 1: Failure (5 cases)
(6) age $>$ 27: Success (3 cases /1 false positive)
(7) Else: Failure (2 cases).

As it can be seen, in this experiment, the rule set produced fewer rules (only 7 rules). These rules were valid for 98% of cases (i.e. 82 cases) in the data set.

5. Conclusions

This study has made a short review of the different ANN rule extraction techniques. Of these techniques some, notably "hybrid neural networks", in which the goal is to extract useful explanations from individual neural networks, have been actively developed. As yet, little evidence exists about research and development in the explanation of several combined neural networks.

This study also has described a novel approach, namely "RDC-ANNE" that is designed to extract useful explanations from several combined neural network classifiers. The primary experimental results revealed that this approach can be used to identify the regions in the data space that have positive impact on the system performance, extract useful explanations from several combined neural networks and enhance the overall utility of current neural network models.

In summary, there are still many challenges to be overcome. Each technique has its own advantages and disadvantages under different circumstances. It is clear, though, that integrating hybrid AI modules into computerised patient records and complex clinical data can provide a great chance for improving quality of care.

References

1. Andrews, R., Diederich, J., and Tickle, A. B. (1995), A Survey and Critique of Techniques for Extracting Rules from Trained Artificial Neural Networks. *Knowledge Based Systems* 8: 373–389.
2. Andrews, R. and Geva, S. (1994), Rule extraction from a constrained error back propagation MLP. *Proceedings of the 5th Australian Conference on Neural Networks*: 9–12.

3. Andrews, R. and Geva, S. (1995), Inserting and extracting knowledge from constrained error back propagation networks. *Proceedings of the 6th Australian Conference on Neural Networks.*
4. ANZDATA (2000), Data Dictionary: ANZDATA Registry Database.
5. Ashutosh, K., Lee, H., and Mohan, C. K. *et al.* (1992), Prediction criteria for successful weaning from respiratory support: statistical and connectionist analyses. *Crit. Care Med.* 20: 1295–1301.
6. Baxt, W. G. (1991), Use of an artificial neural network for the diagnosis of myocardial infarction. *Ann. Intern. Med.* 115: 843–848.
7. Baxt, W. G. (1995), Application of artificial neural networks to clinical medicine. *Lancet* 346: 1135–1137.
8. Castro, J. L., Mantas, C. J., and Benitez, J. M. (2003), Interpretation of artificial neural networks by means of fuzzy rules. *IEEE Transactions on Neural Networks* 13(1): 101–116.
9. Craven, M. (1996), *Extracting Comprehensible Models from Trained Neural Networks.* Ph.D. thesis, Department of Computer Sciences, University of Wisconsin, Madison.
10. Craven, M. W. and Shavlik, J. W. (1994), Using sampling and queries to extract rules from trained neural networks. *Proceedings of the 11th International Conference on Machine Learning*: 37–45.
11. Craven, M. W. and Shavlik, J. W. (1996), Extracting tree-structured representations from trained networks. *Advances in Neural Information Processing Systems* 8: 24–30.
12. Das, G., Lin, K., Mannila, H., Renganathan, G., and Smyth, P. (1998), Rule discovery from time series. *Proceedings of the fourth International Conference on Knowledge Discovery and Data Mining.*
13. Dayhoff, J. E. and DeLeo, J. M. (2001), Artificial neural networks: opening the black box. *Cancer* 91 (8): 1615–1635.
14. Diederich, J., Hild, H., and Bakiri, G. (1995), A comparison of ID3 and back propagation for English text-to-speech mapping. *Machine Learning* 18: 51–80.
15. Doyle, H., Dvorchik, I., Mitchell, S., Marino, I., Ebert, F., McMichael , J., and Fung, J. (1994), Predicting outcomes after liver transplantation. A connectionist approach. *Ann. Surg.* 219(4): 408–415.
16. Fu, L. (1991), Rule learning by searching on adapted nets. *Proceedings of the 9th National Conference on Artificial Intelligence*: 590–595.
17. Fu, L. (1994), Rule generation from neural networks. *IEEE Transactions on Systems, Man, and Cybernetics* 28(8): 1114–1124.
18. Giles, C. L., Miller, C. B., Chen, D., Chen, H., Sun, G. Z., and Lee, Y. C. (1992), Learning and extracting finite state automata with second-order recurrent neural networks. *Neural Computation* 4(3): 393–405.
19. Giles, C. L. and Omlin, C. W. (1993), Extraction, insertion, and refinement of symbolic rules in dynamically driven recurrent networks. *Connection Science* 5(3–4): 307–328.
20. Itchhaporia, D., Snow, P. B., and Almassy, R. J. *et al.* (1996), Artificial neural networks: current status in cardiovascular medicine. *J. Am. Coll. Cardiol.* 28:

2515–2521.
21. Krishnan, R. (1997), A systematic method for decompositional rule extraction from neural networks. *Proceedings of the NIPS'96 Workshop on Rule Extraction from Trained Artificial Neural Networks*: 38–45.
22. LaPuerta, P., L'Italien, G. J., and Paul, S. *et al.* (1998), Neural network assessment of perioperative cardiac risk in vascular surgery patients. *Med. Decis. Making* 18: 70–75.
23. Liberati, D., Setti, E., and Pappaletterra, M. (1994), The application of neural networks in predicting the success of kidney transplants, *AEI Autom. Energ. Inform.* 3(7): 67–70.
24. Masuoka, R., Watanabe, N., Kawamura, A., Owada, Y., and Asakawa, K. (1990), Neurofuzzy systems — fuzzy inference using a structured neural network. *Proceedings of the International Conference on Fuzzy Logic and Neural Networks*: 173–177.
25. Matis, S., Doyle, H., Marino, I., Murad, R., and Uberbacher, E. (1995), Use of Neural Networks for Prediction of Graft Failure following Liver Transplantation. *Proceeding of the Eighth Annual IEEE Symposium on Computer-Based Medical Systems*: 0133.
26. Mitra, S. (1994), Fuzzy MLP based expert system for medical diagnosis. *Fuzzy Sets and Systems* 65(2–3): 285–296.
27. Mobley, B. A., Leasure, R. and Davidson, L. (1995), Artificial neural network predictions of length of stay on a post-coronary care unit. *Heart Lung* 24: 251–256.
28. Omlin, C. W., Giles, C. L. and Miller, C. B. (1992), Heuristics for the extraction of rules from discrete time recurrent neural networks. *Proceedings of the International Joint Conference on Neural Networks* 1: 33–38.
29. Opitz, D. W. and Shavlik, J. W. (1996), Actively searching for an effective neural network ensemble. *Connection Science* 8: 337–353.
30. Ortiz, J., Ghefter, C. G. M. and Silva, C. E. S. (1995), One-year mortality prognosis in heart failure: a neural network approach based on echocardiographic data. *J. Am. Coll. Cardiol.* 26: 1586–1593.
31. Pantazopoulos, D., Karakitsos, P., Iokim-Liossi, A., Pouliakis, A., Botsoli-Stergiou, E. and Dimopoulos, C. (1998), Back propagation neural network in the discrimination of benign from malignant lower urinary tract lesions. *Journal of Urology* 159(5): 1619–1623.
32. Quinlan, J. (1996), Bagging, boosting, and C4.5. *Proceedings of the Thirteenth National Conference on Artificial Intelligence*: 725–730.
33. Setiono, R. (1997), Extracting rules from neural networks by pruning and hidden-unit splitting. *Neural Computation* 9(1): 205–225.
34. Shadabi, F., Cox, R., Sharma, D., and Petrovsky, N. (2004), Use of Artificial Neural Networks in the Prediction of Kidney Transplant Outcomes. *Proceedings of the 8th International Conference on Knowledge-Based Intelligent Information & Engineering Systems (KES 04)* 3: 566–572.
35. Shadabi, F., Cox, R., Sharma, D., and Petrovsky, N. (2005), A Hybrid Decision Tree — Artificial Neural Networks Ensemble Approach for Kidney Transplantation Outcomes Prediction. *9th International Conference on*

Knowledge-Based Intelligent Information & Engineering Systems (KES 05)

36. Shavlik, J. W., Mooney, R., and Towell, G. G. (1991), Symbolic and neural learning algorithms: an experimental comparison. *Machine Learning* 6: 111–143.
37. Sheppard, D., McPhee, D., Darke, C., Shrethra, B., Moore, R., Jurewitz, A., and Gray, A. (1999), Predicting Cytomegalovirus disease after renal transplantation: an artificial neural network approach. *International Journal of Medical Informatics* 54: 55–76.
38. Sima, J. (1995), Neural expert systems. *Neural Net* 8 (2): 261–271.
39. Thrun, S. B. (1994), Extracting provably correct rules from artificial neural networks. Technical Report IAI-TR-93-5, Institute for Informatik III, Universitat Bonn, Germany.
40. Thrun, S. B., Bala, J., Bloedorn, E., Bratko, I., Cestnik, B., Cheng, J., Jong, K. D., Dzeroski, S., Hamann, R., Kaufman, K., Keller, S., Kononenko, I., Kreuziger, J., Michalski, R. S., Mitchell, T., Pachowicz, P., Roger, B., Vafaie, H., de Velde, W. V., Wenzel, W., Wnek, J., and Zhang, J. (1991), The MONK's Problems: A Performance Comparison of Different Learning Algorithms. Tech. report CMU-CS-91-19, Computer Science Department, Carnegie Mellon University.
41. Tickle, A. B., Andrews, R., Golea, M., and Diederich, J. (1998), The truth will come to light: directions and challenges in extracting the knowledge embedded within trained artificial neural networks. *IEEE Transactions on Neural Networks* 9(6): 1057–1067.
42. Tickle, A. B., Orlowski, M., and Diederich, J. (1996), DEDEC: a methodology for extracting rules from trained artificial neural networks. *Proceedings of the AISB'96 Workshop on Rule Extraction from Trained Neural Networks*: 90–102.
43. Towell, G. and Shavlik, J. (1993), The extraction of refined rules from knowledge based neural networks. *Machine Learning* 13(1): 71–101.
44. Tu, J. V. and Guerriere, M. R. J. (1993), Use of a neural network as a predictive instrument for length of stay in the intensive care unit following cardiac surgery. *AMIA*: 666–672.
45. Wall, R. and Cunningham, P. (2000), Exploring the Potential for Rule Extraction from Ensembles of Neural Networks. *11th Irish Conference on Artificial Intelligence & Cognitive Science (AICS).*
46. Wall, R., Cunningham, P., Walsh, P., and Byrne, S. (2003), Explaining the output of ensembles in medical decision support on a case by case basis, *Artificial Intelligence in Medicine* 28: 191–206.
47. Witten, I. H. and Frank, E. (2000), Data Mining: Practical Machine Learing Tools and Techniques with Java Implementations. Morgan Kaufmann, San Diego, CA.
48. Zhou, Z. H., Chen, S. F., and Chen, Z. Q. (2000), A statistics based approach for extracting priority rules from trained neural networks. *Proceedings of the IEEE-INNS-ENNS International Joint Conference on Neural Networks* 3: 401–406.
49. Zhou, Z. H. and Jiang, Y. (2003), Medical diagnosis with C4.5 rule preceded

by artificial neural network ensemble. *IEEE Transactions on Information Technology in Biomedicine* 7(1): 37–42.
50. Zhou, Z. H. and Jiang, Y. (2004), NeC4.5: neural ensemble based C4.5. *IEEE Transactions on Knowledge and Data Engineering* 16(6): 770–773.
51. Zhou, Z. H., Jiang, Y., and Chen, S. F. (2003), Extracting symbolic rules from trained neural network ensembles. *AI Communications* 16(1): 3–15.

Mining Multiple Models

Graham J. Williams

School of Information Sciences and Engineering, University of Canberra, Canberra ACT 2601, Australia
E-mail: graham.williams@togaware.com

Data mining is much more than simply building statistical models from large collections of data. In particular, this paper records a core task of mining as exploring through the space of models that are built in a data mining project. The idea was first introduced through the concept of multiple inductive learning (MIL) (Williams, 1988, 1991) and further developed in practice as mining the data mine (Williams and Huang, 1997). Many data mining advances that have since emerged have further developed the idea: multiple modelling, ensemble learning, bagging and boosting all help the data miner explore different ideas and look for different insights in modelling. In this paper we review these ideas and a number of data mining projects that highlight the significant role played by mining the data mine.

Keywords: data mining; ensembles; hot spots; multiple models; health.

1. Introduction

A data miner is engaged in the activity of aggregating very large collections of data to explore for new insights and understandings that will provide improvements for some process of interest. Application areas include customer relationship management, fraud prevention and control, and risk rating, to list but a few. Data mining is commonly defined as the non-trivial extraction of novel, implicit, and actionable knowledge from large databases (Fayyad *et al.*, 1996).

The tools deployed by a data miner include common statistical modelling approaches as well as modelling approaches developed from research into machine learning and artificial intelligence. Traditionally, this means building decision trees or logistic regression models or neural networks.

Data underlies data mining and comes in many shapes and sizes. For data mining, the data is generally characterised by its sheer size. Its size is one of the key differentiators from traditional research in machine learning and statistics. Each entity (which may also be referred to as a record in data base terminology, or a training instance in machine learning terminology, or a sample in statistical terminology) might be described by anywhere from 10 to 20,000, or more, features, and we may have from 20 to 200 million, or more, entities. Generally, small sets of entities arise in situations where we have very many features describing such entities, as is typical in genomics research, image data, and text mining. Datasets with fewer features but many more entities are typical in industry and government describing clients or customers. Thus the data is often of at least megabytes in size, usually in the gigabytes, and less frequently, in the terabytes.

Traditional approaches to modelling and data mining tend to deal with flat data in the form of a single row of data representing a single entity, with no relational data explicitly allowed. That is, links between entities must be captured in some other way, and repeated data needs to be aggregated in some way so that all entities have the same signature (the same number of features describing each entity).

Complex relationships, then, are generally not mined in data mining. In the administrative medical domain, for example, the entities that exist include patients and doctors, but also receptionists, pathologists, specialists, insurance claims officers, *et c.* Complex relationships exist among all these entities but generally remain too complex to be handled by today's data mining technology. Instead, the complex relationships need to be re-represented in a simpler, flatter form.

Whilst statistics provides many of the traditional tools deployed for modelling in data mining, the data miner spends much more time through other phases of a data mining project, which includes business understanding, data understanding, data preparation and cleaning, modelling, evaluation, and deployment (Fayyad *et al.*, 1996). Common wisdom, indeed, is that modelling is just a small portion of the overall task (often less than 10% of the overall effort in any data mining project).

More broadly, data mining is about deploying multiple technologies to enable data exploration, data analysis, and data visualisation of very large databases at a high level of abstraction, generally without a well defined, specific hypothesis in mind, to extract knowledge from the data in any, and in many, ways.

The technology deployed in data mining comes from a diverse arena of research disciplines, beginning with databases, quickly drawing in machine learning and statistics, and encompassing high performance computing, computational mathematics, intelligent systems, visualisation, and web services. Together, these technologies deliver a rich, if sometime diverse, toolbox that a data miner deploys to deliver knowledge from data by analysing relationships in information.

In this paper we explore what might happen *after* we have built a model. Indeed, we review the idea of building multiple models and then exploring these models for the insights we need to deliver from data mining. Projects deploying such an approach are briefly described.

2. Mining Models

Making the simple observation that, in building decision tree models, for example, choices between different splits sometimes only marginally differentiate variables, Williams (1988) introduced the idea of building multiple decision trees and combining them into a single model. This began the data mining approach of exploring through a much richer space of models to identify and extract more information than otherwise would have been. It also later eventuated, from theoretical studies by many others, that ensemble learning was a good approach to model building and data mining (Hastie *et al.*, 2001).

The original work of Williams (1991) used the Australian Resources Information System (ARIS) database (Walker *et al.*, 1985). This consisted of some 11,000 entities, each recording extensive geographical information about a 700 square kilometre region of Australia. In particular, the rangeland regions of Australia were used in the study (8,000 entities). For each region 40 features were selected describing dominant soil type, vegetation, moisture indicators, and distance to nearest seaport.

The output variable for the study was a measure of the viability of the pastoral use of the land (for sheep and cattle grazing). Some 106 entities had been manually assessed by pastoral experts as to their viability, and this small dataset (although, at the time, regarded as reasonably sized) was used for model building. A version of the ID3 algorithm for decision tree induction (Quinlan, 1986), using the information-theoretic cost function, was used.

In building models in this domain, the decision tree algorithm only marginally chose one variable over another, to result in sometimes quite different looking trees. This fact was taken advantage of, rather than thought of as a problem, so as to produce multiple models, each giving different, but useful, insights into the domain. The final model developed for this domain consisted of multiple decision trees, with conflicts between the models being resolved logically. The system was called MIL for Multiple Inductive Learning.

The key outcome of this research was the idea of building multiple models and combining them. Follow-on research took this idea further in the context of data mining with the realisation that model building was really the starting point to achieving the goals of data mining. Williams and Huang (1997) introduced the concept of mining the knowledge mine, and hot spots data mining.

The basic idea is that of building models that can be decomposed into smaller units that effectively describe different regions of a dataset (or population). Converting decision trees to rule sets is common practice, starting with the C4.5 tool (Quinlan, 1993). Rules, generally in the form of a conjunction of conditions, can then be used to symbolically describe these different regions of a dataset. Indeed, we can think of each set of conditions, each rule, as a nugget! A nugget captures some collection of entities, and our task in data mining is to determine which nuggets are the most interesting. The generation of the nuggets can be left to a variety of approaches, but a common one we introduced is to combine clustering with tree building where the cluster identifier becomes the target variable in the tree building. We call this hot spots data mining.

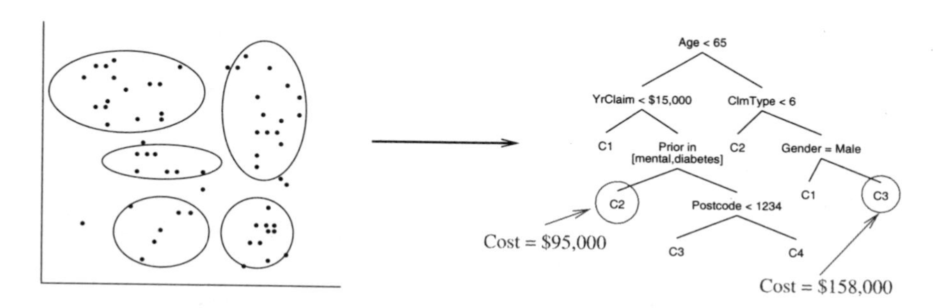

Fig. 1. The Hot Spots data mining process.

We can picture the hot spots data mining process as in Figure 1. Working from left to right, we start with a dataset (a 2-dimensional dataset in this case) that has no specific target variables. Thus we have an unsupervised learning problem. By clustering the data in some way, for example using traditional k-means, we can end up with a collection of clusters. For a very large dataset (e.g., millions of entities) we may indeed end up with quite a number of clusters (upwards of 100 or even 1000).

Each cluster can then be considered as a class, and entities will then belong to one class or another. This class can then be used as a target variable for a supervised learning problem. Using a decision tree builder we produce a set of rules (from a single decision tree each rule corresponds to a single path from the root node to a leaf node — providing a conjunction of conditions).

Each of these rules (or paths, or perhaps we can call them "nuggets") can then be considered independent of any other nuggets. The concept is to then explore through this space of nuggets searching for any that are interesting by some measure.

The nuggets might be as simple as the following:

Nugget 1 Age is between 28 and 35 **and** Weeks ≥ 10
Nugget 2 Weeks < 10 **and** Benefits $>$ \$350

We note again that they are simple conjunctions of conditions.

For each nugget we collect together summary data about those entities that make up the subset of the dataset described by the nugget. This might include things like the size of the nugget and average values of various features over those entities in the nugget, or measures of how far the nugget average is from the population average, and so on.

A simple example might be the following table where perhaps we have a total of just 280 nuggets and we might collect various summary items as in:

Nugget	Size	Age	Gender	Services	Benefits	Weeks	Hoard	Regular
1	9000	30	F	10	30	2	1	1
2	150	30	F	**24**	**841**	4	2	4
3	1200	65	M	7	220	20	1	1
4	80	45	F	**30**	**750**	10	1	1
5	90	10	M	12	**1125**	10	**5**	2
6	800	55	M	8	550	7	1	**9**
⋮								
280	30	25	F	15	450	15	2	6
All	40,000	45		8	30	3	1	1

Specific nuggets are then scored as to their interestingness based on a number of measures. For example, the bold entries in the table indicate values that are found to be more than two standard deviations from the population values. Thus we add scores to these nuggets. By this we produce a ranking of nuggets which can then be explored by domain experts in order, looking for new insights.

Over many different data mining projects, these ideas have repeatedly shown themselves to provide more insights into the relationships between entities and features, with respect to some target variable. In the following section we briefly review a number of these projects.

3. Applications in Health

Australia has a universal health care system that has been providing primary medical care to patients since 1975. For administrative purposes (i.e., to make payments to the doctors) data is collected for each transaction performed. Since the introduction of Medicare in 1975, over a terabyte of data has been collected, and mostly never analysed.

This tremendous resource, that can tell quite a story about the changing health of Australians, started being used with data mining in the early 1990s. Over the years it has been used for identifying inappropriate provider practices, and for identifying public fraud committed against Medicare.

The mining of the knowledge mine approach was successfully deployed to identify a particular type of fraud being committed by a group of patients against Medicare. The particular group ranked highly on a number of disjoint features, and in combination this led the domain experts to follow up on their insurance claims, and eventually determine that they were fraudulent.

Another major piece of health data mining was made possible with the creation of the Queensland Linked Dataset (Williams *et al.*, 2002b). This was the culmination of a project between CSIRO Data Mining, the Commonwealth Department of Health and Ageing and the Queensland Department of Health, bringing together a large collection of health care data for the purpose of data mining.

The Medicare program (MBS), as mentioned above, together with the Pharmaceutical Benefits Scheme (PBS), provide universal health care insurance for Australians. These schemes cost several tens of billions of dollars each year and data relating to virtually every non-hospital medical activity in Australia since 1975 is recorded. But a significant gap in this data was information relating to hospitalisations of patients, which was a State rather than a Commonwealth responsibility. This project brought together these datasets for the first time in Australia.

The resulting dataset consisted of 5 years of MBS and PBS transactions and 4 years of Queensland hospital admissions data for all patients in

Queensland. The data was carefully de-identified so as to preserve patient privacy and confidentiality. The dataset consisted of records for 1.1 million individuals who were hospitalised in Queensland between 1995 and 1999, and there were 3 million hospital records in the data. For these patients there are 100 million MBS transactions and 60 million PBS transactions. For hospital records there are nearly 60 variables recorded, nearly 20 for MBS and 15 for PBS. Overall these data account for 500MB, 8GB, and 4GB respectively.

We have deployed this dataset in a number of data mining tasks, but the early work explored building multiple models and from these models exploring for significant, if rare, relationships between interactions with the medical system, and, for example, episodes in hospital. Indeed, this early work led to initial discoveries in the dataset of relationships between multiple drug prescriptions and hospitalisation for specific conditions (Williams *et al.*, 2002a).

4. Summary

In this paper we review the idea of modelling as one step along the path to data mining, where the aim is to gain insights into the world we are modelling, and with these insights, to take action to improve our business processes or our understanding of how things work. In particular, we have presented the genesis of the idea of building multiple models and illustrated applications where this has demonstrated useful outcomes.

References

1. Fayyad, U., Piatetsky-Shapiro, G., Smyth, P. and Uthurusamy, R., eds (1996), *Advances in knowledge discovery and data mining*, The MIT Press, Cambridge, MA.
2. Hastie, T., Tibshirani, R. and Friedman, J. (2001), *The elements of statistical learning: Data mining, inference, and prediction*, Springer Series in Statistics, Springer-Verlag, New York.
3. Quinlan, J. R. (1986), Induction of decision trees, *Machine Learning* 1(1), 81–106.
4. Quinlan, J. R. (1993), *C4.5: Programs for machine learning*, Morgan Kaufmann.
5. Walker, P. A., Cocks, K. D. and Young, M. D. (1985), Regionalising continental data sets, *Cartography* 14(1), 66–73.
6. Williams, G., Baxter, R., Kelman, C., Rainsford, C., He, H., Gu, L., Vickers, D. and Hawkins, S. (2002a), Estimating episodes of care using linked medical claims data, in *Proceedings of the 15th Australian Joint Conference on Ar-*

tificial Intelligence (AI02) Canberra, Lecture Notes in Artificial Intelligence, Vol. 2557, pp. 660–671, Springer-Verlag, Canberra, Australia.

7. Williams, G. J. (1988), Combining decision trees: Initial results from the MIL algorithm, in J. S. Gero and R. B. Stanton, eds, *Artificial Intelligence Developments and Applications*, pp. 273–289, Elsevier Science Publishers B.V. (North-Holland).
8. Williams, G. J. (1991), Inducing and combining decision structures for expert systems, Ph.D. thesis, Australian National University. `http://papers.togaware.com/gjwthesis.pdf`.
9. Williams, G. J. and Huang, Z. (1997), Mining the knowledge mine: The Hot Spots methodology for mining large, real world databases, in A. Sattar, ed., *Advanced Topics in Artificial Intelligence, Vol. 1342 of Lecture Notes in Computer Science*, pp. 340–348, Springer-Verlag. `http://papers.togaware.com/ai97.pdf`.
10. Williams, G., Vickers, D., Baxter, R., Hawkins, S., Kelman, C., Solon, R., He, H. and Gu, L. (2002b), The Queensland linked data set, Technical report, CSIRO Mathematical and Information Systems.

PART F

Abstracts without Papers

Ranked Set Sampling: A Simple Idea of Great Use

Zehua Chen

Department of Statistics and Applied Probability, National University of Singapore, Singapore 117543
E-mail: stachenz@nus.edu.sg

Ranked set sampling s a simple idea of great use. It was proposed half a century ago. The last fifteen years or so have witnessed considerable development in the research and applications of ranked set sampling. In this talk, we give an overview on ranked set sampling and discuss its essence. We present some novel applications of ranked set sampling in areas of clinical trials, genetic quantitative trait loci mappings and others. By doing so, we wish to provide the audience with a philosophical view on ranked set sampling and shed some light on a broader range of its applications.

Properties of Nearest-neighbour Classifiers

Peter Hall

CMA, Mathematical Sciences Institute, Australian National University,
Canberra ACT 0200, Australia
E-mail: peter.hall@maths.anu.edu.au

The k-th nearest neighbour rule is arguably the simplest and most intuitively appealing nonparametric classification procedure. However, relatively little is known about the manner in which this method is influenced by the value of k, or about properties of empirical selectors for k. We shall discuss these and related issues.

Large Covariance Matrices: Estimation and Inference in High Dimensions

Iain M. Johnstone

CMA, Mathematical Sciences Institute, Australian National University,
Canberra ACT 0200, Australia
and
Department of Statistics, Stanford University,
Stanford CA 94305, USA
E-mail: imj@stanford.edu

Principal components analysis (PCA) and canonical correlation analysis (CCA) are among the workhorses of applied multivariate data analysis. Modern data sets are apt to have a large number of variables, and so a mode of approximation in which the number of variables is comparable to the number of observations is sometimes relevant. The talk will survey several topics relating to PCA and CCA where this type of approximation has led to new results.

Bootstrapping in Clustered Populations

Alan H. Welsh

CMA, Mathematical Sciences Institute, Australian National University, Canberra ACT 0200, Australia
E-mail: alan.welsh@maths.anu.edu.au

We consider the problem of making analytic inferences about the parameters of a simple super-population model which generates clustered populations. We suppose that we have a finite population generated by this model in which we know the cluster structure in the sense that we know to which cluster every unit in the population belongs. From this finite population, we suppose that we select a sample through a simple, noninformative two-stage sampling scheme in which (i) we select a sample of clusters and then (ii) within each selected cluster, we select a sample of units. We assume that the sampling at both stages is ignorable given knowledge of the cluster structure. We discuss estimating the parameters of the model and then explore the use of the bootstrap for estimating the variances of estimators of the super-population parameters.

On the Linear Aggregation Problem in the General Gauss-Markov Model

Hans Joachim Werner

Department of Statistics, Faculty of Economics, University of Bonn, D-53113 Bonn, Germany
E-mail: hjw.de@uni-bonn.de

We consider the linear aggregation problem in the general, possibly singular, Gauss-Markov model. For the true underlying micro-relations, which explain the micro-behaviour of the individuals, no restrictive rank conditions are assumed. We introduce several estimators for certain linear transformations of the systematic part of the corresponding macro-relations and discuss their properties. [This research includes some joint work with Fikri Akdeniz (Adana, Turkey).]

AUTHOR INDEX

SUBJECT INDEX